NONSTEADY DUCT FLOW

Wave-Diagram Analysis

NONSTEADY DUCT FLOW

Wave-Diagram Analysis

BY

GEORGE RUDINGER

Principal Physicist
Cornell Aeronautical Laboratory, Inc.
Buffalo, New York

DOVER PUBLICATIONS, INC.
New York

Published in Canada by General Publishing Company, Ltd., 30 Lesmill Road, Don Mills, Toronto, Ontario.
Published in the United Kingdom by Constable and Company, Ltd., 10 Orange Street, London WC 2.

This Dover edition, first published in 1969, is an unabridged and corrected republication of the work originally published in 1955 by D. Van Nostrand Company, Inc., under the title *Wave Diagrams for Nonsteady Flow in Ducts.* The author has added a new preface, a second appendix and an updated bibliography.

Library of Congress Catalog Card Number: 68-20953

Manufactured in the United States of America
Dover Publications, Inc.
180 Varick Street
New York, N. Y. 10014

PREFACE TO THE DOVER EDITION

It is about twelve years since this work was published by Van Nostrand, and its reception then indicated that it filled a definite need. It is now out of print, and I am most grateful to Dover Publications for giving me the opportunity to prepare a new edition. The aim of the presentation remains the same: to provide a consistent set of computing procedures for the numerical analysis of nonsteady duct flow. Experimental results are therefore not presented, but reference is made to them in various places to indicate the occasional need for more refined procedures.

A number of misprints have been eliminated, and the discussion of a few topics has been changed to make it clearer or to give a more realistic representation of the physical phenomena. About three dozen references have been added, but, as in the original edition, these should be considered typical rather than exhaustive.

There is a new Appendix, which presents some new material which has become available since the publication of the original edition. It contains sections on improved boundary conditions for wave reflections from open-ended duct configurations, the problem of the uniqueness of solutions for the passage of strong shock waves through a contraction of a duct, and the use of digital computers. Computing procedures that are most convenient for manual computations are, in general, not the most suitable ones for digital computers. Since such procedures are adequately described in other texts, only a brief discussion is presented here to indicate the nature of the difficulties and the steps that may be taken to overcome them. Even with the enormously increased availability of digital computers, it will still often be necessary to prepare some preliminary wave diagrams without recourse to large machines, so that the system presented in this text should continue to have a useful function.

Finally, I wish to thank the many collegues who made valuable suggestions for the improvement of the presentation of some topics and who pointed out misprints and minor inaccuracies in the original text.

George Rudinger

Buffalo, N.Y.
February, 1968

PREFACE TO THE FIRST EDITION

The foundations for the theoretical study of pressure waves of large amplitude were laid during the first half of the last century, and attempts to utilize nonsteady flow in gas turbines and jet engines date back to the early years of this century. In spite of this long history, the field of nonsteady gas flow became established as an important branch of fluid dynamics only in recent years.

The enormous difficulties that confront any attempt to solve the general system of partial differential equations of fluid dynamics are too well known to require any elaboration. In the case of nonsteady flows through ducts, an essential simplification is obtained by treating the flow as quasi-one-dimensional, but even then, the equations cannot be solved analytically except in cases of extreme simplicity. One is, thus, forced to solve specific problems by methods of graphical and numerical approximations which, in accordance with mathematical terminology, are referred to generally as the *method of characteristics*. The resulting plots representing the propagation of waves in space and time are conveniently named *wave diagrams*. Procedures for their construction have often been published in reports or journals that are not easily available, as can be seen from a survey of the bibliography. Furthermore, many techniques, such as the handling of the boundary conditions, have often been described in general terms only, and the details have been left for the user to work out for himself. Probably as a result of this situation, there has been a general tendency of authors to re-derive the methods of solution for their particular problems from the fundamental differential equations of fluid dynamics. Since the computing procedures are by no means unique, a great variety of them can now be found scattered throughout the literature. They vary greatly in convenience of operation, and appealing procedures, sometimes, become rather difficult to handle when extended to cover more complicated flow fields.

For these reasons, a need had been felt for some time to prepare a consistent set of computing procedures to cover most of the problems of nonsteady flow that ordinarily may be encountered. Such a system of procedures had been accumulated by the author in the course of a number of years. Some of these are similar to those found in earlier publications and some are, probably, new. As is typical for a field in a phase of rapid growth, derivations must sometimes be based on assumptions for which experimental verification is not yet available, and the tentative nature of the resulting procedures is stressed.

The text is so arranged that it can be used either as a convenient manual of required procedures or for study by a newcomer to the field of nonsteady flow. The multitude of required procedures may be quite confusing in the beginning, and the only way to become proficient in the preparation of wave diagrams is to practice until, at least, the more common techniques have become thoroughly familiar. Many of these are illustrated in a number of examples throughout the text. In addition, several completely worked-out wave diagrams are included. The examples demonstrate many types of problems that can be solved by means of wave diagrams. Familiarity with problems of steady, compressible flow in ducts will be assumed.

Throughout the text, emphasis is placed on the practical aspects of carrying out the computations, and a number of useful auxiliary charts and tables have been added to speed up the procedures.

This text was prepared under the auspices of Project SQUID which is sponsored jointly by the Office of Naval Research (Department of the Navy), the Office of Scientific Research (Department of the Air Force), and the Office of Ordnance Research (Department of the Army).

The author is greatly indebted to many members of the Aerodynamic Research Department of Cornell Aeronautical Laboratory for valuable criticisms and suggestions regarding the presentation of the subject matter—in particular, to Dr. J. V. Foa (now at Rensselaer Polytechnic Institute), Mr. H. R. Lawrence, and Dr. G. H. Markstein. Appreciation is also expressed to Mr. D. Feigenbaum for organizing the computation of Table 1 in Chapter XI with the aid of I.B.M. equipment. Special thanks are due to Mrs. Arleene Dowd for working out the examples in Chapter IX.

George Rudinger

Buffalo, N. Y.
November, 1954

To

ELLEN

CONTENTS

NONSTEADY DUCT FLOW

Wave-Diagram Analysis

I

INTRODUCTION

The first theoretical study of sound waves of large amplitude was apparently undertaken by Poisson [1] (1807) who derived a relation indicating that the velocity of wave propagation is the sum of the speed of sound and the disturbance velocity produced by the wave. It was pointed out later by Stokes [2] (1848) that Poisson's result implies a change of wave form as the wave progresses. The general properties of waves of finite amplitude were studied in 1859 by Earnshaw [3] and, in more general form, by Riemann.[4] A discussion of this early work can be found in references 5 and 81. All these investigations were aimed at a basic understanding of wave phenomena and were not concerned with the solution of specific problems.

Application of nonsteady flow was not attempted until several decades later. The shock tube was invented by Vieille [6] in 1899, and efforts to utilize nonsteady flow in gas turbines and jet engines date back to the beginning of this century; these engines are briefly reviewed in references 82 and 96. All early efforts of utilizing nonsteady flow were, apparently, based on intuitive insight. The first analysis of a problem of practical importance seems to have been made by Kobes [67] (1910) who investigated the wave phenomena in the long pressure lines of air brakes. Capetti [44] (1923) tried to study wave phenomena in Diesel engines and published a method of analysis in which the effects of variable duct area and of wall friction were already treated. This work seems to have been largely overlooked. Advances continued to be slow until World War II, but, by now, work on nonsteady flow covers a wide field as can be judged from the following list of applications:

Nonsteady-flow jet engines (such as pulsejets or intermittent ramjets);
Nonsteady-flow gas turbines and gas generators (such as the Brown-Boveri "Comprex");
Exhaust systems for internal combustion engines;
Internal ballistics of guns;
Pulse starting methods of supersonic wind tunnels;
Stabilization of shock waves in supersonic diffusers;
Nonsteady flow in gas lines;
Wave phenomena resulting from combustion processes;
Use of shock tube techniques for:
Creation of supersonic and hypersonic flows;
Fundamental gas dynamics studies;
Study of properties of shock waves;
Study of the kinetics of chemical reactions.

Specific references for such applications are given in the bibliography in Chapter X which contains many representative references but is by no means complete.

The problems considered here deal with the propagation of disturbances—pressure waves—in ducts. In general, reflected waves are created whenever a wave passes through a region of gradually or discontinuously changing flow conditions or when an end of the duct is reached. Wave interactions and continued reflections build up a flow pattern of ever-increasing complexity and, in spite of simplifying assumptions, analytical solutions can, in general, not be obtained.

The fundamental partial differential equations that describe the flow of compressible fluids are too complicated to be dealt with directly. The usual practice is to reduce these equations to a manageable form by omitting those terms that are of small magnitude for the problems under investigation. In the problems treated here, the cross-sectional dimensions of the duct can be considered small compared to the length of the duct. One is thus naturally led to the simplification of quasi-one-dimensional flow already familiar from steady-flow theory. Under these conditions, all flow variables are assumed to be uniformly distributed over any section of the duct, and one has to deal only with time and one space coordinate as independent variables. In a coordinate system of these two variables, one may follow the propagation of gas particles and pressure waves graphically. Such a plot is called a *wave diagram*, and the method of computation is generally known as the *method of characteristics*, a mathematical term that is taken over from the theory of partial differential equations. The computing procedures can, however, be derived without specific reference to the theory of characteristics * by following essentially the reasoning already used by Riemann.[4] This line of approach will be taken in the following.

Many of the procedures described in the following chapters are illustrated by numerical examples. In addition, complete wave diagrams for the solution of a number of more extensive problems are presented in Chapter IX. The procedures used are not unique, but they have been found convenient to use regardless of the complexity of the problem under investigation. A brief discussion of alternative procedures is presented in Chapter VIII.d. Certain wave diagram procedures involve the use of steady-flow relations. Although familiarity with these will be assumed, a number of such relations are collected in the Appendix for convenient reference. Various charts and tables that are useful during the construction of wave diagrams are presented in Chapter XI.

* A general discussion of the theory of characteristics can be found, for instance, in references 13, 17, 47, 99, and 130.

II

LIST OF SYMBOLS

a	ft/sec	speed of sound
$\mathfrak{A}$	$= \dfrac{a}{a_0}$	nondimensional form of a
A	ft^2	cross-sectional area
c_p	Btu/lb, °R	specific heat at constant pressure
c_v	Btu/lb, °R	specific heat at constant volume
d	ft	diameter of the duct
D	$= M\left(1+\dfrac{\gamma-1}{2}M^2\right)^{-\frac{\gamma+1}{2(\gamma-1)}}$	Mach number function
f	lb/slug	sum of body and dissipative forces per unit mass
$\mathfrak{F}$	$= \dfrac{fL_0}{{a_0}^2}$	nondimensional form of f
g	$= 32.2$ ft/sec^2	acceleration due to gravity
H	$= 778$ ft lb/Btu	heating value of fuel
$I(\tau)$		pressure ratio defined on page 149 (see also table in Appendix II.A)
J	ft lb/Btu	mechanical equivalent of heat
k, K_1, K_2, K_3		constant parameters defined in Chapter VI.h
L	ft	length
M		Mach number (always taken as positive)
M_S		shock Mach number (always taken as positive)

p	lb/ft^2	pressure
P	$= \dfrac{2}{\gamma - 1}\mathcal{A} + \mathcal{U}$	Riemann variables
Q	$= \dfrac{2}{\gamma - 1}\mathcal{A} - \mathcal{U}$	Riemann variables
q	Btu/lb	heat added per unit weight
R	ft lb/slug, °R	gas constant
$\mathcal{R}$	$= \dfrac{ud}{\nu}$	Reynolds number
s	Btu/lb, °R	specific entropy
S	$= \dfrac{s}{c_p(\gamma - 1)} = \dfrac{gJs}{\gamma R}$	nondimensional form of s
t	sec	time
T	°R	temperature
u	ft/sec	flow velocity relative to the duct
$\mathcal{U}$	$= \dfrac{u}{a_0}$	nondimensional form of u
$\hat{\mathcal{U}}, \hat{\mathcal{U}}'$		nondimensional velocities with which the flow enters and leaves a shock wave, respectively
v	ft/sec	velocity of a piston in the duct
$\mathcal{V}$	$= \dfrac{v}{a_0}$	nondimensional form of v
v_f	ft/sec	burning velocity (always taken as positive)
$\mathcal{V}_f$	$= \dfrac{v_f}{a_0}$	nondimensional form of v_f
w	ft/sec	shock velocity relative to the duct
$\mathcal{W}$	$= \dfrac{w}{a_0}$	nondimensional form of w
w_f	ft/sec	velocity of a flame front relative to the duct

$\mathcal{W}_f$ $= \dfrac{w_f}{a_0}$ .... nondimensional form of w_f

x ft. .... distance coordinate

Z $= \tanh \dfrac{\gamma - 1}{4} (S_L - S_R)$ .... (see Chapter VI.f)

α .... air/fuel ratio

β .... angle between the direction of the duct and an axis of rotation

γ $= \dfrac{c_p}{c_v}$ .... ratio of specific heats

η_c .... combustion efficiency

λ .... friction coefficient

ν ft^2/sec .... kinematic viscosity

ξ $= \dfrac{x}{L_0} = \dfrac{x}{a_0 t_0}$ .... nondimensional form of x

ρ slugs/ft^3 .... density

τ $= \dfrac{a_0 t}{L_0} = \dfrac{t}{t_0}$ .... nondimensional form of t

$\phi(\tau)$ .... defined on page 277 in Appendix II.A.1

ψ slugs/sec ft .... mass flow removed per unit length

Ψ $= \dfrac{L_0 a_0 \psi}{\gamma A p_0}$ .... nondimensional form of ψ

ω sec^{-1} .... angular velocity

Subscripts

0 .... reference conditions

s .... stagnation conditions

e .... conditions at the end of the duct

E .... conditions in the external region which surrounds the open end of a duct

L,R .... conditions on the left, or right, side of a surface of discontinuity in the flow

u,b .... unburned and burned gas on the two sides of a flame front

1, 2, 3, $\cdots$ and other subscripts	points or regions in a wave diagram as defined from time to time

A prime (′) will be used to denote the value of a variable on the high-pressure side of a shock wave.

A star (*) will be used to indicate that $M = 1.0$ at this station.

Double subscripts denote mean values, e.g., $\mathfrak{U}_{1,2} = \dfrac{\mathfrak{U}_1 + \mathfrak{U}_2}{2}$.

Operators

$\dfrac{D}{Dt} = \dfrac{\partial}{\partial t} + u\dfrac{\partial}{\partial x}$, $\dfrac{D}{D\tau} = \dfrac{\partial}{\partial \tau} + \mathfrak{U}\dfrac{\partial}{\partial \xi}$	derivative in the direction of a particle path (substantial derivative)
$\dfrac{\delta_\pm}{\delta t} = \dfrac{\partial}{\partial t} + (u \pm a)\dfrac{\partial}{\partial x}$, $\dfrac{\delta_\pm}{\delta \tau} = \dfrac{\partial}{\partial \tau} + (\mathfrak{U} \pm \mathfrak{A})\dfrac{\partial}{\partial \xi}$	derivative in the direction of a characteristic
$\Delta(\ )$	finite change of a variable in the direction of a particle path
$\Delta_\pm(\ )$	finite change of a variable in the characteristic directions

III

FUNDAMENTALS OF WAVE DIAGRAM CONSTRUCTION

III.a. Assumptions

A great variety of nonsteady-flow problems can be solved by means of wave diagrams provided that the following assumptions can be made:

1. The flow through the duct is treated as quasi-one-dimensional.
2. The gas follows the ideal-gas law and the values of the specific heats are constant.* This assumption does not exclude the possibility that two or more different gases are involved in a problem. Whenever two gases are in contact, it will be assumed that no diffusion across the interface takes place. Occasionally, it may be desirable to make allowance for the changes of the specific heats during combustion. In such cases, a discontinuous change will be assumed (see Chapter VI.h).
3. The configuration of the duct must be prescribed. The duct shape may also vary in time and such variations may

* Procedures based on other equations of state or variable specific heats are indicated, for instance, in references 16–18, and 21. In general, however, the complications that would be introduced by such refinements do not seem to be warranted in view of other simplifying assumptions that must be made. Kantrowitz[27] pointed out that the temperature-sensitive part of the specific heats, which is the heat capacity due to molecular vibration or rotation, is unable in most cases to follow rapid changes of state; such lag effects should, therefore, be considered simultaneously with the variations of the specific heats. For these reasons, one can expect only qualitative results if the problems studied involve extreme temperatures or pressures. These would occur, for instance, if shock Mach numbers above 5, say, must be dealt with. Extension of the procedures to nonequilibrium flow would follow the same approach that has been used for the analysis of supersonic steady flow (see, for instance, reference 128).

be prescribed either explicitly as a function of time or implicitly as a function of the flow conditions, for instance, the instantaneous pressure.

4. The conditions governing the flow at the ends of the duct must be given.
5. The initial conditions of the gas in the duct—state and flow velocity—must be given. (Strictly speaking, it is only necessary that the flow conditions be prescribed along a given curve in the wave diagram which must not be a characteristic (see Chapter IV.a).)
6. The mode of any heat addition to the flow must be prescribed.
7. Particulars must be given about any body and dissipative forces acting on the flow and about any loss of mass flow through the walls of the duct.

It happens sometimes that pieces of the required information are not well known, for example, particulars about a combustion process. It will then be necessary to replace these by "reasonable" assumptions. It must be emphasized that a wave diagram cannot supply such missing information. One may, however, check assumptions made by comparing experimental observations with results obtained from wave diagrams, provided these are not insensitive to a variation of the assumptions.

III.b. Selection of Flow Variables

A problem is completely solved once the state of the gas—described by two state parameters—and its flow velocity are known at all times and at all points of the duct. In principle, one could thus choose any three *independent* parameters as the dependent variables of a given problem. However, certain combinations of parameters prove to be more convenient than others, and it is, therefore, important that a judicious choice be made. Experience has shown that the following variables are particularly convenient:

The flow velocity relative to the duct.... u
The local speed of sound............... a
The specific entropy.................. s

Since small disturbances in a gas propagate with the speed of sound, it is not surprising that the latter rather than the temperature T assumes a major importance in any study of wave phenomena. Since the pressure p is prescribed in many cases, it could be selected as a suitable second state parameter. However, the specific entropy is felt to be even more convenient. Isentropic flow phenomena play an important part in nonsteady-flow studies, and in such cases, the number of dependent variables is then reduced from three to two. Whenever the problems investigated involve heat addition to the flow or dissipative forces acting on the fluid, the entropy rise is readily calculated from the assumed mechanism of the process.

The graphical procedures are greatly facilitated by the use of nondimensional quantities which are obtained in the following manner: A reference state, indicated by the subscript zero, is selected. Since only entropy differences will be required, the value of s_0 can always be set equal to zero without loss of generality. The speed of sound at this reference state, a_0, is then used to make both a and u nondimensional. Thus, we define

$$\left.\begin{aligned} \mathcal{A} &= \frac{a}{a_0} \\ \mathcal{U} &= \frac{u}{a_0} \\ S &= \frac{s}{c_p(\gamma - 1)} = \frac{gJs}{\gamma R} \\ \mathcal{A}_0 &= 1.0 \\ S_0 &= s_0 = 0 \end{aligned}\right\} \qquad \text{(III.b.1)}$$

where c_p, γ, and R are the specific heat at constant pressure, the ratio of specific heats, and the gas constant, respectively. Since the customary units for s and c_p are Btu/lb, °R, while R is expressed in ft lb/slug, °R, the acceleration due to gravity, g, and the mechanical equivalent of heat, J, appear in the relation for S. The reason for making the entropy nondimensional in the manner shown is that the foregoing combinations of parameters are the ones that will be frequently required.

The reference state may be one of the states that occur in a given problem, but this is not necessary. Problems are sometimes encountered in which more than one gas is involved and a reference state must be chosen for each. Some care must then be exercised since, for instance, equal reference speeds of sound do not imply equal reference temperatures. This problem will be further discussed in Chapter VI.f.

Both a and u are velocities and any combination of them could also serve as a flow variable. Indeed, it will be seen that the combinations, in nondimensional form,

$$\left.\begin{aligned} P &= \frac{2}{\gamma - 1}\mathfrak{a} + \mathfrak{u} \\ Q &= \frac{2}{\gamma - 1}\mathfrak{a} - \mathfrak{u} \end{aligned}\right\} \qquad \text{(III.b.2)}$$

are of the utmost importance. These quantities are, generally, referred to as *Riemann variables*.

The independent variables of the problems are the distance and time coordinates, x and t, respectively. These are made nondimensional by selecting an arbitrary reference length L_0 or a reference time t_0, and by using the already chosen value of a_0:

$$\left.\begin{aligned} \xi &= \frac{x}{L_0} = \frac{x}{a_0 t_0} \\ \tau &= \frac{a_0 t}{L_0} = \frac{t}{t_0} \\ L_0 &= a_0 t_0 \end{aligned}\right\} \qquad \text{(III.b.3)}$$

It is usually convenient, although not necessary, that L_0 or t_0 be a quantity of particular significance in the problem at hand.

III.c. Basic Equations

1. *The Continuity Equation* expresses the principle of conservation of mass. Consider a volume element of the duct $A\,dx$ (shaded in Fig. III.c.1). Fluid is entering and leaving this

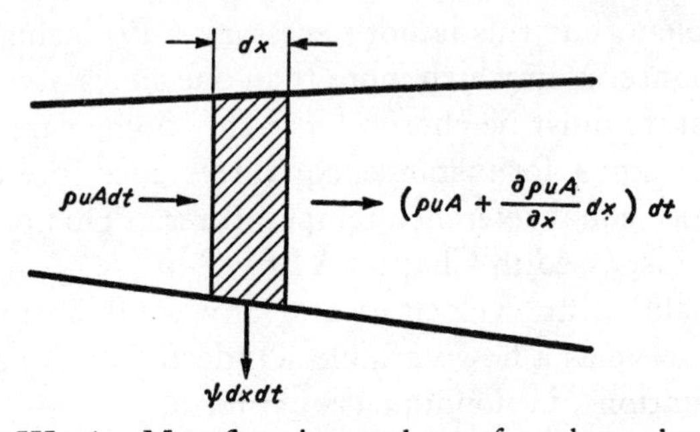

FIG. III.c.1. Mass flows into and out of a volume element

element and may also leave the duct through the walls (for instance, porous or perforated walls). The mass flow that is leaving the duct through the walls per unit length will be denoted by ψ. The mass that is entering and leaving the element during the interval dt is indicated in the figure. During this time, the mass contained in the element increases from $\rho A\,dx$ to $\rho A\,dx + \frac{\partial \rho A}{\partial t}\,dt\,dx$ where ρ is the density. Here the area cannot be taken out of the differentiation process since it may be variable with time. The excess of the inflow over the outflow must equal the increase of mass contained in the element and the continuity equation thus becomes

$$\frac{\partial \rho A}{\partial t} + \frac{\partial \rho u A}{\partial x} + \psi = 0 \qquad \text{(III.c.1)}$$

2. *The Momentum Equation* expresses Newton's law that the acceleration of a particle equals the sum of the forces acting on it divided by its mass. Consider a fluid element between two adjacent sections (shaded in Fig. III.c.2). The mass of this element is given by $\rho A\,dx$ and the forces acting on it are indicated in the figure. All body and dissipative forces that one may wish to consider are lumped into a resultant force which per unit mass will be denoted by f. Particulars about

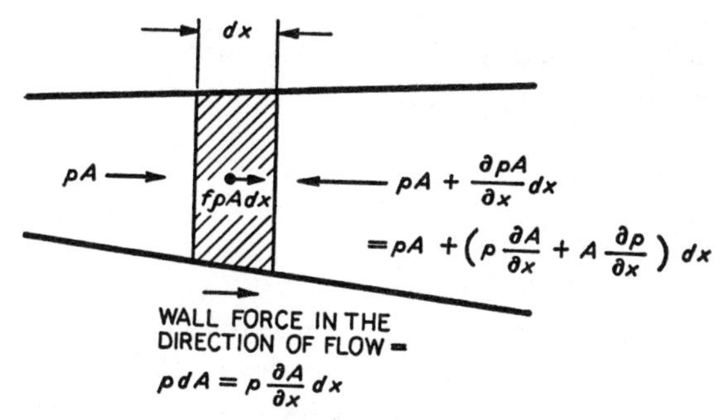

FIG. III.c.2. Forces acting on a fluid element

f must, of course, be given in any specific case (see Chapter V.c.2.2 and V.d). The acceleration of a particle is given by the substantial derivative $\frac{Du}{Dt}$, and the momentum equation thus becomes

$$\frac{Du}{Dt} = \frac{\partial u}{\partial t} + u\frac{\partial u}{\partial x} = -\frac{1}{\rho}\frac{\partial p}{\partial x} + f \tag{III.c.2}$$

3. *Entropy Conditions.* As a fluid element is moving along the duct, its energy content may vary as result of combustion, interaction with neighboring elements, and other phenomena. These energy changes take place with or without simultaneous changes of the specific entropy of the particles. For any given problem of nonsteady flow, the manner in which the entropy of an element varies is readily calculated from the assumptions

made (see Chapter V.c.2). The methods of wave diagram construction may, therefore, be derived in a completely general form if the substantial derivative $\frac{Ds}{Dt}$ is prescribed either explicitly as a function of x and t or implicitly as a function of the prevailing flow conditions. The instantaneous value of the entropy of any particle can then always be obtained from the prescribed conditions for $\frac{Ds}{Dt}$ by integrating along the particle path in the wave diagram. Depending on the assumptions made, one may distinguish three types of flow:

(1) The entropy of all particles is the same and constant—isentropic flow: $\frac{\partial s}{\partial t} = 0, \frac{\partial s}{\partial x} = 0$. (Such flows have also been called homentropic.)

(2) Each particle maintains a constant value of its entropy, but different particles may have different entropies. Mathematically, this condition is expressed by the vanishing of the substantial derivative: $\frac{Ds}{Dt} = \frac{\partial s}{\partial t} + u \frac{\partial s}{\partial x} = 0.$ These flows will be referred to as multi-isentropic. (Such flows sometimes are called isentropic.)

(3) In the most general case, the entropy of each particle varies with time in some prescribed manner (depending, for example, on the mode of heat addition).

4. *The Equation of State* for an ideal gas is given by

$$p = \rho RT \qquad \text{(III.c.3)}$$

Since the state of the gas will mostly be expressed in terms of entropy and the speed of sound, alternative forms of the equation of state are required. The speed of sound is given by

$$a^2 = \gamma \frac{p}{\rho} = \gamma RT \qquad \text{(III.c.4)}$$

and the relation between entropy, pressure, and temperature is given by

$$s - s_1 = c_p \ln \frac{T}{T_1} - \frac{R}{gJ} \ln \frac{p}{p_1} \tag{III.c.5}$$

where subscript 1 indicates some state from which entropy changes are measured. Note that Eq. (III.c.5) implies constant specific heats since it is derived by integrating the second law of thermodynamics on the basis of this assumption (see, for instance, reference 85). Eqs. (III.c.3 to 5) may be combined to yield any needed relation between state variables. In terms of the nondimensional variables defined in the preceding chapter, the following two relations will be frequently required

$$\frac{p}{p_1} = \left(\frac{\mathcal{A}}{\mathcal{A}_1}\right)^{\frac{2\gamma}{\gamma-1}} e^{-\gamma(\mathcal{S}-\mathcal{S}_1)} \tag{III.c.6}$$

$$\frac{\rho}{\rho_1} = \left(\frac{\mathcal{A}}{\mathcal{A}_1}\right)^{\frac{2}{\gamma-1}} e^{-\gamma(\mathcal{S}-\mathcal{S}_1)} \tag{III.c.7}$$

If state 1 is chosen as the reference state, these relations are somewhat simplified because then $\mathcal{A}_1 = \mathcal{A}_0 = 1.0$ and $\mathcal{S}_1 = \mathcal{S}_0 = 0$ (see Eqs. (III.b.1)).

III.d. Derivation of the Characteristics Relations

The continuity and momentum equations (Eqs. III.c.1 and 2) may be written in the form

$$\frac{\partial \ln \rho}{\partial t} + u \frac{\partial \ln \rho}{\partial x} + \frac{\partial u}{\partial x} = -u \frac{\partial \ln A}{\partial x} - \frac{\partial \ln A}{\partial t} - \frac{a^2}{\gamma p A} \psi \tag{III.d.1}$$

$$\frac{\partial u}{\partial t} + u \frac{\partial u}{\partial x} = -\frac{a^2}{\gamma} \frac{\partial \ln p}{\partial x} + f \tag{III.d.2}$$

where use has been made of Eq. (III.c.4). The derivatives of

p and ρ in these equations may be expressed in terms of derivatives of a and s, since (see the preceding chapter)

$$d \ln p = \frac{2\gamma}{\gamma - 1} d \ln a - \frac{gJ}{R} ds = \frac{2\gamma}{\gamma - 1} \frac{da}{a} - \frac{gJ}{R} ds$$

$$d \ln \rho = \frac{2}{\gamma - 1} d \ln a - \frac{gJ}{R} ds = \frac{2}{\gamma - 1} \frac{da}{a} - \frac{gJ}{R} ds$$

Substituting this into Eqs. (III.d.1 and 2), one obtains

$$\frac{2}{\gamma - 1} \frac{\partial a}{\partial t} + \frac{2}{\gamma - 1} u \frac{\partial a}{\partial x} + a \frac{\partial u}{\partial x}$$

$$= -au \frac{\partial \ln A}{\partial x} - a \frac{\partial \ln A}{\partial t} + a \frac{gJ}{R} \frac{Ds}{Dt} - \frac{a^3 \psi}{\gamma p A} \qquad \text{(III.d.3)}$$

$$\frac{\partial u}{\partial t} + u \frac{\partial u}{\partial x} + \frac{2}{\gamma - 1} a \frac{\partial a}{\partial x} = a^2 \frac{gJ}{\gamma R} \frac{\partial s}{\partial x} + f \qquad \text{(III.d.4)}$$

If Eq. (III.d.4) is added to and subtracted from Eq. (III.d.3), it is easily verified that the result is given by

$$\frac{\partial}{\partial t}\left(\frac{2}{\gamma - 1} a \pm u\right) + (u \pm a) \frac{\partial}{\partial x}\left(\frac{2}{\gamma - 1} a \pm u\right)$$

$$= -au \frac{\partial \ln A}{\partial x} - a \frac{\partial \ln A}{\partial t} + a \frac{gJ}{R}\left(\frac{Ds}{Dt} \pm \frac{a}{\gamma} \frac{\partial s}{\partial x}\right) \pm f - \frac{a^3 \psi}{\gamma p A} \qquad \text{(III.d.5)}$$

The left-hand side of this equation represents the derivative of the parameters $\frac{2}{\gamma - 1} a \pm u$ in a direction in the x,t-plane such that

$$\frac{dx}{dt} = u \pm a \qquad \text{(III.d.6)}$$

It will be convenient to use special symbols for these derivatives:

$$\left.\begin{aligned}\frac{\delta_+}{\delta t} &= \frac{\partial}{\partial t} + (u + a)\frac{\partial}{\partial x}\\ \frac{\delta_-}{\delta t} &= \frac{\partial}{\partial t} + (u - a)\frac{\partial}{\partial x}\end{aligned}\right\} \tag{III.d.7}$$

With the aid of these and the substantial derivative

$$\frac{D}{Dt} = \frac{\partial}{\partial t} + u\frac{\partial}{\partial x}$$

it is possible to eliminate the $\frac{\partial s}{\partial x}$ terms of Eq. (III.d.5) since

$$a\frac{\partial}{\partial x} = \frac{\delta_+}{\delta t} - \frac{D}{Dt} = -\frac{\delta_-}{\delta t} + \frac{D}{Dt} \tag{III.d.8}$$

One obtains then

$$\begin{aligned}\frac{\delta_\pm}{\delta t}\left(\frac{2}{\gamma - 1}a \pm u\right) = &-au\frac{\partial \ln A}{\partial x} - a\frac{\partial \ln A}{\partial t}\\ &+ a\frac{gJ}{\gamma R}\frac{\delta_\pm s}{\delta t} + (\gamma - 1)a\frac{gJ}{\gamma R}\frac{Ds}{Dt} \pm f - \frac{a^3\psi}{\gamma p A}\end{aligned} \tag{III.d.9}$$

By multiplying these equations with L_0/a_0^2, they may be transformed into a nondimensional form and written in terms of the variables defined by Eqs. (III.b.1, 2, and 3). The nondimensional form of f will be denoted by

$$\mathfrak{F} = \frac{fL_0}{a_0^2} \tag{III.d.10}$$

The pressure in the last term of Eq. (III.d.9) may be expressed in terms of nondimensional speed of sound and of entropy with the aid of Eq. (III.c.6). This leads to a nondimensional form of ψ which is denoted by

$$\Psi = \frac{L_0 a_0 \psi}{\gamma p_0 A} \tag{III.d.11}$$

Eq. (III.d.9) finally becomes

$$\frac{\delta_+ P}{\delta\tau} = -\mathcal{A}\mathcal{U}\frac{\partial \ln A}{\partial\xi} - \mathcal{A}\frac{\partial \ln A}{\partial\tau} + \mathcal{A}\frac{\delta_+ S}{\delta\tau} + (\gamma - 1)\mathcal{A}\frac{DS}{D\tau} + \mathfrak{F} - \Psi\mathcal{A}^{\frac{\gamma-3}{\gamma-1}} e^{\gamma S} \tag{III.d.12}$$

$$\frac{\delta_- Q}{\delta\tau} = -\mathcal{A}\mathcal{U}\frac{\partial \ln A}{\partial\xi} - \mathcal{A}\frac{\partial \ln A}{\partial\tau} + \mathcal{A}\frac{\delta_- S}{\delta\tau} + (\gamma - 1)\mathcal{A}\frac{DS}{D\tau} - \mathfrak{F} - \Psi\mathcal{A}^{\frac{\gamma-3}{\gamma-1}} e^{\gamma S} \tag{III.d.13}$$

Since there are three dependent variables—$\mathcal{A}$, $\mathcal{U}$, and S—a third relation is required to complete the system of equations. This is provided by the entropy conditions which must be prescribed for any problem (see Chapter III.c.3) and can be stated only as a general function F so that

$$\frac{DS}{D\tau} = F(\mathcal{A}, \mathcal{U}, S, \xi, \tau) \tag{III.d.14}$$

Eqs. (III.d.12, 13, and 14) form the basis for a solution of nonsteady-flow problems by means of a step-by-step procedure. These equations indicate how the parameters P, Q, and S vary in the ξ,τ-plane along curves that are given by

$$\left.\begin{aligned} \frac{d\xi}{d\tau} &= \mathcal{U} + \mathcal{A} \quad \text{for } P \\ \frac{d\xi}{d\tau} &= \mathcal{U} - \mathcal{A} \quad \text{for } Q \\ \frac{d\xi}{d\tau} &= \mathcal{U} \qquad\quad \text{for } S \end{aligned}\right\} \tag{III.d.15}$$

In agreement with the terminology used in the theory of partial differential equations, the three types of curves defined above are called characteristics, or characteristic curves, of the fundamental partial differential equations, and the wave diagram is also often referred to as a characteristic diagram.

Eqs. (III.d.12, 13, and 14) are essentially wave equations since they indicate how P, Q, and S, respectively, propagate in the characteristic directions. For this reason, curves of direction $\mathcal{U} + \mathcal{A}$ may also be referred to as P-waves, and curves of direction $\mathcal{U} - \mathcal{A}$ as Q-waves. The third family, the curves of direction $\mathcal{U}$, are, of course, the paths of fluid elements already referred to. Thus, it seems natural to refer to the plot showing propagation of characteristics in the ξ,τ-plane as a wave diagram.

The right-hand sides of Eqs. (III.d.12 and 13) contain a number of terms each of which represents the effect of certain conditions prescribed for a problem: Changes of area, both along the duct and with time, the effects of entropy gradients, body and dissipative forces, and of mass removal along the duct. Only in extremely rare cases will all these effects have to be considered simultaneously so that usually the procedures are somewhat simplified. Of particular importance is the case of isentropic flow in ducts of constant cross section, without any body forces acting on the flow and without any mass removal from the duct. In this case, all terms on the right-hand side of Eqs. (III.d.12 and 13) become zero so that the values of P and Q remain constant along their respective characteristics. For $S =$ const., Eq. (III.d.14) is not required and the problem reduces to one with two dependent variables only. The resulting simplification of the procedures for this case will become clear in Chapter V.

IV

THE WAVE DIAGRAM

IV.a. Step-by-step Procedure for Constructing Wave Diagrams

The system of partial differential equations (III.d.12, 13, and 14) can be solved only by some numerical procedure. Through every point of the wave diagram passes one curve from each of the three families of characteristics. Let these curves be approximated by short pieces of straight lines and let $\Delta_+ P$,

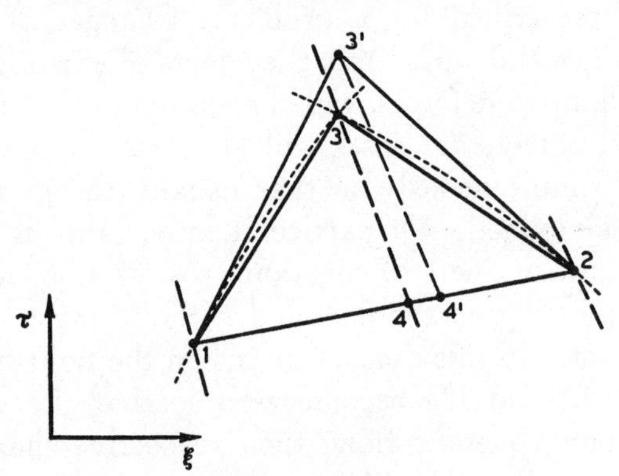

FIG. IV.a.1. Determination of a new point 3 of the wave diagram from known points 1 and 2

$\Delta_- Q$, and ΔS denote small but finite changes of the corresponding variables along their respective characteristics.

In Fig. IV.a.1, let 1 and 2 represent two points for which $\mathfrak{a}$, $\mathfrak{u}$, and S have already been determined by previous calculations. Once these variables are known, the Riemann variables

P and Q are, of course, also known (Eqs. III.b.2), and any other variables that may be of interest are then easily obtained.

The P-wave through 1 and the Q-wave through 2 intersect at a new point 3. The characteristics are indicated by dotted lines, but for the purpose of a step-by-step procedure, these must be approximated by a sequence of short straight sections, the slope of which is then a mean between the values at each end. (See, however, the remarks later in this chapter.) For instance, the correct slope of the P-wave through points 1 and 3 is given by $\mathcal{U}_1 + \mathcal{A}_1$ and $\mathcal{U}_3 + \mathcal{A}_3$, respectively, and the slope of the corresponding straight line section may be approximated by the arithmetic mean

$$(\mathcal{U} + \mathcal{A})_{1,3} = \frac{(\mathcal{U} + \mathcal{A})_1 + (\mathcal{U} + \mathcal{A})_3}{2}$$

As indicated here, double subscripts will be used to denote the mean of the values of some variable at two neighboring points in the wave diagram.

A particle path through 3 is also indicated in the figure (dashed line), and its intersection with a line connecting 1 and 2 determines a point 4. From the flow conditions at points 1 and 2, the corresponding values at 4 can be obtained by interpolation. At point 3, the following conditions must then be satisfied.

$$S_3 = S_4 + \Delta S = S_4 + \left(\frac{DS}{D\tau}\right)_{3,4} \Delta\tau$$

$$= S_4 + \left(\frac{DS}{D\tau}\right)_{3,4} (\tau_3 - \tau_4) \qquad \text{(IV.a.1)}$$

$$P_3 = P_1 + \Delta_+ P = P_1 + \left(\frac{\delta_+ P}{\delta\tau}\right)_{1,3} \Delta_+\tau$$

$$= P_1 + \left(\frac{\delta_+ P}{\delta\tau}\right)_{1,3} (\tau_3 - \tau_1) \qquad \text{(IV.a.2)}$$

$$Q_3 = Q_2 + \Delta_- Q = Q_2 + \left(\frac{\delta_- Q}{\delta\tau}\right)_{2,3} \Delta\tau$$

$$= Q_2 + \left(\frac{\delta_- Q}{\delta\tau}\right)_{2,3} (\tau_3 - \tau_2) \qquad \text{(IV.a.3)}$$

The derivatives in these equations are given by Eqs. (III.d.14, 12, and 13), respectively. From P_3 and Q_3, one obtains, then, $\mathcal{A}_3$ and $\mathcal{U}_3$ simply by solving Eqs. (III.b.2):

$$\left.\begin{aligned} \mathcal{A}_3 &= \frac{\gamma - 1}{4} (P_3 + Q_3) \\ \mathcal{U}_3 &= \tfrac{1}{2} (P_3 - Q_3) \end{aligned}\right\} \qquad \text{(IV.a.4)}$$

As soon as these variables have been calculated, one may go on to the next point in the wave diagram.

Since the mean values indicated in the foregoing relations cannot be found until the flow conditions at 3 have been determined, one must carry out an iteration procedure, using at first some approximations for the location and flow conditions at 3. Since the iteration must be continued until Eqs. (IV.a.1 to 3) are all satisfied, it is not important in what order or how these approximations are selected. Indeed, an "educated" guess can sometimes save a number of iteration steps.

Essentially, one proceeds as follows: First, one obtains an approximate location of point 3 either by guessing, or as the intersection of the tangents to the characteristics at 1 and 2, given by $\mathcal{U}_1 + \mathcal{A}_1$ and $\mathcal{U}_2 - \mathcal{A}_2$, respectively. This point is marked in Fig. IV.a.1 as 3′. Knowing the directions $\mathcal{U}_1$ and $\mathcal{U}_2$, one can then estimate where a particle path through 3′ would intersect the line connecting 1 and 2. The flow conditions at this point 4′ are obtained by interpolation between 1 and 2. One should check whether a particle path through 3′ plotted with a slope corresponding to $\mathcal{U}_{3',4'}$ is in reasonable agreement with the original estimate and, if necessary, obtain a better estimate for 4′. With the aid of such preliminary values as $\tau_{3'}$ and $S_{4'}$, one can calculate $S_{3'}$ from Eq. (IV.a.1). In the next step, $P_{3'}$ and $Q_{3'}$ are computed from Eqs. (IV.a.2 and 3). In these calculations one must use either the values of $\mathfrak{a}$ and $\mathfrak{u}$ at points 1 and 2 or mean values based on estimates for $\mathfrak{a}_{3'}$ and $\mathfrak{u}_{3'}$. Whenever this calculation involves the product of two or more variables, one should, strictly, use the mean values of the complete terms rather than the product of the mean values of the individual factors. For instance, it would be more accurate to use $\left(\mathfrak{a}\mathfrak{u}\,\dfrac{\partial \ln A}{\partial \xi}\right)_{\text{mean}}$ than $\mathfrak{a}_{\text{mean}} \times \mathfrak{u}_{\text{mean}} \times \left(\dfrac{\partial \ln A}{\partial \xi}\right)_{\text{mean}}$. However, the latter procedure is faster and entirely adequate unless the steps between consecutive characteristics are rather large. $P_{3'}$ and $Q_{3'}$ determine new values $\mathfrak{a}_{3''}$ and $\mathfrak{u}_{3''}$. In general, these will not agree with the estimated values, and the entire procedure is then repeated, using the results of the first iteration as approximations. This leads to a second approximation, and the process is repeated until the results do not change any further within the accuracy of the calculations. In this manner, the whole network of characteristics in the ξ,τ-plane may be constructed, provided one can start the diagram from known conditions. It is seen from the foregoing that these initial conditions must be prescribed along a line in the wave diagram that is everywhere different from a characteristic. (If points 1 and 2 would lie on the same characteristics, the method would obviously break down.)

These general procedures of wave diagram construction are quite slow. Fortunately, all of the described steps are often not necessary and, furthermore, considerable simplifications are usually permissible. It must be realized that, in trying to solve a nonsteady-flow problem, certain simplifying assumptions must be made which may make the results uncertain by amounts that often considerably exceed the errors that are introduced by simplified procedures. Frequently, the characteristics are only slightly curved, at least in some part of the wave diagram. Consequently, the mean slope between two consecutive points along a wave differs only insignificantly from the slope at either one of them. Whenever this is the case, a simplification is possible by plotting the wave segments that connect two neighboring points with a slope that corresponds to either one of them. It is immaterial whether the slope at the earlier or later point is used, since this simplification is only permissible when there is little difference between the two slopes. If the steps are sufficiently small, the improvement of accuracy that is obtainable by the iteration process is often meaningless in view of other limitations of wave diagrams. Only the first step as described above need then be carried out, and the values thus obtained are taken as the final ones. In case of doubt, one can always carry out the iteration for a few points and check that the results will not be significantly changed.

The increments of the Riemann variables and entropy along their respective characteristics are made up of a number of terms that depend on the problem under consideration (see Eqs. III.d.12, 13, and 14). These terms will be discussed individually in Chapter V, and it will also be shown that, for the computation of some of them, certain short-cuts may be used. It is only occasionally necessary that several terms must be considered simultaneously. This slows down the work but does not introduce any complications.

In almost all problems, the region in which the procedures may be applied is bounded by the ends of the duct or by discontinuities, such as contact surfaces or shock waves. For points located at the end of the duct, one of the Riemann variables cannot be computed in the preceding manner, and one of the equations (IV.a.2 or 3) is then replaced by an appropriate boundary condition. The latter must also prescribe the entropy level in the case of inflow. Across a discontinuity where the flow variables undergo sudden changes, the fundamental partial differential equations (III.c.1 and 2) do not apply. Such a discontinuity, therefore, divides the wave diagram into two parts. Each must be prepared separately and the two diagrams must be properly matched at their boundary. A great variety of boundary and matching conditions may occur in various problems, and these will be considered in Chapter VI.

IV.b. Types of Pressure Waves

From the manner in which points of the wave diagram are constructed (see Chapter IV.a), it can be seen that the effect of any disturbance of local flow conditions can be felt at other points only after the P- and Q-waves that pass through the origin of the disturbance arrive there. The characteristics are thus signals which carry information about local flow disturbances to other parts of the duct. Actual pressure waves are spread over a certain time and are therefore made up of an entire family of characteristics. The P- and Q-waves are defined also in the absence of actual flow disturbances, i.e., for steady flow. However, if the flow is steady, the Riemann variables remain constant at any point of the duct, and in this trivial case the P- and Q-waves become two families of parallel curves. As a result of flow disturbances, the characteristics will no longer remain parallel but will either diverge or converge.

Consider a steady isentropic flow through a duct of constant cross section. In this case, all flow variables have constant values throughout the wave diagram and the characteristics are straight lines. Suppose that somewhere, say at the left side of the duct, a disturbance is created. As a result, the characteristics coming from the left—the P-waves—will have modified values, whereas the Q-waves are not affected. Along a P-wave, the value of P is then constant as a property of characteristics and Q is constant because of the assumption that no disturbances originate at the right side of the duct. Therefore, $\mathfrak{U}$ and $\mathfrak{A}$ are also constant along the P-wave, and one thus obtains the important conclusion that disturbance waves which travel into an undisturbed region—also called *simple waves*—have characteristics that are straight lines. Note that this conclusion holds only in the case of isentropic flow in a duct of constant cross section.

The Q-waves, on the other hand, cross P-waves for which the value of P varies from wave to wave. Therefore, $\mathfrak{U} - \mathfrak{A}$ is not constant along the Q-waves which become curved lines. However, they remain nonintersecting since they cross a P-wave at constant slope (the value of $\mathfrak{U} - \mathfrak{A}$ remains constant along any P-wave). Entirely analogous results would have been found if the disturbance waves had come from the right instead of the left side of the duct.

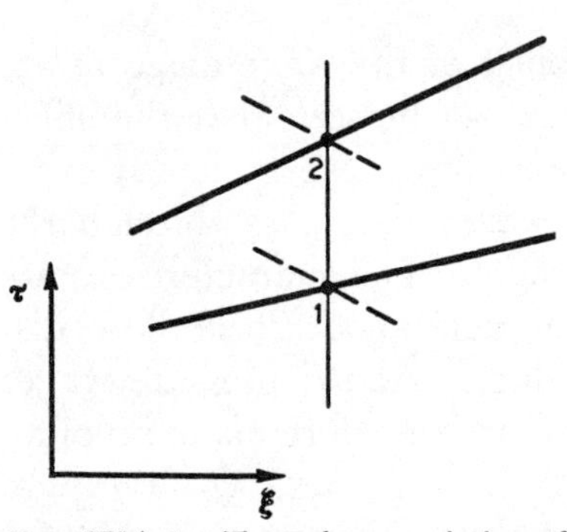

FIG. IV.b.1. Two characteristics of a simple P-wave ($P_1 \neq P_2$, $Q_1 = Q_2$)

A pressure wave moving into a steady-flow region will cause either a decrease or an increase of pressure. One can therefore make a distinction between *expansion* (or *rarefaction*) and *compression waves*. More complex wave forms may be composed of a succession of expansion and compression elements. To see how these two types of waves appear in a wave diagram, consider again isentropic flow in a duct of constant cross section and assume that a disturbance wave is coming from the left side of the duct. In Fig. IV.b.1, let 1 and 2 be two points in the wave diagram located at the same section of the duct. In view of the foregoing, the P-waves through these points are then straight but not parallel lines. Although 1 and 2 lie on different Q-waves (dotted lines), both points have the same value of Q because of the assumption that no disturbance comes from the right side of the duct.

If it is assumed that the disturbance is an expansion wave, the pressure and, therefore, also the speed of sound at 2 are lower than at 1. The following relations then apply:

$$\mathfrak{a}_2 < \mathfrak{a}_1$$

$$\frac{2}{\gamma - 1}\mathfrak{a}_2 - \mathfrak{u}_2 = \frac{2}{\gamma - 1}\mathfrak{a}_1 - \mathfrak{u}_1$$

It follows immediately that $\mathfrak{u}_2 < \mathfrak{u}_1$ and therefore $\mathfrak{u}_2 + \mathfrak{a}_2 < \mathfrak{u}_1 + \mathfrak{a}_1$ and $P_2 < P_1$. Thus, any P-wave propagates with a slower velocity than that of preceding waves or, in other words, an expansion wave is composed of diverging characteristics, and the values of the corresponding Riemann variable decrease.

The same reasoning would show that, in a compression wave, the characteristics converge and the corresponding Riemann variable increases.

In a compression wave, the later wave elements travel in a medium in which the speed of sound is higher than for earlier elements and in which the flow velocity is increased in the direction of wave travel. The later wave elements thus tend to catch up with earlier ones and the wave "steepens." Eventually, the characteristics will meet, and at this point the value of the Riemann variables, and therefore also the pressure, increase discontinuously—a *shock wave* is formed in the flow field. Once the characteristics meet, the method of wave diagram construction as described so far breaks down. A shock wave is no longer a characteristic but becomes a variable boundary in the wave diagram. Particles that pass through a shock wave also undergo an entropy rise which may have to be taken into account. This introduces further complications in the construction of a wave diagram. The handling of problems involving shock waves will be described in Chapter VI.e.

In the case of an expansion wave, the divergence of the characteristics leads to a continuing "flattening out" of the wave.

In this case, no discontinuity appears and the flow remains isentropic. However, if one follows the characteristics of an expansion wave backward, it seems to have originated at a point of discontinuity, namely, where the backward projections of the characteristics meet. This is, of course, not necessarily the actual origin of the wave since diverging characteristics could also have been created by a later disturbance.

In general, the backward projections of all the characteristics that make up an expansion wave do not originate from a

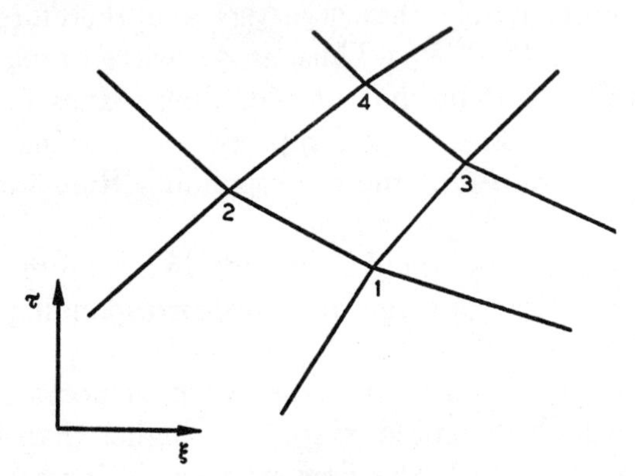

FIG. IV.b.2. Interaction of two pressure waves

single point; if they do, the wave is called a *centered expansion wave*. A wave of this type would, for example, be created by an instantaneous decrease of pressure at some point of the duct. Such conditions may arise as a result of the interaction of discontinuities, such as shock waves, or shocks and contact surfaces (see Chapter VII). There are also instances where the disturbance caused by a rapid pressure drop may be closely approximated by a centered expansion wave (see the examples in Chapter IX).

A *centered compression wave* in which all characteristics converge to a single point represents a special case of shock formation; it has acquired some importance in connection with facilities for short-duration tests at extremely high flow velocities.[122, 127]

Until now, only pressure waves coming from one side of the

duct have been considered. Usually, however, there will be pressure waves coming from both directions, and it is important to study their interaction. In Fig. IV.b.2., let a compression wave coming from the left be indicated by the P-waves through points 1 and 2 that lie on the same Q-wave.

By the same method that was used before in connection with Fig. IV.b.1, one may show that for a compression wave the conditions $P_2 > P_1$ and $\mathcal{U}_2 + \mathcal{A}_2 > \mathcal{U}_1 + \mathcal{A}_1$ are satisfied. Let a second Q-wave intersect the two P-waves at points 3 and 4. From Eqs. (IV.a.4) and from the consideration that the Riemann variables remain constant along their respective characteristics, one obtains immediately

$$\mathcal{U}_3 + \mathcal{A}_3 = \frac{1 + \gamma}{4} P_1 - \frac{3 - \gamma}{4} Q_3$$

and

$$\mathcal{U}_4 + \mathcal{A}_4 = \frac{1 + \gamma}{4} P_2 - \frac{3 - \gamma}{4} Q_3$$

The second terms on the right-hand side are equal and, for $P_2 > P_1$, it follows that $\mathcal{U}_4 + \mathcal{A}_4 > \mathcal{U}_3 + \mathcal{A}_3$. In other words, the converging characteristics in a compression wave remain converging irrespective of the type of the wave crossed.

Quite generally, one may say: If pressure waves are traveling simultaneously in both directions, their property of converging or diverging characteristics is maintained, but their classification as compression or expansion waves must now be based on the behavior of the pressure along a characteristic of the crossing wave and not at a fixed section of the duct. Two crossing pressure waves of similar type will always increase or decrease the pressure at a section of the duct, depending on their type; the result of the crossing of an expansion wave by a compression wave will be a decrease or increase of pressure, depending on which wave is the stronger one. It is thus seen that, in general, there is no unique relation between the wave phenomena and the pressure history at one point of the duct.

The discussion, so far, has been limited to isentropic flows in ducts of constant cross section. In more general flows, neither

P nor Q remains constant along its respective characteristics. Any modification of the P (or Q) values that is caused by a disturbance will also modify the Q (or P) values, an effect which is called *partial wave reflection*. The reflected waves, in turn, are reflected, and so on until a complex flow pattern is built up in a short time.

Another property of the characteristics should be discussed here. Suppose that in a duct of constant cross section a wave travels into a region of steady flow or rest. In this region, the Riemann variables are constant and their derivatives are zero. Within the pressure wave, the variations of P (if it is a P-wave) depend only on the manner in which the wave was created. Thus, when the first characteristic of this wave passes a given station in the duct, the values of $\frac{\partial P}{\partial \tau}$ and $\frac{\partial P}{\partial \xi}$ may change abruptly from zero to some finite value. Since at this station, the derivatives of Q are zero, it follows then that the derivatives of the flow variables also undergo a discontinuous change. Characteristics are thus recognized as lines along which discontinuities of the derivatives of the flow variables can propagate. In the derivation of the fundamental differential equations it is required only that the flow variables be continuous and no restriction has to be placed on their derivatives. It is now seen that solutions of the differential equations are pieced together in the wave diagram whereby discontinuities of the derivatives of the flow variables may appear along the characteristic lines that separate regions where the solution is continuous. This type of discontinuity does not introduce any difficulties into the wave diagram procedures. Whenever discontinuities are discussed in the following chapters, they always refer to sudden changes of the flow variables and not their derivatives. The fundamental differential equations do not apply across such discontinuities and special procedures are required to take care of them (Chapter VI).

V

GENERAL FLOW PROBLEMS

V.a. Isentropic Flows in a Duct of Constant Cross Section

By far the simplest procedures are obtained for the case of isentropic flow in ducts of constant cross section in the absence of body forces and mass removal through the walls. In this case, Eqs. (IV.a.2 and 3) simplify to

$$\left.\begin{aligned} P_3 &= P_1 \\ Q_3 &= Q_2 \end{aligned}\right\} \tag{V.a.1}$$

The solution for point 3 is thus obtained in the simplest possible manner, and wave diagrams may be constructed rather quickly.

For this particular case, an interesting property of the characteristics should be pointed out. Consider the closed loop in a wave diagram that is formed by two P- and two Q-waves (Fig. V.a.1). The Riemann variables then satisfy the con-

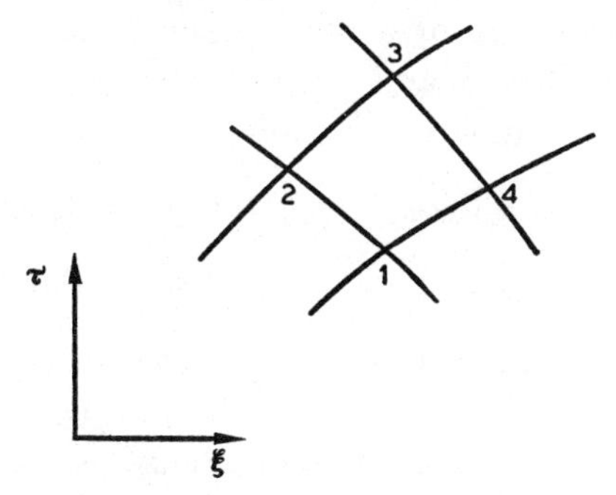

FIG. V.a.1. Closed loop formed by two P- and two Q-waves

ditions

$$P_1 = P_4 \quad \text{or} \quad \frac{2}{\gamma - 1}\,\mathfrak{a}_1 + \mathfrak{u}_1 = \frac{2}{\gamma - 1}\,\mathfrak{a}_4 + \mathfrak{u}_4$$

$$Q_2 = Q_1 \quad \text{or} \quad \frac{2}{\gamma - 1}\,\mathfrak{a}_2 - \mathfrak{u}_2 = \frac{2}{\gamma - 1}\,\mathfrak{a}_1 - \mathfrak{u}_1$$

$$P_3 = P_2 \quad \text{or} \quad \frac{2}{\gamma - 1}\,\mathfrak{a}_3 + \mathfrak{u}_3 = \frac{2}{\gamma - 1}\,\mathfrak{a}_2 + \mathfrak{u}_2$$

$$Q_4 = Q_3 \quad \text{or} \quad \frac{2}{\gamma - 1}\,\mathfrak{a}_4 - \mathfrak{u}_4 = \frac{2}{\gamma - 1}\,\mathfrak{a}_3 - \mathfrak{u}_3$$

The sum of these equations yields

$$\mathfrak{u}_2 - \mathfrak{u}_1 = \mathfrak{u}_3 - \mathfrak{u}_4 \quad \text{or} \quad \mathfrak{u}_4 - \mathfrak{u}_1 = \mathfrak{u}_3 - \mathfrak{u}_2 \qquad \text{(V.a.2)}$$

and, similarly, if the second and fourth equations are subtracted from the sum of the first and third, one obtains

$$\mathfrak{a}_2 - \mathfrak{a}_1 = \mathfrak{a}_3 - \mathfrak{a}_4 \quad \text{or} \quad \mathfrak{a}_4 - \mathfrak{a}_1 = \mathfrak{a}_3 - \mathfrak{a}_2 \qquad \text{(V.a.3)}$$

These results show that, although neither $\mathfrak{u}$ nor $\mathfrak{a}$ is constant along any characteristic, the change of either $\mathfrak{u}$ or $\mathfrak{a}$ from one characteristic to another of the same family remains constant, provided this change is measured between points that lie on a crossing characteristic.

Numerical Example: Assume that Fig. V.a.1 (not drawn to scale) represents part of a wave diagram and that the flow variables for points 2 and 4 have already been determined. Suppose that the numerical values found are

$$\mathfrak{a}_2 = 1.040 \qquad \mathfrak{a}_4 = 1.060$$

$$\mathfrak{u}_2 = 0.200 \qquad \mathfrak{u}_4 = 0.600$$

The Riemann variables that are required in the computations for point 3 are, therefore, $P_2 = 5.400$ and $Q_4 = 4.700$. In an actual wave diagram, these values would, of course, be already known from the preceding calculations. From Eqs. (V.a.1) and Eqs. (IV.a.4), one obtains the solution (using a value $\gamma = 1.4$)

$$\mathcal{A}_3 = (P_3 + Q_3)/10 = 1.010$$

$$\mathcal{U}_3 = (P_3 - Q_3)/2 = 0.350$$

The location of point 3 is found as the intersection of the P-wave through 2 and the Q-wave through 3 drawn with the slopes $\mathcal{U}_2 + \mathcal{A}_2 = 1.240$ and $\mathcal{U}_4 - \mathcal{A}_4 = -0.460$, respectively. If one wishes to use mean velocities, one would use slopes corresponding to $(\mathcal{U} + \mathcal{A})_{2,3} = (1.240 + 1.360)/2 = 1.300$ and $(\mathcal{U} - \mathcal{A})_{3,4} = -(0.460 + 0.660)/2 = -0.560$, respectively. It depends on the accuracy desired whether or not the more accurate location of point 3 obtained by the second method is really required. (See also the example in Chapter IX.b.1 and Fig. IX.b.3.)

V.b. Ducts of Variable Cross Section

1. *Rigid Ducts.* The terms to be considered are now

$$\left.\begin{aligned} \Delta_+ P &= -\mathcal{A}\mathcal{U}\frac{\partial \ln A}{\partial \xi}\Delta_+ \tau \\ \Delta_- Q &= -\mathcal{A}\mathcal{U}\frac{\partial \ln A}{\partial \xi}\Delta_- \tau \end{aligned}\right\} \qquad \text{(V.b.1)}$$

If it is possible to select a duct or approximate the actual duct by one whose shape may be expressed in the form

$$\frac{A}{A_{\xi_1}} = e^{k(\xi - \xi_1)}$$

where k is a constant, then $\partial \ln A/\partial \xi$ = const., and one variable factor is eliminated from the calculations.

The values of $\Delta_\pm \tau$ in Eqs. (V.b.1) must be determined by iteration although, as pointed out in Chapter IV.a, this can be avoided if it is permissible to plot the characteristics with a slope that corresponds to the beginning of the wave segments; the values of $\Delta_{\pm}\tau$ can then be obtained directly from the wave diagram.

By means of an alternative procedure, the increments of the Riemann variables may be computed directly without having first to determine $\Delta_{\pm}\tau$. Considering that the area is not a function of time, one obtains with the aid of Eqs. (III.d.7)

$$\frac{\partial \ln A}{\partial \xi} = \frac{1}{\mathcal{U} + \mathcal{A}} \frac{\delta_{+} \ln A}{\delta \tau} = \frac{1}{\mathcal{U} - \mathcal{A}} \frac{\delta_{-} \ln A}{\delta \tau}$$

The increments of the Riemann variables along their respective characteristics then become

$$\left.\begin{aligned} \Delta_{+}P &= -\frac{\mathcal{A}\mathcal{U}}{\mathcal{U} + \mathcal{A}} \Delta_{+} \ln A \\ \Delta_{-}Q &= -\frac{\mathcal{A}\mathcal{U}}{\mathcal{U} - \mathcal{A}} \Delta_{-} \ln A \end{aligned}\right\} \qquad \text{(V.b.2)}$$

Eqs. V.b.2 become highly inaccurate when the flow velocity becomes near sonic so that either $\mathcal{U} + \mathcal{A}$ or $\mathcal{U} - \mathcal{A}$ approaches zero. The P- and Q-waves then tend to become parallel to the τ-axis, and $\Delta_{+} \ln A$ or $\Delta_{-} \ln A$ also approach zero. In such regions, the calculations should be based on Eqs. (V.b.1) which do not lead to indeterminate forms.

On the basis of Eqs. (V.b.2), one may simplify the procedures still further. The actual duct shape is approximated by a series of constant-area sections that are joined by discontinuous steps. The wave diagram is thus divided into a number of strips parallel to the τ-axis. Each strip is assigned a value of $\ln A$ and, although it is convenient to select constant steps of $\ln A$, this is not necessary. Within each strip, the simple rules for constant area ducts then apply, and the Riemann variables change only when the characteristics enter a new strip. This procedure is quite fast, but the increments of $\ln A$ should be reasonably small. What constitutes "reasonably small" steps is, of course, determined by the demands on the accuracy, but steps of $\Delta_{\pm} \ln A = 0.1$ or 0.2 should, in general, be permissible.

This simplified procedure may be illustrated with reference to Fig. V.b.1. Let the areas of the strips be denoted by A, A', A'', $\cdots$, and let 1 and 2 be two points whose flow variables have already been determined. The next point 3 lies then at

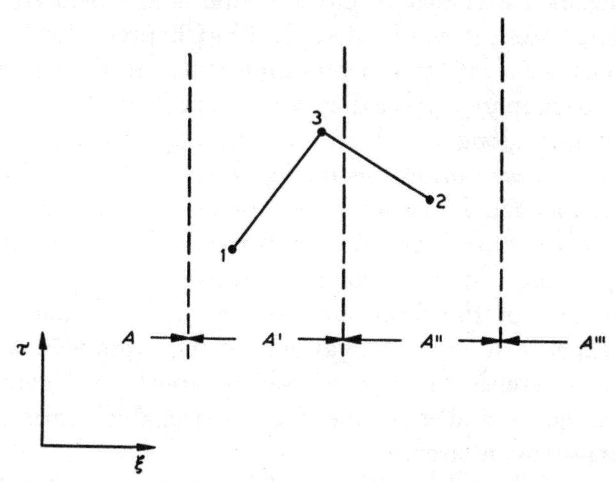

FIG. V.b.1. Duct of gradually varying cross section; the changes are approximated by small discontinuous steps

the intersection of the P-wave through 1 and the Q-wave through 2; assume, for instance, that it lies in the same strip as 1. In view of the foregoing, P_3 and Q_3 are then determined by

$$P_3 = P_1$$

$$Q_3 = Q_2 + \Delta_- Q = Q_2 - \frac{\mathfrak{A}_2 \mathfrak{U}_2}{\mathfrak{U}_2 - \mathfrak{A}_2} (\ln A' - \ln A'')$$

If a point should fall exactly on the boundary line between two strips, it is immaterial within the accuracy of the procedure whether it is considered to be situated on the left or right side of the line as long as the changes of the Riemann variables are computed in a consistent manner.

In the described procedure, no reflected waves are drawn from the points where a wave passes from one strip to the next. If such reflections are to be included in the wave diagram, the problem becomes one of matching the wave diagrams on either side of the discontinuity. This approach is preferable whenever a change of cross section is approximated by a single large step; it will be described in Chapter VI.g.1.

By means of any of the procedures described, it is always possible to prepare a wave diagram irrespective of how fast the area changes along the duct. In the derivation of the procedures, the assumption was made, however, that the flow may be treated as one-dimensional, and this assumption will not be satisfied if the cross section changes too rapidly. The limits of validity of the one-dimensional treatment depend not only on the direction of the flow but also on the wave phenomena. Some experimental investigations of this problem can be found in references 41 and 59 which report good agreement between theory and experiment for "reasonable" variations of the cross-sectional area.

Numerical Example: With reference to Fig. V.b.1 (not drawn to scale) assume that $A''/A' = 1.15$ and, therefore, $\Delta_{\pm} \ln A = \pm \ln A''/A' = \pm 0.14$. Assume further that $\gamma = 1.4$, and that the calculations for points 1 and 2 have already been performed with the results $\mathfrak{A}_1 = 1.060$, $\mathfrak{U}_1 = 0.600$, ($P_1 = 5.900$), and $\mathfrak{A}_2 = 1.040$, $\mathfrak{U}_2 = 0.200$ ($Q_2 = 5.000$). If wave segments are plotted with a slope corresponding to the "early" points (points 1 and 2 in this example), then the location of point 3 is already determined. If, however, one wishes to use mean slopes, one must estimate in which strip point 3 is located. In case of doubt, a guess must be made and its correctness verified after completion of the calculations and plotting of the wave segments. In this example, it appears that point 3 should be located in the same strip as 1, and in view of the preceding data one can thus compute

$$P_3 = P_1 = 5.900$$

$$Q_3 = Q_2 - \frac{\mathfrak{A}_2 \mathfrak{U}_2}{\mathfrak{U}_2 - \mathfrak{A}_2} (\ln A' - \ln A'') = 4.653$$

and, therefore, $\mathfrak{A}_3 = 1.055$ and $\mathfrak{U}_3 = 0.624$.

Spherical and Cylindrical Waves. The procedures of this Chapter may also be used to treat spherical or cylindrical waves for which $\partial \ln A/\partial\xi = n/\xi$, where $n = 1$ for cylindrical and $n = 2$ for spherical symmetry. Then, Eqs. (V.b.1) become

$$\Delta_+ P = -\frac{n}{\xi} \mathfrak{A}\mathfrak{U}\, \Delta_+ \tau$$

$$\Delta_- Q = -\frac{n}{\xi} \mathfrak{A}\mathfrak{U}\, \Delta_- \tau$$

These expressions become indeterminate at the origin where both $\mathfrak{U}$ and ξ are equal to zero. One method to treat this singularity was described by Roberts.[99] Near the center, $\mathfrak{U}$ may be assumed to be proportional to ξ, so that $\partial\mathfrak{U}/\partial\xi = \mathfrak{U}/\xi$; furthermore, the second term on the left-hand side of Eq. (III.d.3) may be neglected. Consequently, this equation takes the form

$$\frac{2}{\gamma - 1} \frac{\partial \mathfrak{A}}{\partial \tau} = -(n+1)\frac{\mathfrak{A}\mathfrak{U}}{\xi}$$

from which $\mathfrak{U}/\xi$ may be substituted into the foregoing expressions for the Riemann variables. Near the center, the equations

$$\Delta_+ P = -\frac{n}{n+1} \frac{2}{\gamma - 1} \frac{\partial \mathfrak{A}}{\partial \tau} \Delta_+ \tau$$

$$\Delta_- Q = -\frac{n}{n+1} \frac{2}{\gamma - 1} \frac{\partial \mathfrak{A}}{\partial \tau} \Delta_- \tau$$

thus should be used.

2. *Duct Area Variable with Time.* Problems of this kind are rare.[125] If the area changes in a manner that the cross section remains uniform over the length of the duct, the terms to be

considered are

$$\left.\begin{aligned} \Delta_+P &= -\mathfrak{a}\,\frac{\partial \ln A}{\partial \tau}\,\Delta_+\tau \\ \Delta_-Q &= -\mathfrak{a}\,\frac{\partial \ln A}{\partial \tau}\,\Delta_-\tau \end{aligned}\right\} \qquad \text{(V.b.3)}$$

It is here again possible to express the time derivative in terms of the derivatives in the characteristic directions (Eqs. III.d.7) since $\partial A/\partial \xi = 0$.

One obtains

$$\left.\begin{aligned} \Delta_+P &= -\mathfrak{a}\Delta_+ \ln A \\ \Delta_-Q &= -\mathfrak{a}\Delta_- \ln A \end{aligned}\right\} \qquad \text{(V.b.4)}$$

The strip method described in Chapter V.b.1 may thus be used again with the only difference that the strips are now parallel to the ξ-axis and the increments of the Riemann variables must be computed from Eqs. V.b.4.

If the changing area is prescribed as a function of both τ and ξ, or implicitly as a function of the flow conditions, such as the pressure, the simple strip method is no longer applicable since the derivatives with respect to τ and ξ cannot be expressed exclusively in terms of the derivative in a characteristic direction. In this case, the terms may be combined to the substantial derivative of ln A, but this will, in general, be of no help in the calculations. The increments of the Riemann variables may be computed from the equations

$$\left.\begin{aligned} \Delta_+P &= -\mathfrak{a}\left[\mathfrak{u}\,\frac{\partial \ln A}{\partial \xi} + \frac{\partial \ln A}{\partial \tau}\right]\Delta_+\tau \\ \Delta_-Q &= -\mathfrak{a}\left[\mathfrak{u}\,\frac{\partial \ln A}{\partial \xi} + \frac{\partial \ln A}{\partial \tau}\right]\Delta_-\tau \end{aligned}\right\} \qquad \text{(V.b.5)}$$

In general, these equations must be evaluated by iteration. It is probably best to estimate the area at a new point in the wave diagram, then calculate the flow parameters at this point in the usual manner, and finally check whether the estimated area is correct for the computed flow conditions. If this is not the case, a better estimate is made and the procedure is then repeated.

An important class of problems of this kind is represented by the pulsating flow of blood through the blood vessels. However, the wave processes then result from the elasticity of the vessel walls, while the fluid may be treated as incompressible, and the cross-sectional area of the vessel replaces the fluid density as one of the dependent variables. Although calculating procedures analogous to those described here may be derived, the actual form of the equations is different. Such equations are given, for instance, in references 126, 132, and 133.

V.c. Flows with Variable Entropy

1. *Multi-isentropic Flows.* These flows are characterized by the condition that the entropy of each gas particle remains constant ($DS/D\tau = 0$), although different particles may have

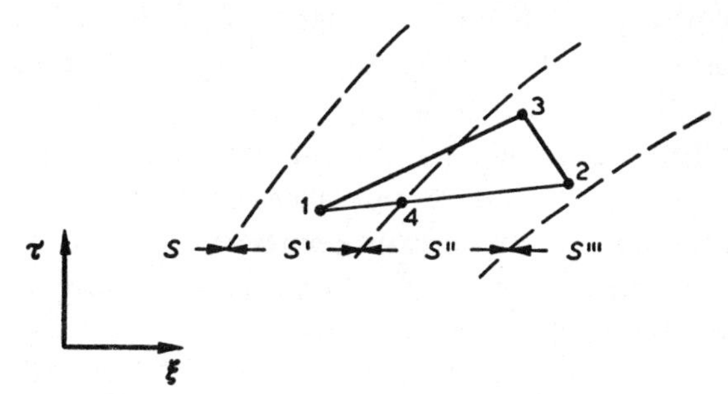

Fig. V.c.1. Multi-isentropic flow; regions of entropy gradients are approximated by a succession of strips of constant entropy

different entropy values. Flows of this type are encountered, for instance, when a compression wave develops into a shock wave that is gradually growing stronger.

The terms controlling the variation of the Riemann variables are

$$\left.\begin{aligned}\Delta_+ P &= \mathfrak{a}\Delta_+ S\\ \Delta_- Q &= \mathfrak{a}\Delta_- S\end{aligned}\right\} \qquad \text{(V.c.1)}$$

These equations can be solved only by means of an iteration procedure. However, unless the entropy gradient is large, the variations in $\mathfrak{a}$ are so small that it is almost always permissible to use the value of $\mathfrak{a}$ at the beginning of the characteristic increment rather than the mean of the values at both ends. The resulting procedure is similar to the one described in Chapter V.b.1. A number of particles is selected, and it is assumed that the entropy has constant values in the strips between them. In Fig. V.c.1. several strips are indicated with entropy values S, S', S'', etc. Again let 1 and 2 be points for which all variables have already been obtained. If the next point 3 is located in the same strip as, say, 2, its Riemann variables are given by

$$P_3 = P_1 + \Delta_+ P = P_1 + \mathcal{A}_{1,3}(S_3 - S_1) \simeq P_1 + \mathcal{A}_1(S_3 - S_1)$$

$$Q_3 = Q_2$$

The only difference between this procedure and the one described in Chapter V.b lies in the fact that now the boundaries of the strips are not fixed but must be constructed as the diagram progresses. This is done by simple interpolation; for example, $\mathcal{U}_4$ is obtained from the known values $\mathcal{U}_1$ and $\mathcal{U}_2$.

The entropy increments from one strip to the next may be constant, but this is not at all necessary. Usually, entropy values are obtained at certain points in the course of wave diagram construction, and it is then easiest to form strips from the particle paths through these points.

2. *Nonisentropic Flows*

2.1. *Flows with Heat Addition.* The manner in which the entropy of the gas elements varies must be prescribed as part of the information required to construct a wave diagram. The terms that control the variations of the Riemann variables are now

$$\begin{aligned} \Delta_+ P &= \mathcal{A}\left[(\gamma - 1)\frac{DS}{D\tau}\Delta_+\tau + \Delta_+ S\right] \\ \Delta_- Q &= \mathcal{A}\left[(\gamma - 1)\frac{DS}{D\tau}\Delta_-\tau + \Delta_- S\right] \end{aligned} \qquad \text{(V.c.2)}$$

where $DS/D\tau$ is a prescribed function of the flow variables and coordinates (Eq. III.d.14). In this case, no short-cuts are possible and Eqs. (V.c.2) must be solved by means of the iteration procedure described in Chapter IV.a. In problems of this kind, the variables often change so rapidly that it is necessary to use the mean values of $\mathfrak{a}$ and, perhaps, also of $DS/D\tau$ rather than the values at either end of a wave segment.

The rate of specific entropy changes of a gas may be expressed in terms of the heat that is added to it as

$$\frac{Ds}{Dt} = \frac{1}{T}\frac{Dq}{Dt}$$

where q is the amount of heat added (measured in Btu/lb). If, in this equation, the temperature is expressed by the speed of sound and nondimensional variables are introduced (Eqs. III.b.1 and 3), one obtains

$$\frac{DS}{D\tau} = \frac{1}{\mathfrak{a}^2}\frac{gJt_0}{a_0^2}\frac{Dq}{Dt} \tag{V.c.3}$$

The last factor of this equation has been left in dimensional form as being more convenient for practical computations.

The most important problems of nonisentropic flows are those concerned with combustion of some fuel-air mixture. If the heating value of the fuel is denoted by H (in Btu/lb), the air/fuel ratio by α, and the combustion efficiency by η_c, the amount of heat released per unit weight of the mixture is given by

$$q = \frac{H\eta_c}{\alpha + 1} \tag{V.c.4}$$

At the present time, the rate of heat release of a burning gas mixture as function of the flow conditions is not well known. Burning proceeds throughout the mixture in a complicated and certainly not one-dimensional manner. It thus becomes necessary to select some one-dimensional model on which the wave diagram procedures can be based. If one follows the particle lines in a wave diagram, each line may be considered as representing the motion of an infinitesimal layer of gas. The rate at which heat is released per unit weight in this layer is given by Dq/Dt, and it is for this quantity that some assumption must be made.

One may assume, for instance, that the rate of heat release is constant while burning is taking place so that

$$\frac{Dq}{Dt} = \frac{q}{t_c}$$

where t_c denotes the combustion time. No matter what assumption is made, one must remember that the total amount of heat released cannot exceed the value of q computed from Eq. (V.c.4). Combustion of any layer must stop after a time interval t_c which is determined by the equation

$$q = \int_{t_i}^{t_i+t_c} \frac{Dq}{Dt}\, dt = \frac{a_0^2}{gJ} \int_{\tau_i}^{\tau_i+\tau_c} \mathfrak{a}^2 \frac{DS}{D\tau}\, d\tau \qquad \text{(V.c.5)}$$

where t_i is the time at which the layer is ignited. Eq. (V.c.5) can be solved for t_c directly if the rate of heat release is prescribed as a function of time only. Otherwise, the integration must be carried out simultaneously with the construction of the wave diagram for as many layers as may be necessary.

The combustion process slightly changes the composition of the gas. However, in view of the simplifying assumptions that must be made for the rate of heat release, one is, in general, quite justified to neglect the resulting small variations of the gas constant and specific heats.

A simplification of the procedures is possible if one assumes that $\partial S/\partial \xi = 0$ within the combustion region. Since then $DS/D\tau = \delta_{\pm} S/\delta\tau$ (see Eqs. (III.d.7)), Eqs. (V.c.2) reduce to

$$\left.\begin{aligned} \Delta_+ P &= \gamma \mathfrak{a} \Delta_+ S \\ \Delta_- Q &= \gamma \mathfrak{a} \Delta_- S \end{aligned}\right\} \qquad \text{(V.c.6)}$$

It is now possible to divide the wave diagram into strips parallel to the ξ-axis and assign a fixed entropy value to each strip. The procedure becomes then quite analogous to that described in Chapters V.b and V.c.1.

Other assumptions that may be made for the combustion process are: Instantaneous combustion either in a whole region or in an advancing flame front. In either case, the problem involves discontinuities and will be treated in Chapters VI.h. and VII.j.

Numerical Example: Consider a combustible air/fuel mixture ($\gamma = 1.4$), burning uniformly in a duct of constant cross section,

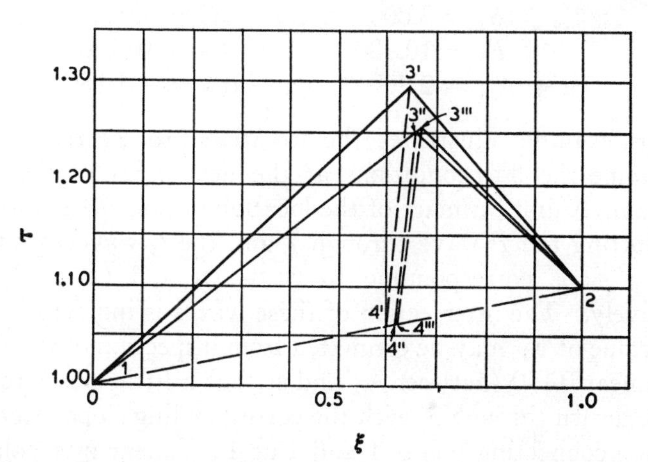

FIG. V.c.2. Determination of a new point 3 from known points 1 and 2, showing steps of iteration for the numerical example

and suppose that $H = 19{,}000$ Btu/lb, $\alpha = 60$, and $\eta_c = 80\%$. If the combustion time is of the order of one millisecond, the average rate of heat release in the mixture is approximately given by (see Eq. V.c.4) $q/t_c = 250{,}000$ Btu/lb sec. If one selects as reference values the initial speed of sound in the mixture, say, $a_0 = 1120$ ft/sec, and $t_0 = 10^{-3}$ sec, the initial rate of entropy rise in terms of nondimensional variables becomes approximately equal to 5 (from Eq. V.c.3). It seems, therefore, reasonable that one might prepare a wave diagram in which $DS/D\tau = 5.0$ during the entire combustion process. This assumption implies that the reduction of the rate of entropy increase due to increasing values of $\mathcal{A}$ is just compensated by a simultaneous increase of the rate of heat release as combustion proceeds (see Eq. V.c.3).

Let 1 and 2 represent two points of the wave diagram (Fig. V.c.2) for which the flow conditions have already been obtained, say,

$$\begin{aligned} \mathcal{A}_1 &= 2.050 & \mathcal{A}_2 &= 2.060 \\ \mathcal{U}_1 &= 0.150 & \mathcal{U}_2 &= 0.180 \\ S_1 &= 3.000 & S_2 &= 3.040 \\ P_1 &= 10.400 & Q_2 &= 10.120 \\ (\mathcal{U} + \mathcal{A})_1 &= 2.200 & (\mathcal{U} - \mathcal{A})_2 &= -1.880 \end{aligned}$$

This example illustrates the iteration process required to determine the flow conditions at the next point of the wave diagram. A first estimate of the location of point 3 is obtained by plotting the P-wave through 1 and the Q-wave through 2 with a slope corresponding to $(\mathcal{U} + \mathcal{A})_1$ and $(\mathcal{U} - \mathcal{A})_2$, respectively. The intersection of these waves is indicated by 3′. The value of $\mathcal{U}_{3'}$ may be estimated from inspection of the figure to be near 0.170 (between $\mathcal{U}_1$ and $\mathcal{U}_2$ but a little closer to $\mathcal{U}_2$). A line drawn through 3′ with the corresponding slope intersects the line connecting points 1 and 2 at 4′. Linear interpolation of $\mathcal{U}_{4'}$ between $\mathcal{U}_1$ and $\mathcal{U}_2$ yields $\mathcal{U}_{4'} = 0.168$, indicating that there is no need to correct the guess of 0.170. The value of $S_{4'}$ can then also be obtained by interpolation between S_1 and S_2; this yields $S_{4'} = 3.024$. One obtains thus from Eq. (IV.a.1)

$$S_{3'} = S_{4'} + \frac{DS}{D\tau}(\tau_{3'} - \tau_{4'})$$

$$= 3.024 + 5.0 \times 0.240 = 4.224$$

where the τ values are taken from Fig. V.c.2.

Once $S_{3'}$ is found, it is possible to compute $P_{3'}$ and $Q_{3'}$ with the aid of Eqs. (V.c.2). For instance, $P_{3'} = P_1 + \mathfrak{a}_1[0.4 \times 5.0 \times (\tau_{3'} - \tau_1) + (S_{3'} - S_1)] = 14.119$. Since $\mathfrak{a}_{3'}$ is not yet known, one can use only $\mathfrak{a}_1$ instead of $\mathfrak{a}_{1,3'}$ in this calculation. Similarly, one obtains $Q_{3'} = 13.362$ and, therefore, $\mathfrak{a}_{3'} = 2.748$ and $\mathfrak{u}_{3'} = 0.378$. These values could be somewhat improved by recalculating $P_{3'}$ and $Q_{3'}$ using $\mathfrak{a}_{1,3'}$ and $\mathfrak{a}_{2,3'}$, respectively. It is better, however, to correct first the location of point 3 by replotting the P- and Q-waves with slopes corresponding to $(\mathfrak{u} + \mathfrak{a})_{1,3'} = 2.663$ and $(\mathfrak{u} - \mathfrak{a})_{2,3'} = -2.125$. Their intersection leads to an improved position $3''$. One can now plot the particle path through $3''$, using the mean velocity $\mathfrak{u}_{3',4'} = 0.273$ which determines point $4''$. A new interpolation yields $\mathfrak{u}_{4''} = 0.169$ and $S_{4''} = 3.025$. With the aid of the latter value, one obtains, similar to before, $S_{3''} = 3.975$. The next step is the calculation of $P_{3''}$ and $Q_{3''}$, using now the values $\mathfrak{a}_{1,3'}$ and $\mathfrak{a}_{2,3'}$, respectively. This yields $P_{3''} = 13.939$, $Q_{3''} = 13.089$, and, therefore, $\mathfrak{a}_{3''} = 2.703$ and $\mathfrak{u}_{3''} = 0.425$.

The entire process is now repeated. First, a further improved location $3'''$ is obtained which, in this case, almost coincides with $3''$. If the particle path is then plotted with a slope corresponding to $\mathfrak{u}_{3'',4''} = 0.297$, the resulting point $4'''$ may be considered identical with $4''$. Since $\tau_{3''} - \tau_{4''}$ is almost exactly equal to $\tau_{3'''} - \tau_{4'''}$, no noticeable change can be found between $S_{3''}$ and $S_{3'''}$. The calculation of $P_{3'''}$ and $Q_{3'''}$, using now $\mathfrak{a}_{1,3''}$ and $\mathfrak{a}_{2,3''}$, respectively, yields $P_{3'''} = 13.920$ and $Q_{3'''} = 13.076$, or $\mathfrak{a}_{3'''} = 2.700$ and $\mathfrak{u}_{3'''} = 0.422$. These values are already so close to those obtained before that one could consider them as final. As a check, one should, however, recompute the Riemann variables with the aid of $\mathfrak{a}_{1,3'''}$ and $\mathfrak{a}_{2,3'''}$, respectively. One obtains thus, finally, $\mathfrak{a}_3 = 2.699$, $\mathfrak{u}_3 = 0.422$, and $S_3 = 3.975$.

An example involving a different assumption about the combustion process is presented in Chapter IX.f.

2.2. *Flows with Wall Friction.* The flow phenomena involving wall friction and boundary layers are highly complex even in steady flow (see, for instance, the brief survey by Prandtl [94]). Consider a steady flow entering a duct of constant cross section without boundary layer so that the flow velocity is uniform over the entire cross section. Following the flow into the duct, one observes then the formation of a boundary layer of gradually increasing thickness in which the flow velocity is reduced to zero at the wall. At first, the effect of the boundary layer on the core of the flow, where the velocity is still uniform, is the same as if the duct were converging. It may require a distance of many duct diameters, depending on the Reynolds number of the flow, before the boundary layer reaches the center of the duct when fully developed pipe flow is established.

A fully developed pipe flow may be considered as quasi-one-dimensional so that the action of wall friction is assumed to be uniformly distributed over the entire cross section. It is seen from Eq. (III.c.2) that in steady, incompressible flow, the friction force per unit mass f is related to the easily observable pressure gradient through the relation

$$f = \frac{1}{\rho}\frac{\partial p}{\partial x}$$

It is customary to express the friction effects by means of a friction coefficient λ defined by

$$f = -\frac{\lambda}{2d}u^2 \qquad \text{(V.c.7)}$$

where d is the diameter of the duct. In compressible flow, the simple relation between f and the pressure gradient no longer applies since heating of the gas due to friction causes an expansion and, therefore, a further pressure drop. However, Eq. (V.c.7) holds also in compressible flow and it has been found that λ is practically independent of the flow Mach number (see, for instance, p. 279 of reference 94).

The nondimensional coefficient λ is a function of the surface conditions in the duct and of the Reynolds number $\mathcal{R} = ud/\nu$ where ν is the kinematic viscosity. A number of relations have been proposed for λ, and for smooth pipes, one may use, for instance,

$$\frac{1}{\lambda^{1/2}} = 2.0 \log_{10} (\mathcal{R}\lambda^{1/2}) - 0.8 \qquad \text{(V.c.8)}$$

for all values of $\mathcal{R}$ in the turbulent region.

So far, this discussion has been concerned with steady flows only. If one considers nonsteady flow, the first question that arises is to what extent the value of λ is the same in steady and in nonsteady flow. According to Schultz-Grunow,[80] the steady-flow values of λ are good approximations as long as no flow separation occurs. In steady flow, separation and even reversed flow near the walls may occur if deceleration (due to widening of the duct) becomes too large. Similar effects may appear in nonsteady flow in a constant-area duct due to deceleration in time instead of space. This may occur quite frequently in nonsteady-flow devices, particularly when the flow is brought to rest or completely reversed.

From this brief review, it becomes apparent that friction effects in nonsteady flow depend not only on the instantaneous flow conditions but also on the history of the flow. Apparently, no work has been done so far in which the history of the flow is taken into account. As far as the initial growth of the boundary layer is concerned, investigations seem to have been restricted to the problem of decay of shock waves in long ducts of constant cross section (see, for instance, references 62, 88, and 102).

For the case of fully developed pipe flow, several investigators [38, 42, 44, 47] have published procedures for the inclusion of wall friction in a wave diagram. In view of the uncertainty of the friction coefficient, constant values of λ were used.

It should be pointed out that, in most practical applications, the friction losses are quite small and relatively large errors in their evaluation would be difficult to detect. This indicates, however, that a wave diagram in which friction effects are included, would, in general, not yield any results that are significantly different from those obtained when friction effects were neglected (at a considerable saving of time). On the other hand, in those rare cases in which friction effects may be of real importance, the existing methods are almost certainly inadequate. Until further studies become available, the inclusion of friction effects in a wave diagram will, therefore, only rarely be profitable. However, for the sake of completeness, a brief outline of procedures that are consistent with those described in other chapters are presented in the following. In view of what was stated above, they apply strictly only if one deals with fully developed pipe flow without flow separation.

The nondimensional form of Eq. (V.c.7) becomes

$$\mathfrak{F} = -\frac{\lambda L_0}{2d}\mathfrak{U}^2\frac{\mathfrak{U}}{|\mathfrak{U}|} \tag{V.c.9}$$

according to Eq. (III.d.10) where the last factor is here introduced to provide the correct sign for $\mathfrak{F}$ regardless of the direction of the flow.

The work done by friction per unit mass and unit time is given by $-fu$. In the absence of heat transfer through the walls of the duct, this power is converted into heat that is added to the flow. The rate at which the specific entropy of a gas particle increases as a result of friction is, therefore, given by

$$\frac{Ds}{Dt} = \frac{1}{T}\frac{Dq}{Dt} = -\frac{fu}{gJT} = -\frac{fu\gamma R}{a^2 gJ}$$

In nondimensional form, this relation becomes

$$\frac{DS}{D\tau} = -\frac{\mathfrak{U}\mathfrak{F}}{\mathfrak{A}^2} \tag{V.c.10}$$

From Eqs. (III.d.12 and 13), it is seen that the changes of the

Riemann variables are now given by

$$\Delta_+ P = \left[\mathfrak{F} + (\gamma - 1)\mathfrak{a}\frac{DS}{D\tau}\right]\Delta_+\tau + \mathfrak{a}\Delta_+ S$$

$$\Delta_- Q = -\left[\mathfrak{F} - (\gamma - 1)\mathfrak{a}\frac{DS}{D\tau}\right]\Delta_-\tau + \mathfrak{a}\Delta_- S$$

Substituting Eq. (V.c.10), one obtains the final relations

$$\left.\begin{aligned} \Delta_+ P &= \mathfrak{F}\left[1 - (\gamma - 1)\frac{\mathfrak{u}}{\mathfrak{a}}\right]\Delta_+\tau + \mathfrak{a}\Delta_+ S \\ \Delta_- Q &= -\mathfrak{F}\left[1 + (\gamma - 1)\frac{\mathfrak{u}}{\mathfrak{a}}\right]\Delta_-\tau + \mathfrak{a}\Delta_- S \end{aligned}\right\} \quad \text{(V.c.11)}$$

where $\mathfrak{F}$ must be computed from Eq. (V.c.9).

V.d. Flows with Body Forces

1. *Flows in Gravitational Fields.* The effects of gravity on a flow are usually negligible. It is only in cases of high accelerating forces or extremely large dimensions that the acceleration f must be taken into consideration. The value of f may be either constant or a prescribed function of x and t. The nondimensional form $\mathfrak{F}$ is defined by Eq. (III.d.10), and the increments of the Riemann variables in this case are given simply by

$$\left.\begin{aligned} \Delta_+ P &= \mathfrak{F}\Delta_+\tau \\ \Delta_- Q &= -\mathfrak{F}\Delta_-\tau \end{aligned}\right\} \quad \text{(V.d.1)}$$

These equations may be solved either by iteration or, if the characteristic net is fine enough, by means of the simplified procedure where all characteristics are plotted with a slope corresponding to the beginning point of each wave segment.

Numerical Example: Assume that the question arises whether or not to consider acceleration effects in a given problem. In order to be significant, such effects should modify the Riemann variables by at least 0.001 (about 0.02%) at each step of the wave diagram calculations. One might thus write Eqs. (V.d.1) in the form

$$|\mathfrak{F}\Delta_{\pm}\tau| = \left|\frac{f\Delta_{\pm}t}{a_0}\right| \simeq 0.001$$

If the phenomena take place in a duct that is several feet long, a wave takes a few milliseconds to travel the entire length since the speed of sound is of the order of 1000 ft/sec. Each step in the wave diagram would, therefore, correspond to only a fraction of a millisecond. For instance, if one assumes time increments of 10^{-4} second, f must be at least of the order of 10^4 ft/sec^2 (or over 300 g) to produce noticeable effects.

2. *Flows in Centrifugal Fields.* Flow problems of this kind may become of interest, for example, in the study of gas flows through helicopter blades or in connection with whirling arm tests.

For the construction of a wave diagram one must know the centrifugal acceleration f that is acting at any point of the duct. In a completely general case, the duct would be revolving around a skew axis with the angular velocity ω. Let β be the angle between the two directions (see Fig. V.d.1) and let the x-coordinate along the duct be measured from point A which is nearest to the axis of rotation. The centrifugal acceleration at a point B is then given by $\omega^2 r$ where r is the distance of B from the axis of rotation. As far as the flow in the duct is concerned, only the component in the direction of the duct is important. If σ denotes the angle between the directions of the duct and the centrifugal acceleration, one obtains $f = \omega^2 r \cos\sigma$, where $r\cos\sigma$ must now be expressed in terms of known parameters. It is immediately seen from Fig. V.d.1 that

$$r = \frac{AC}{\cos\epsilon} = \frac{x\sin\beta}{\cos\epsilon}$$

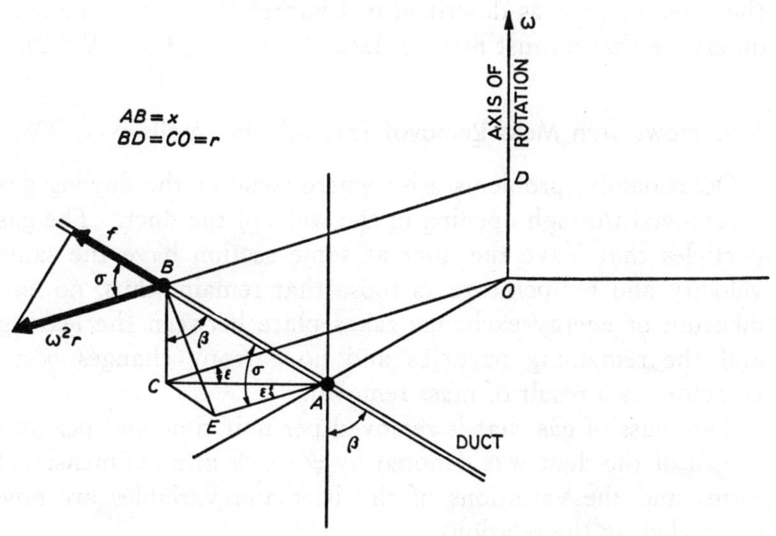

FIG. V.d.1. Isometric view of a duct revolving around a skew axis

If point B is projected on a line AE that is parallel to BD, one also sees that

$$\cos\sigma = \frac{AE}{AB} = \frac{AC\cos\epsilon}{x} = \sin\beta\cos\epsilon$$

Finally, the required component of the centrifugal acceleration is thus given by

$$f = \omega^2 x \sin^2\beta$$

In nondimensional form, this becomes (see Eq. (III.d.10))

$$\mathfrak{F} = \left(\frac{\omega L_0 \sin\beta}{a_0}\right)^2 \xi \tag{V.d.2}$$

It is interesting to note that f is independent of the distance between the duct and the axis of rotation.

The variation of the Riemann variables is again given by Eqs. (V.d.1), and the wave diagram construction proceeds in the same manner as described in Chapter V.d.1 with the only difference that $\mathfrak{F}$ must now be determined from Eqs. (V.d.2).

V.e. Flows with Mass Removal Through the Walls

Occasionally, problems arise where some of the flowing gas is removed through opening in the walls of the duct. The gas particles that leave the duct at some section have the same velocity and temperature as those that remain; thus, no momentum or energy exchange takes place between the leaving and the remaining particles and no entropy changes occur therefore as a result of mass removal.

The mass of gas that is removed per unit time and per unit length of the duct was denoted by ψ (or Ψ in nondimensional form) and the variations of the Riemann variables are now controlled by the relations

$$\left.\begin{aligned} \Delta_+ P &= -\Psi \mathfrak{A}^{\frac{\gamma-3}{\gamma-1}} e^{\gamma S} \Delta_+ \tau \\ \Delta_- Q &= -\Psi \mathfrak{A}^{\frac{\gamma-3}{\gamma-1}} e^{\gamma S} \Delta_- \tau \end{aligned}\right\} \qquad \text{(V.e.1)}$$

The value of ψ may be prescribed not only as a function of x but also as a function of the flow conditions, particularly pressure. Eqs. (V.e.1) may be solved either by iteration or by means of the simplified procedure where all characteristics are plotted with a slope corresponding to the beginning point of each wave segment.

The case of mass addition to the flow ($\psi < 0$) introduces difficulties that are not found when only mass removal is considered. Since it is hardly feasible to inject gas into a nonsteady flow in such a manner that the injected particles have the same velocity and temperature as the instantaneous values of the main flow, mass addition would always be associated with mixing phenomena. No wave diagram procedures have been worked out for this case since mixing of nonsteady flows is as yet too poorly understood.

VI

BOUNDARIES AND DISCONTINUITIES

VI.a. General Methods

It has already been pointed out (Chapter IV.a) that the wave diagram procedures described in Chapters III, IV, and V may be applied only within regions that are bounded either by the ends of the duct or by discontinuities.

The ends of a duct can only be reached by either P- or Q-waves, depending on whether the right or the left end of the

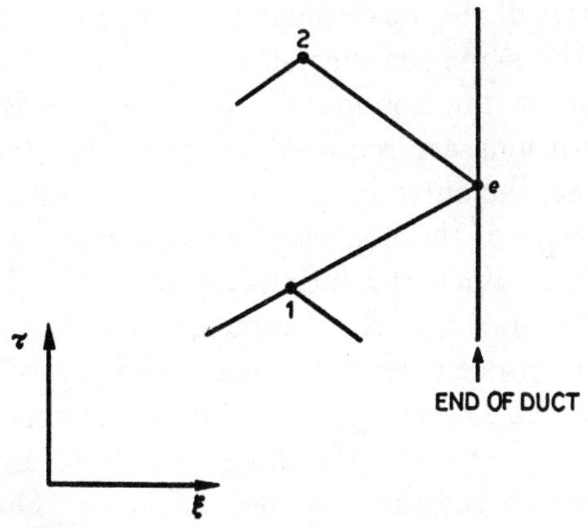

FIG. VI.a.1. Reflection of a characteristic from an end of the duct

duct, respectively, is being considered. In Fig. VI.a.1 which represents the conditions at the right end of a duct, let 1 denote a point for which all variables have already been determined. The P-wave through 1 reaches the end of the duct at a point e

and, in order to compute the flow variables for this point ($\mathfrak{U}_e$, $\mathfrak{A}_e$, and S_e), one requires three conditions:

One relation is provided by the property of the Riemann variables: For points located at the end of the duct, one may always compute

$$\left.\begin{aligned} P_e &= P_1 + \Delta_+ P \\ &= \frac{2}{\gamma - 1}\,\mathfrak{A}_e + \mathfrak{U}_e \quad \text{for the right end, or} \\ Q_e &= Q_1 + \Delta_- Q \\ &= \frac{2}{\gamma - 1}\,\mathfrak{A}_e - \mathfrak{U}_e \quad \text{for the left end.} \end{aligned}\right\} \qquad \text{(VI.a.1)}$$

A second equation is provided by a mathematical formulation of the flow conditions at the end of the duct. In general, this takes the form of some relation between $\mathfrak{U}_e$, $\mathfrak{A}_e$ and, possibly, other parameters.

If the gas is leaving the duct at point e, the particle path through e is part of the wave diagram and S_e can, therefore, be computed in the same manner as the entropy of other points that do not lie on the boundary. In other words, no further boundary conditions are required for outflow. In the case of inflow, however, the entropy level of the incoming gas must be prescribed as part of the boundary conditions.

There are thus sufficient relations to obtain the flow variables at the end of the duct. If the computation of S_e and P_e (or Q_e) by means of the procedures described in the preceding chapters should require the knowledge of the flow variables at point e, one must use an iteration procedure, but the handling of the boundary problem is otherwise not affected. Once the flow variables at e are determined, one may proceed to the next point (2 in Fig. VI.a.1) by means of the ordinary wave diagram procedures.

Whether or not waves propagate beyond the end of a duct is of no concern here since it is fully implied in the boundary conditions. However, in the case of discontinuities, one is interested in the flow and wave phenomena on both sides of the discontinuity surface. In effect, one prepares two wave diagrams, and the flow conditions at their junction must be matched in such a manner that they are compatible with the properties of the discontinuity surface. For any point, one must determine three flow variables on either side of the discontinuity surface, or six unknowns in all. Two of the required six relations are always provided by the known Riemann variables P_L and Q_R (analogous to the Eqs. (VI.a.1)), where subscripts L and R are used to indicate whether a variable refers to the left or right side of the discontinuity, respectively. Furthermore, the entropy level is known at least on one side of the discontinuity. The remaining three relations represent the fundamental continuity, energy, and momentum equations, and depending on the type of the discontinuity these take various forms. The variety of possible procedures will become evident in the following chapters.

VI.b. Closed End of a Duct

The boundary condition at a closed end of a duct is simply expressed by the fact that the flow velocity must be zero. Thus, the gas layer at the end of the duct always remains there, and its entropy level is governed by the conditions that must already be prescribed for points in the interior of the duct.

With the boundary condition $\mathcal{U}_e = 0$, only $\mathcal{A}_e$ remains to be computed. This is done with the aid of one of the Eqs. (VI.a.1), from which one obtains

$$\left.\begin{aligned} \mathcal{A}_e &= \frac{\gamma - 1}{2} P_e \quad \text{for the right end, or} \\ \mathcal{A}_e &= \frac{\gamma - 1}{2} Q_e \quad \text{for the left end.} \end{aligned}\right\} \qquad \text{(VI.b.1)}$$

From Eqs. (IV.a.4), it follows that the Riemann variables for the reflected waves equal those of the arriving waves, or

$P_e = Q_e$ for either end of the duct. Therefore, if P_e increases with time, Q_e also increases, and vice versa, which shows that the reflected waves are of the same type—compression or expansion waves—as the arriving waves (see Chapter IV.b).

These procedures are extremely simple; they are illustrated, for instance, in Example IX.b.1, in the calculation of points 1, 7, 13, 19, etc.

VI.c. Duct Closed by a Moving Piston

This case is very similar to the one treated in the preceding chapter. The piston velocity will be denoted by v (in non-dimensional form, $\mathcal{V} = v/a_0$). The gas layer that is in contact with the piston initially remains in contact with it so that its velocity $\mathcal{U}_e$ is always given by $\mathcal{V}$ (see, however, the remarks about the escape velocity below). Since its entropy level is determined by the conditions prescribed for the interior of the duct, the speed of sound may be computed by substituting $\mathcal{V}$ for $\mathcal{U}_e$ into one of the Eqs. (VI.a.1). This leads to

$$\left.\begin{aligned} \mathcal{A}_e &= \frac{\gamma - 1}{2}(P_e - \mathcal{V}) \quad \text{for the right end, or} \\ \mathcal{A}_e &= \frac{\gamma - 1}{2}(Q_e + \mathcal{V}) \quad \text{for the left end.} \end{aligned}\right\} \qquad \text{(VI.c.1)}$$

Obviously, for $\mathcal{V} = 0$, one obtains the equations that apply in the case of a closed end.

The value of the Riemann variables after reflection is given by

$$\left.\begin{aligned} Q_e &= \frac{2}{\gamma - 1}\mathcal{A}_e - \mathcal{V} \\ &= P_e - 2\mathcal{V} \quad \text{for the right end, or} \\ P_e &= \frac{2}{\gamma - 1}\mathcal{A}_e + \mathcal{V} \\ &= Q_e + 2\mathcal{V} \quad \text{for the left end.} \end{aligned}\right\} \qquad \text{(VI.c.2)}$$

Since the speed of sound cannot be negative, there exists a limiting velocity, $\mathcal{V}_{max}$ for which $\mathcal{A}_e$ and, therefore, also the pressure and density become zero. $\mathcal{V}_{max}$ is the greatest velocity at which the gas can still remain in contact with the piston, whereas at still greater velocities the piston separates from the gas and then ceases to have any effect on the flow phenomena. This "escape velocity" is easily calculated. If the gas is initially at rest, the Riemann variables for this state are given by $P_0 = Q_0 = \dfrac{2}{\gamma - 1}\, \mathcal{A}_0$. If the piston moves away from the gas, it produces an expansion wave traveling in the opposite direction. It has already been shown that expansion waves produce isentropic changes of state (Chapter IV.b), and the Riemann variables of the characteristics that reach the piston are thus also given by $\dfrac{2}{\gamma - 1}\, \mathcal{A}_0$. If this value is substituted in one of the Eqs. (VI.c.1), one obtains for the escape velocity ($\mathcal{A}_e = 0$)

$$|\mathcal{V}_{max}| = \frac{2}{\gamma - 1}\, \mathcal{A}_0 \qquad \text{(VI.c.3)}$$

For air at room temperature, this corresponds to a velocity of the order of 5000 ft/sec.

The piston motion may be given as part of the information required to solve a problem. In this case, the path of the piston can be plotted in the ξ,τ-plane before starting the construction of the wave diagram. Whenever, in the course of the computations, a characteristic terminates at the piston, the required boundary condition is then given by the piston velocity at the particular instant. In a different group of problems, the motion of the piston is not explicitly prescribed, but the latter moves under the influence of the forces that act on it. The resultant of these forces is primarily given by the difference of the pressures on both sides of the piston multiplied by the piston area, but other forces such as friction may also have to be considered. The piston acceleration is thus obtained when the resultant force is divided by the mass of the piston. The path of the piston in the wave diagram must now be obtained from an integration of the acceleration, and since the piston motion, in turn, affects the wave phenomena in the duct (on both sides of the piston!), this integration can only proceed simultaneously with the construction of the wave diagram.

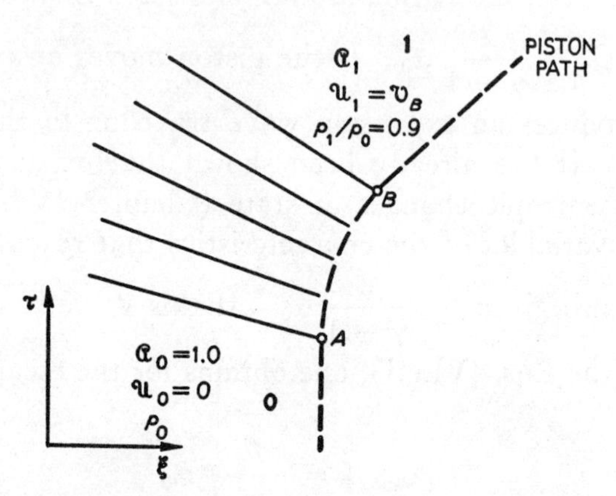

FIG. VI.c.1. Expansion wave created behind an accelerated piston

Numerical Example: A piston in a duct starting from rest accelerates until it reaches such a velocity that the pressure in the gas behind it is reduced by 10 percent. What piston velocity must be reached if the gas is (a) air and (b) helium, and if the initial gas temperature is 530°R?

The wave diagram corresponding to this problem is sketched in Fig. VI.c.1, and it is convenient to select the initial conditions as reference conditions. The heavy dashed line represents the piston path and indicates that the motion starts at point *A* and the final velocity is reached at *B*. Several characteristics of the expansion wave that is created by the piston motion are also shown. The first significant characteristic is the *Q*-wave through *A* which travels with the velocity

$(\mathfrak{U} - \mathfrak{A})_0 = -1.0$. The following points along the piston path are all characterized by the conditions that $P = P_0$ and that $\mathfrak{U}$ is equal to the instantaneous piston velocity. Only the final point B is of interest here, and the corresponding Q-wave travels with the velocity $(\mathfrak{U} - \mathfrak{A})_1$ where $\mathfrak{U}_1 = \mathfrak{V}_B$ is the required final piston velocity. One has, therefore, the relation

$$P_1 = \frac{2}{\gamma - 1} \mathfrak{A}_1 + \mathfrak{V}_B = P_0 = \frac{2}{\gamma - 1}$$

The problem statement and the condition that changes of state across the expansion wave are isentropic combine to

$$\frac{p_1}{p_0} = \mathfrak{A}_1^{\frac{2\gamma}{\gamma-1}} = 0.9$$

The actual value of the final piston velocity is, therefore, obtained as

$$v_B = \frac{2}{\gamma - 1} (1 - 0.9^{\frac{\gamma-1}{2\gamma}}) a_0$$

For air of the prescribed temperature, the speed of sound is given by $a_0 = 1128$ ft/sec; for helium, it is 2.93 times greater (see Table of Properties of Some Important Gases in Chapter XI). The answer to the stated problem is, therefore, that the piston must reach a velocity of 85 ft/sec in air or 208 ft/sec in helium in order to reduce the pressure in the gas behind it by 10 percent.

VI.d. Open End of a Duct

1. *General Considerations.* When pressure waves reach an open end of a duct they are partly transmitted and partly reflected. Thus, a nonsteady external flow field is created and the boundary conditions should be represented by the fact that the internal and external flows must be identical at their junction, that is, at the open end section. It would be ex-

tremely difficult to analyze in detail the three-dimensional wave phenomena that occur in the vicinity of the exit. This difficulty is avoided by making some assumptions which allow wave reflections to be determined without detailed knowledge of the external flow patterns.

The wave phenomena near the exit tend to produce a pressure there that would establish itself if the flow remained steady. How fast this equilibrium pressure is approached depends directly on the dimensions of the duct cross section. If the time required by waves to travel across the duct section is sufficiently small compared to the time in which the flow conditions are significantly changed by the arriving waves, one may consider the flow at the exit as quasi-steady. Customarily, the boundary conditions at an open end are, therefore, approximated by those which apply in steady flow, and the short pressure fluctuations produced at the exit by the arriving pressure waves are neglected. Results obtained from this assumption are in good agreement with experimental observations. Occasionally, however, the delay in establishing the steady-flow boundary conditions does become important. More refined boundary conditions, which should then be used, are described in Chapter VII.e and Appendix II.A.

The use of the steady-flow boundary conditions leads to simple procedures which depend, however, on whether inflow or outflow takes place, and whether the flow is subsonic or supersonic. Furthermore, cases may have to be studied where the gas in the external region is not at rest but in relative motion to the duct. This variety of conditions requires that great care be exercised in selecting the proper boundary conditions in every case. Furthermore, the boundary conditions may change at a time that is not known beforehand, for instance, when the flow direction reverses from outflow to inflow. One should, therefore, always check that the computed flow conditions are compatible with the boundary condition used.

2. *Outflow.* According to the remarks in Chapter VI.a, only one boundary condition is required in the case of outflow. As

long as the flow velocity at the exit remains subsonic, and the gas in the external region is at rest, the pressures at the exit and in the external region are equal in steady flow. With the aid Eq. (III.c.6), the boundary condition $p_e = p_E$ is, therefore, represented by

$$\mathcal{A}_e = \left(\frac{p_E}{p_0}\right)^{\frac{\gamma-1}{2\gamma}} e^{\frac{\gamma-1}{2} S_e} \quad \text{for subsonic outflow.} \qquad \text{(VI.d.1)}$$

The exit velocity follows then directly from Eqs. (VI.a.1).

It should be observed that, in the case of isentropic flow, Eq. (VI.d.1) yields a constant value for $\mathcal{A}_e$ and from Eqs. (IV.a.4), it follows that the sum $P_e + Q_e$ is then also constant. Any variation of P_e is, therefore, associated with a variation of Q_e in the opposite sense, which shows that a compression wave is reflected as an expansion wave, and vice versa.

Since only the pressure of the external region enters into Eq. (VI.d.1), wave reflection is independent of the density or the type of gas that is present there. This may seem surprising, but it is a direct consequence of the use of steady-flow boundary conditions.

Frequently, the state of the gas in the external region is selected as reference state, so that $p_E = p_0$ and $S_E = 0$. If $S_e = 0$ also applies, Eq. (VI.d.1) reduces to $\mathcal{A}_e = 1.0$, a particularly convenient boundary condition.

Both a compression wave that travels toward the exit and an expansion wave that travels in the opposite direction accelerate the flow toward the exit. The described procedure applies as long as the flow remains subsonic. Once the flow velocity at the exit is sonic, the value of $\mathcal{U}_e - \mathcal{A}_e$ at the right end of the duct (or $\mathcal{U}_e + \mathcal{A}_e$ at the left end) becomes zero, so that no expansion wave can travel back into the duct. In other words, although a compression wave can accelerate the flow indefinitely, acceleration by an expansion wave cannot produce higher than sonic velocities.

The condition $p_e = p_E$ need no longer be satisfied for sonic exit velocities and p_e may be larger than p_E; $\mathfrak{a}_e$ is then obtained from Eq. (VI.a.1) and the condition of sonic flow. In this case the boundary condition may be expressed in the form

$$\left.\begin{aligned} \mathfrak{a}_e &= \frac{\gamma - 1}{\gamma + 1} P_e \quad \text{for the right end, and for } \mathfrak{u}_e = \mathfrak{a}_e, \text{ or} \\ \mathfrak{a}_e &= \frac{\gamma - 1}{\gamma + 1} Q_e \quad \text{for the left end, and for } -\mathfrak{u}_e = \mathfrak{a}_e. \end{aligned}\right\} \tag{VI.d.2}$$

As long as the flow is subsonic, the P- and Q-waves travel in opposite directions since $\mathfrak{u} + \mathfrak{a}$ is always positive, and $\mathfrak{u} - \mathfrak{a}$ is always negative. As soon as the flow velocity exceeds the

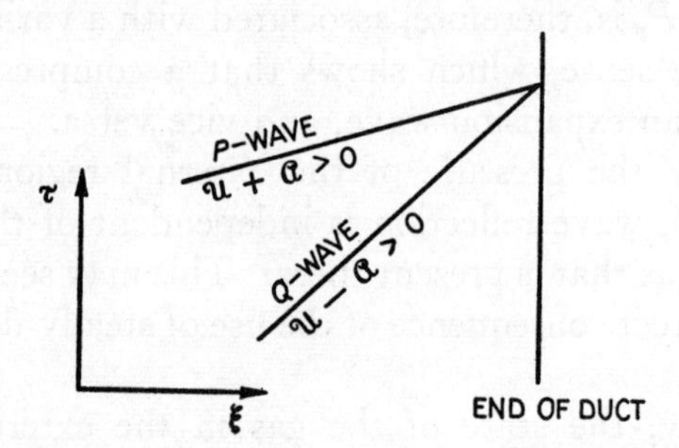

FIG. VI.d.1. Supersonic outflow from a duct

speed of sound, both P- and Q-waves travel in the same direction. It follows that in the case of supersonic outflow, both P- and Q-waves reach the exit, and the flow conditions there can be obtained by means of the standard procedures without recourse to any boundary conditions (see Fig. VI.d.1). This expresses the fact that a supersonic flow cannot be affected by the downstream conditions. The only signals that can travel upstream in a supersonic flow are shock waves of such strength that their propagation velocity is greater than the flow velocity, or the shock Mach number must be greater than the flow Mach number. It is, therefore, possible for the pressure in the external region to be larger than the exit pressure, as long as the Mach number of a shock wave whose pressure ratio is given by p_E/p_e is smaller or, at most, equal to the exit flow Mach number. This is expressed by the condition (see Eq. (VI.e.2))

$$\frac{p_E}{p_e} \leq \frac{2\gamma M_e^2 - \gamma + 1}{\gamma + 1} \quad \text{for supersonic outflow.} \qquad \text{(VI.d.3)}$$

As soon as Eq. (VI.d.3) is no longer satisfied, a shock forms and travels into the duct. Outflow is then, of course, no longer supersonic, and the appropriate boundary condition must be used. The handling of problems that involve shock waves will be discussed in Chapter VI.e.

Numerical examples for the cases discussed here are presented at the end of Chapter VI.d.3.

3. *Isentropic Inflow.* Whenever the flow enters the duct from an external reservoir, it must be accelerated from rest to the entrance velocity. Within the accuracy of the assumption that the flow in the vicinity of the end of the duct can be treated as quasi-steady, the state of the gas in the external region represents the stagnation condition (denoted by subscript s) for the flow in the inlet section of the duct. If the flow is to remain isentropic after it enters the duct, it must further be assumed that no flow separation occurs at the edge of the inlet section. Although this requires that the duct be terminated by a properly shaped short inlet nozzle, the assumption of isentropic inflow is usually also applied to straight open ends. This greatly simplifies wave diagram construction since the entropy variations that are caused by flow separation at the edge of the duct are then neglected. The errors thus introduced become appreciable only if the entropy rise $S_e - S_E$ is greater than about 0.01. Even then, more elaborate procedures (see Chapter VI.d.5) are often not justified in view of other simplifying assumptions that may have to be made.

According to the discussion in Chapter VI.a.1, two boundary conditions are required for inflow. These are provided by the conservation of energy in the flow from the external region to the inlet (assumed to be steady), and by the prescribed entropy level of the gas in the external region. The boundary conditions may thus be expressed by the relations

$$\left.\begin{aligned} \mathcal{U}_e^2 + \frac{2}{\gamma - 1}\,\mathcal{A}_e^2 = \frac{2}{\gamma - 1}\,\mathcal{A}_{e,s}^2 &= \frac{2}{\gamma - 1}\,\mathcal{A}_E^2 \\ S_e &= S_E \end{aligned}\right\} \qquad \text{(VI.d.4)}$$

The first of Eqs. (VI.d.4) together with one of Eqs. (VI.a.1) supplies the two equations which must be solved for $\mathcal{A}_e$ and $\mathcal{U}_e$. The solution of the resulting quadratic equation for $\mathcal{A}_e$ is given by

$$\mathcal{A}_e = \frac{P_e(\text{or } Q_e) + \sqrt{\dfrac{\gamma + 1}{\gamma - 1}\,\mathcal{A}_E^2 - \dfrac{\gamma - 1}{2}\,P_e^2(\text{or } Q_e^2)}}{\dfrac{\gamma + 1}{\gamma - 1}} \qquad \text{(VI.d.5)}$$

where P_e or Q_e must be used, depending on whether wave reflection takes place at the right or left end of the duct, respectively. The value of $\mathcal{U}_e$ follows then from Eqs. (VI.a.1). Of the two roots of the quadratic equation only the one shown can be used since the other root would lead to a sign for $\mathcal{U}_e$ that corresponds to outflow.

In a given problem, $\mathcal{A}_E$ is usually constant and $\mathcal{A}_e$ is thus only a function of the Riemann variable of the waves arriving at the end of the duct. The procedures can, therefore, be considerably speeded up if one prepares an accurate plot of $\mathcal{A}_e$ versus P_e(or Q_e) for the given value of $\mathcal{A}_E$. In order to avoid having to prepare a new chart every time a new value of $\mathcal{A}_E$ must be used, it is preferable to plot $\mathcal{A}_e/\mathcal{A}_E$ versus P_e(or Q_e)$/\mathcal{A}_E$ (see Charts 1a, b, c in Chap. XI). The use of this plot requires only one extra multiplication and one division, and if the state of the gas in the external region is selected as reference state ($\mathcal{a}_E = \mathcal{a}_0 = 1.0$), even this additional computing labor is eliminated.

Flow into the duct may be preceded by outflow, and the outflowing gas may be different or may have a different entropy level from the gas in the external region. Therefore, as soon as the flow reverses, an interface is formed which separates the gas entering the duct from the gas layer that just reached the exit of the duct but did not flow out. The treatment of such interfaces is discussed in Chapter VI.f (see also Example IX.b.3). After flow reversal takes place, some of the gas that has flowed out of the duct may be flowing back, mixed with the gas of the external region. The properties of the inflowing gas may, therefore, not be well defined, particularly at the beginning of inflow. One might allow for this by assuming modified values for the properties of the gas, but no general method of eliminating this uncertainty is available.

Numerical Example: Let the left end of a duct be open to the atmosphere. Suppose waves arrive at the end which are characterized by their Riemann variables $Q_1 = 4.500$, $Q_2 = 5.000$, $Q_3 = 5.750$, $Q_4 = 6.000$, and $Q_5 = 6.500$. Suppose further that the atmospheric conditions had been chosen as reference conditions ($\mathcal{a}_E = \mathcal{a}_0 = 1.0$, $p_E = p_0$) and that the specific entropy of the air in duct is the same as that of the atmosphere ($S = 0$). Determine the flow conditions at the exit of the duct at the times when the prescribed Q-waves arrive there.

In an actual wave diagram, the direction of the flow at the end of the duct is only occasionally in doubt since the trend of the flow conditions is known from preceding work. However, for the purpose of demonstrating the procedures suppose that there are no preconceived notions about the flow conditions. Under such circumstances, it is then best to apply first those conditions which require the simplest procedures, namely, subsonic outflow. The boundary condition for this case becomes $\mathcal{a}_e = 1.000$ (from Eq. (VI.d.1)) and one has, therefore (for $\gamma = 1.4$),

$$\mathcal{U}_e = 5\mathcal{A}_e - Q_e = 5 - Q_e$$

If one substitutes the given Q-values into this relation, one obtains $\mathcal{U}_1 = 0.500$, $\mathcal{U}_2 = 0$, $\mathcal{U}_3 = -0.750$, $\mathcal{U}_4 = -1.000$, and $\mathcal{U}_5 = -1.500$, respectively. It is now necessary to check which of these results satisfy the assumption of subsonic outflow.

Since we are dealing here with the left end of a duct, outflow requires negative values of $\mathcal{U}$. It is thus seen that $\mathcal{U}_1$ cannot be the correct solution since it came out positive. For this case, Chart 1a (or direct evaluation from Eq. (VI.d.5)) yields $\mathcal{A}_1 = 0.983$ from which the velocity of inflow follows as $\mathcal{U}_1 = 5 \times 0.983 - 4.500 = 0.415$. At point 2, the flow reverses direction. As the limiting case of both inflow and outflow, one obtains, of course, the correct answer from either procedure.

At points 3, 4, and 5, the condition of outflow is satisfied, but at 5, the absolute value of the flow velocity is greater than the speed of sound. This again violates the condition for which the calculation was carried out. Here, the flow becomes supersonic and, as pointed out above, no boundary condition is required, but the value of P_5 must be known in order to solve for $\mathcal{A}_5$ and $\mathcal{U}_5$. Suppose $P_5 = 4.250$; one would then obtain $\mathcal{A}_5 = 1.075$ and $\mathcal{U}_5 = -1.125$. In this case, the exit pressure would no longer be atmospheric; one would find instead, $p_5/p_0 = \mathcal{A}_5{}^7 = 1.66$.

At point 4, outflow just becomes sonic so that this represents the transition point where the boundary conditions change from those of subsonic to those of supersonic outflow.

The results of the given example are collected in the table on page 67 in which the velocities and Riemann variables of the reflected waves have also been included.

Inspection of these figures reveals that the given problem represents the case in which flow into a duct is reversed by a compression wave which is so strong that it produces supersonic

RESULTS OF NUMERICAL EXAMPLE

Point	Q	$\mathfrak{a}$	$\mathfrak{u}$	$\mathfrak{u}+\mathfrak{a}$	P
1	4.500	0.983	0.415	1.398	5.330
2	5.000	1.000	0	1.000	5.000
3	5.750	1.000	−0.750	0.250	4.250
4	6.000	1.000	−1.000	0	4.000
5	6.500	1.075	−1.125	−0.050	4.250

outflow. As the flow inside the duct becomes supersonic, the direction of propagation of the P-waves is reversed and they are swept downstream. For instance, the P-wave from point 3 where the flow is still subsonic first enters the duct, but as a result of this interaction with subsequent Q-waves, it is turned around and reaches the exit again at point 5. These phenomena are qualitatively indicated in Fig. VI.d.2.

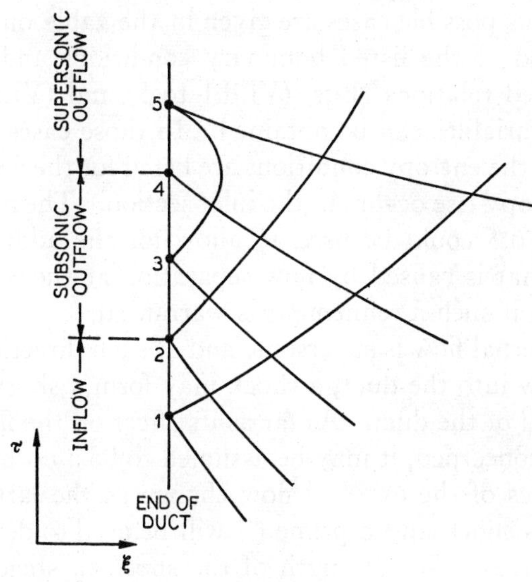

FIG. VI.d.2. Reflection of Q-waves from the left end of a duct (see the numerical example)

The few points discussed here cover a wide range of flow variables. In any actual problem, many more intermediate points would have to be computed, particularly in those regions where the propagation velocities change signs, in order to obtain reliable results.

4. *Effect of External Flows.* The boundary conditions for open ends that were presented in the preceding chapters are based on the assumption that the gas surrounding the end of the duct is at rest (except for disturbances that are produced by the phenomena inside the duct). If the duct is submerged in a steady flow, the boundary conditions must be somewhat modified and, depending on the direction of the internal and external flows relative to the duct, one can recognize essentially four different cases, with further variations depending on whether the external flow is subsonic or supersonic. It must be pointed out, however, that some of the boundary conditions listed below are only approximations even for steady flows.

The various possible cases are given in the table on page 69. With the aid of the listed boundary conditions and the previously stated relations (Eqs. (VI.d.1 to 5) and (VI.a.1)), the other flow variables can be obtained. In those cases that deal with inflow, the entropy conditions are listed for the assumption that no entropy rise occurs in the inlet section. The methods of Chapter VI.d.5 could be used to allow for the additional entropy rise that is caused by flow separation at the edge of the inlet section if such a refinement is warranted.

If the external flow is supersonic and of such direction that it tends to flow into the duct, a shock may form a short distance from the end of the duct. As far as its effect on the flow inside the duct is concerned, it may be assumed to be a normal shock. The variables of the external flow change as the latter passes through this shock and a prime (′) will be used to denote their modified values. The strength of the shock in steady flow is completely determined. In nonsteady flow, the shock moves as a result of pressure waves coming from the duct and its

BOUNDARY CONDITIONS AT AN OPEN END IN THE CASE OF EXTERNAL FLOW

Flow Conditions	M_E	Boundary Conditions	Remarks
	<1.0	$p_e = p_E$	
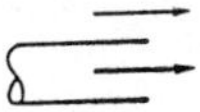	>1.0	?	Too little is known to list any boundary condition for this case.
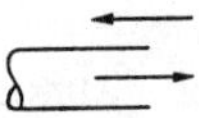	<1.0	$p_e = p_{E,s}$	The external flow must have a stagnation point near the duct exit in order to allow outflow.
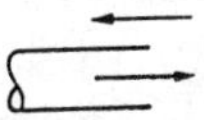	>1.0	$p_e = p_{E,s}'$	The outflowing gas stream is equivalent to a blunt body, requiring a shock to be formed near the duct exit.
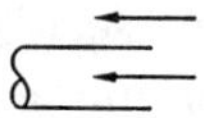	<1.0	$\mathcal{A}_{e,s} = \mathcal{A}_{E,s}$ $S_e = S_E$	
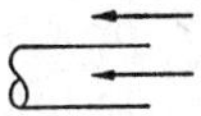	>1.0	$\mathcal{A}_e = \mathcal{A}_E$ $\mathcal{U}_e = \mathcal{U}_E$ $S_e = S_E$ or $\mathcal{A}_{e,s} = \mathcal{A}_{E,s}$ $S_e = S_E'$	*Steady* supersonic inflow (only a shock of sufficient strength could leave the duct but inflow could then no longer be supersonic) A shock forms near the duct inlet and the boundary conditions are then the same as those listed for $M_E < 1.0$ but with the state variables modified by the shock. Which of these two cases applies, follows uniquely from the initial conditions of the problem under investigation and the ensuing wave phenomena.
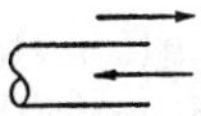	<1.0	$p_{e,s} = p_E(?)$ $S_e = S_E(?)$	These conditions are doubtful and they are offered only as tentative suggestions. For $\mathcal{U}_e = 0$, they lead to the condition $p_e = p_E$ which is in agreement with the first case. In view of the uncertainty of the pressure condition, the assumption of isentropic flow is not likely to increase the error significantly.
	>1.0	?	Too little is known to list any boundary condition for this case.

strength varies accordingly. By using the constant shock strength computed from steady-flow conditions, one may introduce a considerable error into the wave diagram in the case of strong shocks. No methods are as yet available to deal with this situation.

For these and other reasons stated in the table, a number of the listed boundary conditions can only be considered as rough approximations that may be used until more information becomes available.

5. *Nonisentropic Inflow*

5.1. *General Method.* Unless the duct has proper fairing, the inflowing gas stream separates at the edge of the inlet and a flow pattern develops as schematically shown in Fig. VI.d.3. A stationary vortex is formed just inside the inlet section and forces the incoming gas stream to contract (vena contracta) before expanding to fill the entire cross section of the duct. This flow process is not isentropic, and it may become important to make allowance for the entropy rise. For the purpose of wave diagram construction, it will be assumed that stream contraction and subsequent widening occur at the same section of the duct.

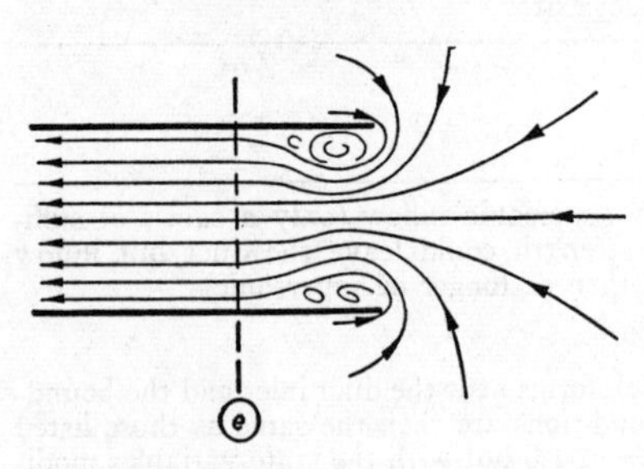

FIG. VI.d.3. Inflow with separation at the inlet section of the duct.

Subscript e will be used to indicate the conditions after the flow fills again the entire cross section. The Riemann variable of a wave arriving at the end of the duct (one of the Eqs. (VI.a.1)) and the conservation of energy in the flow from the external region into the duct (the first of Eqs. (VI.d.4)) provide two relations in which $\mathcal{A}_e$ and $\mathcal{U}_e$ are the only unknowns. These relations are exactly the same as in the case of isentropic inflow (Chapter VI.d.3), and the same method can therefore be used to solve them. This does not mean that a given wave will produce the same inflow velocity regardless of whether the inflow process is treated as isentropic or not. In the nonisentropic flow, the characteristics must pass through a region of variable entropy before they reach the end of the duct, and the resulting modifications of the Riemann variables lead to inflow velocities that are different from those obtained in the isentropic case (see Example IX.b.2). Since the entropy at station e must be known before the Riemann variable of the arriving wave can be computed, while the latter is required to find the flow conditions at e, it is seen that the problem must be solved by iteration: Assume a value for the Riemann variable of the arriving wave, and then compute the corresponding flow conditions at e. The latter determine $\mathcal{S}_e$, which allows an improved value of the Riemann variable to be computed, and so on.

The entropy rise during inflow depends on the configuration of the inlet section and must be computed on the basis of suitable assumptions. Since the boundary conditions are assumed to be the same as those applying in steady flow, one may determine the entropy rise also experimentally from measurements of the drop of stagnation pressure under various flow conditions. The entropy rise $\mathcal{S}_e - \mathcal{S}_E$ is best plotted as function of the speed of sound ratio $\mathcal{A}_e/\mathcal{A}_E$. The latter quantity can be directly obtained from the Riemann variable with the aid of Chart 1 (Chap. XI) so that all the necessary data for the iteration procedure are readily available.

5.2. *Borda Nozzle.* If the walls of the duct are sufficiently thin, one may assume the flow pattern to be the same as if the inlet were sharp-edged (Borda nozzle). In this case, a simple relation may be derived for the entropy rise.

Consider a large pressure reservoir connected to the outside through a Borda nozzle (Fig. VI.d.4). Let p_E and p_e be the pressures inside and outside of this reservoir, respectively. The exhaust jet produces a thrust that is given by $A\rho_e u_e^2$, and this must be equal to the unbalance of the pressures acting on the

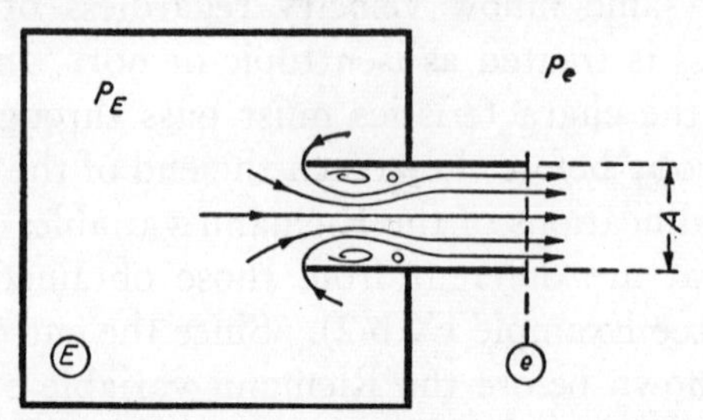

FIG. VI.d.4. Scheme used for deriving entropy rise in a Borda nozzle

walls of the system, which is $A(p_E - p_e)$. One obtains, therefore,

$$p_E = p_e + \rho_e u_e^2 = p_e(1 + \gamma M_e^2)$$

The reservoir temperature and the stagnation temperature in the jet must be equal, or

$$T_E = T_{e,s} = T_e\left(1 + \frac{\gamma - 1}{2} M_e^2\right)$$

The entropy rise is then obtained from Eq. (III.c.5) as

$$s_e - s_E = -c_p \ln\left(1 + \frac{\gamma - 1}{2} M_e^2\right) + \frac{R}{gJ} \ln (1 + \gamma M_e^2)$$

If the entropy is expressed in its nondimensional form, this relation becomes, finally,

$$S_e - S_E = \frac{1}{\gamma} \ln \frac{1 + \gamma M_e^2}{\left(1 + \frac{\gamma - 1}{2} M_e^2\right)^{\frac{\gamma}{\gamma-1}}} \qquad \text{(VI.d.6)}$$

This result can be immediately applied to wave diagrams, where the subscripts e and E have again the previously assigned meanings. For any value of P_e(or Q_e)$/\mathcal{A}_E$, one obtains first $\mathcal{A}_e/\mathcal{A}_E$ from Chart 1; this, together with the known Riemann variable, yields $\mathcal{U}_e/\mathcal{A}_E$ and thus also M_e. The corresponding entropy rise follows from Eq. (VI.d.6). In Charts 2a, b, c (Chapter XI), the values of $S_e - S_E$ are plotted as function of $\mathcal{A}_e/\mathcal{A}_E$ for different values of γ.

A numerical example for flow into a Borda nozzle is presented in Chapter IX.b.2.

VI.e. Shock Waves

1. *General Remarks.** It was pointed out in Chapter IV.b that the merging of the characteristics of one family indicates the formation of a shock wave. Depending on whether the *P*- or the *Q*-waves merge, one may speak of *P*-shocks or *Q*-shocks, respectively.

Since the fundamental differential equations (Eqs. III.c.1 and 2) do not apply across discontinuities, a shock wave divides the wave diagram into two parts which must be treated separately and properly matched at the discontinuity.

Because the changes of the flow variables across a shock wave can be considered as taking place instantaneously, the flow conditions on the two sides of the shock wave can be matched as if the flow were steady. From the continuity, momentum, and energy equations for steady flow, one can derive relations between the flow variables upstream and downstream of a shock wave which may be written in a variety of forms and are known by the collective term "Rankine-Hugoniot equations." Essentially, these are relations between the various quantities that may be used to describe the strength of a shock wave. Let the symbols $\hat{\mathcal{U}}$ and $\hat{\mathcal{U}}'$, respectively, represent the nondimensional velocities with which the flow enters and leaves the shock wave, and let the corresponding state variables be denoted by

* Since shock waves are extensively treated in any textbook on compressible flow (for instance, reference 85), their essential properties will be assumed to be known.

p, p', S, S', and so forth. The following shock relations are listed here for convenient reference.

$$\frac{\rho'}{\rho} = \frac{\hat{u}}{\hat{u}'} = \frac{1 + \dfrac{\gamma+1}{\gamma-1}\dfrac{p'}{p}}{\dfrac{\gamma+1}{\gamma-1} + \dfrac{p'}{p}} \tag{VI.e.1}$$

$$\frac{p'}{p} = \frac{2\gamma}{\gamma+1} M_S^2 - \frac{\gamma-1}{\gamma+1} \tag{VI.e.2}$$

The shock Mach number M_S (always considered as a positive quantity) is defined as the Mach number of a supersonic flow in which the shock would be stationary; it is, therefore, given by

$$M_S = \frac{\hat{u}}{a} \tag{VI.e.3}$$

Other relations of interest are

$$\frac{\hat{u}'}{a'} = \left[\frac{1 + \dfrac{\gamma-1}{2} M_S^2}{\gamma M_S^2 - \dfrac{\gamma-1}{2}}\right]^{1/2} \tag{VI.e.4}$$

$$S' - S = \frac{1}{\gamma(\gamma-1)} \ln\left\{\left(\frac{2\gamma}{\gamma+1} M_S^2 - \frac{\gamma-1}{\gamma+1}\right) \times \left[\frac{1 + \dfrac{\gamma-1}{2} M_S^2}{\dfrac{\gamma+1}{2} M_S^2}\right]^{\gamma}\right\} \tag{VI.e.5}$$

It is seen from Eq. (VI.e.3) that aM_S is the velocity with which a shock wave would propagate in a gas at rest. If the gas is not at rest but flowing with some velocity u, the resulting shock velocity relative to the duct becomes the sum of the two velocities and will be denoted by w, or in nondimensional form by $\mathcal{W} = w/a_0$. Since in a gas rest, a P-shock travels to the right and a Q-shock to the left, the following relations apply:

$$\left.\begin{aligned} \mathcal{W} &= \mathcal{U} + \mathcal{A}M_S \quad \text{for a } P\text{-shock} \\ \mathcal{W} &= \mathcal{U} - \mathcal{A}M_S \quad \text{for a } Q\text{-shock} \end{aligned}\right\} \qquad \text{(VI.e.6)}$$

Let a shock wave pass through a point 1 of the wave diagram (Fig. VI.e.1) and let 1′ indicate the same point after passage of

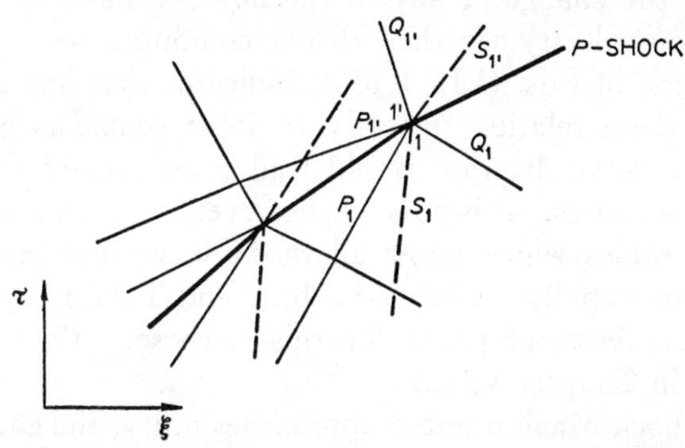

FIG. VI.e.1. Shock wave of gradually varying strength (general case)

the shock wave. Since the velocity with which the flow enters a shock wave is supersonic relative to the shock, the latter meets both P- and Q-waves. The flow conditions ahead of the shock wave are, therefore, determined regardless of the shock strength. This is also seen from Eqs. (VI.e.6) which show that a P- or Q-shock always overtakes the corresponding characteristics. The flow behind the shock wave, relative to the latter, is subsonic. A P-shock is, therefore, overtaken by P-waves and a Q-shock by Q-waves.

Assume that the wave diagram has already been constructed on both sides of the shock wave for all points earlier than 1 and 1′. In the case of a P-shock, the values of P_1, Q_1, S_1, and $P_{1'}$ can then be obtained by means of the procedures described so far. The P- and Q-waves that are carried along in the wave diagram cannot be expected to intersect at points that are located exactly on the shock path and, in general, it will thus be necessary to find the required waves by interpolation. The problem consists then in computing $Q_{1'}$ and $S_{1'}$ from the known variables with the aid of the Rankine-Hugoniot relations. Entirely analogous considerations apply in the case of a Q-shock when $P_{1'}$ and $S_{1'}$ must be determined from known values of P_1, Q_1, S_1, and $Q_{1'}$. The strength of the shock wave is determined by the change of any of the flow variables so that its propagation velocity can then also be computed.

Inspection of Eqs. (VI.e.1 to 6) indicates that any attempt to apply these relations directly to solve problems of shock waves in a wave diagram would lead to extremely time-consuming procedures. It is possible, however, to prepare auxiliary charts or tables which allow all shock wave problems to be solved quite rapidly. Charts 3a, b, c, and Tables 1a, b, c, in Chapter XI were prepared for this purpose. They will be discussed in Chapter VI.e.3.

If the shock Mach number approaches unity, the changes of the flow variables across the shock approach zero. There is, however, an important difference in the manner in which the change of entropy vanishes compared to that of the other variables. The symbol Δ will be used to indicate the change of a variable across the shock wave.

For Mach numbers close to unity, the entropy change given by Eq. (VI.e.5) may be expanded in a power series in terms of $(M_S{}^2 - 1)$ with the result

$$\Delta S = \frac{2}{3(\gamma + 1)^2} (M_S{}^2 - 1)^3 + \text{higher order terms} \tag{VI.e.7}$$

The pressure change across the shock may be obtained directly from Eq. (VI.e.2) as

$$\frac{\Delta p}{p} = \frac{p' - p}{p} = \frac{2\gamma}{\gamma + 1} (M_S{}^2 - 1) \tag{VI.e.8}$$

This shows that, as M_S approaches 1.0, Δp vanishes as the first power of $(M_S^2 - 1)$, while ΔS vanishes as the third power of this quantity. Thus, the entropy change becomes negligible, whereas the change of pressure may still be quite appreciable. A weak shock wave will, therefore, be defined as one for which the entropy change can be neglected within the accuracy required for any particular wave diagram. This not only simplifies the procedures for those points that are actually located on the path of the shock wave, but also for all other points in the wave diagram that lie in the flow behind the shock wave since no entropy changes are introduced.

A detailed discussion of the procedures for weak and strong shocks is presented in the following chapters. Once the solution for a point has been obtained, the continuation of the shock path can be plotted in the wave diagram. If it is desired to use the mean shock velocity between consecutive points along the shock path, the location of any new point can, at first, only be estimated. The procedures then become steps in the required iteration process and must be repeated as often as may be necessary. However, such refinement is, in general, worth while only if $\mathcal{W}$ changes rapidly along the shock path.

2. *Weak Shock Waves.* The procedures will be derived for the case of an isentropic flow in a duct of constant cross section, but since wave phenomena that take place at some distance from the shock wave are of no concern here (it has already been pointed out that the boundary conditions at a shock wave are the same as if the flow were steady), this assumption involves no loss of generality. Let the P-waves through points 1 and 2 (Fig. VI.e.2) meet at a point 3 where a shock wave first appears in the wave diagram. Assume further that the shock path has already been plotted between points 3 and 4 and that two more P-waves meet at the shock at 4.

The value of Q_4 is obtained by means of the standard wave diagram procedures, and the problem is to find $Q_{4'}$. The change of Q due to crossing of the shock wave can be expressed in the form

$$\Delta Q_4 = Q_{4'} - Q_4 = \frac{2}{\gamma - 1}(\mathfrak{a}_{4'} - \mathfrak{a}_4) - (\mathfrak{u}_{4'} - \mathfrak{u}_4)$$

Another Q-wave which does not cross the shock wave is also shown in Fig. VI.e.2, and its intersections with the P-waves through 4 and 4′ define the points 5 and 6. Because of the

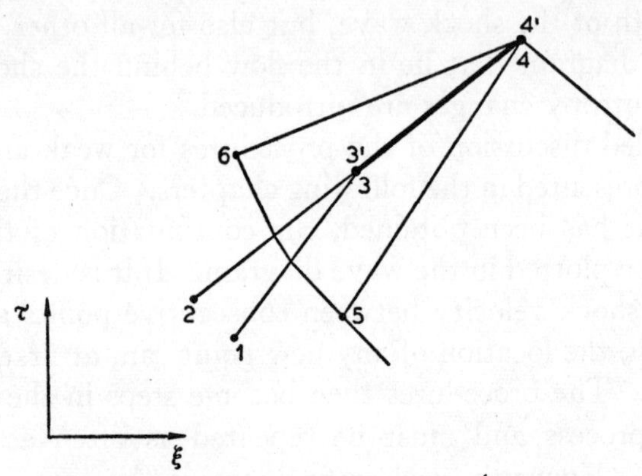

FIG. VI.e.2. Formation of a weak shock wave

assumption of isentropic flow in a duct of constant cross section, the property of characteristics that was expressed by Eqs. (V.a.2 and 3) can now be written as

$$\mathfrak{a}_{4'} - \mathfrak{a}_4 = \mathfrak{a}_6 - \mathfrak{a}_5$$

and

$$\mathfrak{u}_{4'} - \mathfrak{u}_4 = \mathfrak{u}_6 - \mathfrak{u}_5$$

If this is substituted in the foregoing relation for ΔQ_4 one obtains

$$\Delta Q_4 = Q_6 - Q_5 = 0$$

because of the constancy of Q along a Q-wave. In other words, as long as the entropy change across a shock wave can be neglected, the latter behaves like a characteristic since it does not modify the value of the Riemann variable of a crossing characteristic. However, the propagation velocity even of a weak shock wave is not equal to that of a characteristic. It can be obtained as follows: For small isentropic changes of state, Eq. (III.c.6) can be expanded into a Taylor series of the form

$$\frac{\Delta p}{p} = \frac{2\gamma}{\gamma - 1}\frac{\Delta \mathcal{a}}{\mathcal{a}} + \text{higher order terms}$$

If this is substituted into Eq. (VI.e.8), one obtains

$$M_S{}^2 = 1 + \frac{\gamma + 1}{\gamma - 1}\frac{\Delta \mathcal{a}}{\mathcal{a}} + \text{higher order terms}$$

or, approximately,

$$M_S \simeq 1 + \frac{1}{2}\frac{\gamma + 1}{\gamma - 1}\frac{\Delta \mathcal{a}}{\mathcal{a}} \qquad \text{(VI.e.9)}$$

The exact relation between shock Mach number and speed of sound ratio (listed in Table 1, Chap. XI) is better approximated if the factor $(\gamma + 1)/2(\gamma - 1)$ is slightly increased. The easily remembered values 10/3 and 9/4 corresponding to values of γ of 1.4 and 5/3, respectively, give a better fit over a wider range than Eq. (VI.e.9).

The wave diagram procedures for weak shock waves become thus extremely simple: The flow variables P, Q, S and either P' or Q' are given; S is not required as part of the shock wave procedures but may be needed for the calculation of P and Q. At first, one determines $\mathcal{a}$ and $\mathcal{u}$ from P and Q. On the other side of the shock wave, $\mathcal{a}'$ and $\mathcal{u}'$ follow then from P' and $Q' = Q$ in the case of a P-shock, or from Q' and $P' = P$ in the case of a Q-shock. Finally, $\mathcal{w}$ is computed from Eq. (VI.e.6) where the shock Mach number is obtained either from Table 1 for the computed value of $\mathcal{a}'/\mathcal{a}$ or from Eq. (VI.e.9).

The greatest shock strength for which changes of state can still be considered as isentropic depends, of course, on the required accuracy of the wave diagram. The exact procedures for strong shock waves (Chapter VI.e.3) may be used to estimate the errors that result from the neglect of entropy changes. In the case of air, the simplified procedures described here should always be permissible for $\mathfrak{a}'/\mathfrak{a} \leq 1.06$; even for $\mathfrak{a}'/\mathfrak{a} \leq 1.15$, the errors should be unimportant for many problems. These limits correspond, approximately, to shock pressure ratios of 1.5 and 2.5, respectively.

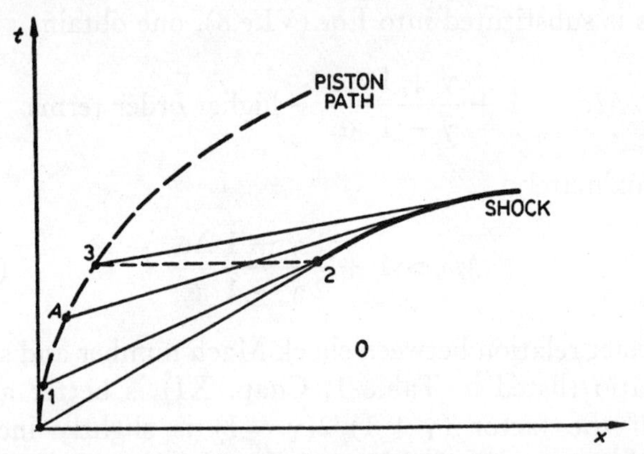

FIG. VI.e.3. Formation of a shock wave ahead of an accelerated piston

Numerical Example: The procedures for weak shock waves should not present any difficulties. They are demonstrated, for instance, in points 21 to 24 of Example IX.b.1.

The following example is chosen to illustrate how, in certain cases, one can obtain exact solutions of nonsteady-flow problems by following the graphical procedures analytically.

Suppose that a piston starting from rest is moving with a constant acceleration b. This creates compression waves which travel ahead of the piston. The leading waves eventually coalesce to form a shock wave the strength of which increases gradually as more and more waves catch up with the shock (Fig. VI.e.3). What is the acceleration for which the shock is formed 10 feet ahead of the piston? Assume that the gas in the duct is air of such initial temperature that the speed of sound is 1140 ft/sec.

Since only analytical relations will be used here, no advantage is gained by using nondimensional variables and the actual quantities are, therefore, shown in the figure. Let the piston start its motion at the origin of the wave diagram. Indicate the initial conditions in the duct by subscript 0 and let subscript A indicate the flow variables at a point on the piston path. The latter is thus given by

$$x_A = \frac{b}{2} t_A{}^2$$

and the piston velocity by

$$v_A = bt_A$$

The characteristic relation $Q_A = Q_0$ can be expressed in the form

$$a_A = a_0 + \frac{\gamma - 1}{2} u_A$$

and, if this is combined with the boundary condition $u_A = v_A$, one obtains the velocity of any P-wave originating at the piston as

$$u_A + a_A = a_0 + \frac{\gamma + 1}{2} bt_A$$

Since Q is constant throughout the wave diagram, all P-waves are straight lines (see Chapter IV.b). The equation of a P-wave in the x,t-coordinate system becomes, therefore,

$$x - x_A = \left(a_0 + \frac{\gamma + 1}{2} bt_A\right)(t - t_A)$$

If point A is taken close to the origin of the coordinate system (point 1), then the intersection of this line with the head of the compression wave, given by $x = a_0 t$, determines point 2 where a shock wave is first formed. The two simultaneous equations for x_2 and t_2 are easily solved. Rigorously, the point of shock formation is obtained by letting t_1 go to zero. If x_1 is expressed in terms of t_1 from the equation of the piston path, and the limit is formed, one obtains

$$x_2 = \lim_{t_1 \to 0} \frac{2a_0^2}{(\gamma + 1)b}\left(1 + \frac{\gamma b}{2a_0} t_1\right) = \frac{2a_0^2}{(\gamma + 1)b}$$

This determines the distance of the point of shock formation from the starting point of the piston. The time elapsed since the start of the motion is given by

$$t_2 = \frac{x_2}{a_0} = \frac{2a_0}{(\gamma + 1)b}$$

During this time, the piston has moved to point 3 which is located at

$$x_3 = \frac{b}{2} t_2^2 = \frac{2a_0^2}{(\gamma + 1)^2 b}$$

The distance from the piston at which the shock wave is formed is, therefore, given by

$$x_2 - x_3 = \frac{2\gamma a_0^2}{(\gamma + 1)^2 b}$$

This equation can now be solved for the acceleration for which the shock forms 10 feet ahead of the piston. Substituting the numerical values also for a_0 and γ, one obtains $b = 616{,}000$ ft/sec^2 or, approximately, 20,000g.

3. *Strong Shock Waves.* If the entropy rise across a shock wave cannot be neglected, the exact Rankine-Hugoniot equations must be used to match the wave diagrams on both sides of the shock. Numerous tables of these relations are available,* but the quantities that are usually tabulated do not include the ones required for the most convenient wave diagram proce-

* See, for instance, references 76–79.

dures. A procedure that is of general usefulness should depend on the strength of the shock only and not on the state and the flow velocity of the gas in which the shock wave is moving. It should also be based on variables that are either directly available in a wave diagram or, at least, easily computed from those.

Clearly, the change of the flow velocity, $\mathfrak{u}' - \mathfrak{u}$, that is produced by a shock wave is independent of the velocity of the gas ahead of the shock wave. Thus the change of the Riemann variables across the shock wave, $\Delta P = P' - P$ and $\Delta Q = Q' - Q$, is invariant to a change of the coordinate system, but these quantities are not functions of the shock strength alone since they depend on the absolute values of $\mathfrak{a}$ and $\mathfrak{u}$. The variables $\Delta P/\mathfrak{a}$ and $\Delta Q/\mathfrak{a}$ are not only invariant to a change of the coordinate system, but are also true measures of the shock strength. It is, therefore, possible to derive the relations between $\Delta P/\mathfrak{a}$ or $\Delta Q/\mathfrak{a}$ and other measures of shock strength such as M_S, $\Delta\mathfrak{u}/\mathfrak{a}$, $\mathfrak{a}'/\mathfrak{a}$, p'/p, and ΔS from the equations for a stationary shock wave (Eqs. VI.e.1 to 5) without loss of generality. Since $\hat{\mathfrak{u}} = -\mathfrak{u}$ for a stationary P-shock and $\hat{\mathfrak{u}} = \mathfrak{u}$ for a stationary Q-shock, one obtains the following relations from the definition of the Riemann variables:

$$\left.\begin{aligned} \frac{\Delta P}{\mathfrak{a}} &= \frac{2}{\gamma - 1}\left(\frac{\mathfrak{a}'}{\mathfrak{a}} - 1\right) + \frac{\mathfrak{u}' - \mathfrak{u}}{\mathfrak{u}} \cdot \frac{\mathfrak{u}}{\mathfrak{a}} \\ &= \frac{2}{\gamma - 1}\left(\frac{\mathfrak{a}'}{\mathfrak{a}} - 1\right) + \left(1 - \frac{\mathfrak{u}'}{\mathfrak{u}}\right) M_S \\ \frac{\Delta Q}{\mathfrak{a}} &= \frac{2}{\gamma - 1}\left(\frac{\mathfrak{a}'}{\mathfrak{a}} - 1\right) - \left(1 - \frac{\mathfrak{u}'}{\mathfrak{u}}\right) M_S \end{aligned}\right\} \text{for a } P\text{-shock} \qquad \text{(VI.e.10)}$$

and, similarly,

$$\left.\begin{aligned} \frac{\Delta P}{\mathfrak{a}} &= \frac{2}{\gamma - 1}\left(\frac{\mathfrak{a}'}{\mathfrak{a}} - 1\right) - \left(1 - \frac{\mathfrak{u}'}{\mathfrak{u}}\right) M_S \\ \frac{\Delta Q}{\mathfrak{a}} &= \frac{2}{\gamma - 1}\left(\frac{\mathfrak{a}'}{\mathfrak{a}} - 1\right) + \left(1 - \frac{\mathfrak{u}'}{\mathfrak{u}}\right) M_S \end{aligned}\right\} \text{for a } Q\text{-shock} \qquad \text{(VI.e.11)}$$

For a P-shock, $\Delta P/\mathfrak{a}$ is known from the preceding work on the wave diagram and, similarly, $\Delta Q/\mathfrak{a}$ is known in the case of a Q-shock. It is seen from Eqs. (VI.e.10 and 11) that $\Delta P/\mathfrak{a}$ for a P-shock is equal to $\Delta Q/\mathfrak{a}$ for a Q-shock and a single graph or table is, therefore, sufficient to solve all problems of shock wave-characteristic interactions. Charts 3a, b, c, represent such graphs for different values of γ. For most wave diagrams, a better accuracy is required than would be obtainable from these charts which cover a wide range of shock strengths. The data are, therefore, presented in Tables 1a, b, c, in Chapter XI in a form that is convenient for actual use.

The measures of shock strength discussed are all positive quantities with the exception of $\Delta\mathfrak{u}/\mathfrak{a}$ which is positive for a P-shock and negative for a Q-shock. Only the absolute value of this quantity, $|\Delta\mathfrak{u}|/\mathfrak{a}$, is entered in Chart 3 and Table 1; the proper sign must be applied in every case.

The shock wave procedures may now be summarized as follows: From the known variables P, Q, S, and P' or Q' compute first $\mathfrak{a}$ and $\mathfrak{u}$; then find $\Delta P/\mathfrak{a}$ or $\Delta Q/\mathfrak{a}$, and for this value obtain $\mathfrak{a}'/\mathfrak{a}$, M_S, and ΔS from Table 1; the first of these yields $\mathfrak{a}'$, and this and P' (or Q') determine $\mathfrak{u}'$; S' is given simply by $S + \Delta S$; and, finally, $\mathfrak{w}$ is calculated from one of Eqs. (VI.e.6).

The values of p'/p and $|\Delta\mathfrak{u}|/\mathfrak{a}$ are also obtainable from Chart 3 and Table 1 and will often be needed.

Numerical Examples: (1) Let a Q-shock in air arrive at a point of the wave diagram where $P = 4.900$, $Q = 5.200$, $S = 0$, and $Q' = 6.600$. Solve for the flow conditions on both sides of the shock wave and find the velocity of the latter if (a) the exact shock relations are used and (b) changes of state across the shock are assumed to be isentropic.

The speed of sound and the flow velocity ahead of the shock wave are immediately found from the Riemann variables as $\mathfrak{a} = 1.010$ and $\mathfrak{u} = -0.150$. The shock strength is, therefore, determined by

$$\frac{\Delta Q}{\mathcal{A}} = \frac{6.600 - 5.200}{1.010} = 1.386$$

For this value, one obtains from Table 1a the quantities $M_S = 1.480$, $\mathcal{A}'/\mathcal{A} = 1.144$, $\Delta S = 0.047$, and $p'/p = 2.39$ so that all required variables can now easily be calculated.

If changes of state across the shock wave are assumed to be isentropic, one has simply $P' = P$, so that both Riemann variables are now known behind the shock. The results for the exact and the approximate calculations are compared in the following table. The exact pressure ratio is taken from Table 1a while the approximate value is computed from the relation $p'/p = (\mathcal{A}'/\mathcal{A})^7$. The approximate shock Mach number is calculated first from Eq. (VI.e.9) and then also with the aid of the improved empirical relation given there.

COMPARISON OF EXACT AND APPROXIMATE SOLUTIONS FOR THE SHOCK WAVE OF EXAMPLE (1)

Variable	Exact	Approximate		
		*		†
$\mathcal{A}'$	1.155		1.150	
$\mathcal{U}'$	−0.825		−0.850	
S'	0.047		0	
M_S	1.480	1.416		1.462
$\mathcal{W}$	−1.645	−1.580		−1.627
p'/p	2.39		2.48	

* From Eq. (VI.e.9).

† From $M_S = 1 + \frac{10}{3} \cdot \frac{\Delta\mathcal{A}}{\mathcal{A}}$.

It is seen that the errors of the approximate calculation are quite noticeable, but they may not yet be considered excessive in some practical applications.

(2) Consider a shock wave moving in a gas at rest. What shock strength is required to produce sonic flow behind the shock when the gas is (a) air, $\gamma = 1.4$, and (b) helium, $\gamma = 5/3$?

This problem is easily solved with the aid of Tables 1a and 1b, respectively. Since the gas ahead of the shock wave is at rest, $\Delta\mathfrak{u}$ represents already the velocity $\mathfrak{u}'$ behind the shock. It is, therefore, only required to find the table entries for which the condition $|\Delta\mathfrak{u}|/\mathfrak{a} = \mathfrak{a}'/\mathfrak{a}$ is satisfied. The results are given in the table below.

STRENGTHS OF SHOCK WAVES THAT PRODUCE SONIC FLOW

Variable	Air	Helium
$\lvert\Delta\mathfrak{u}\rvert/\mathfrak{a} = \mathfrak{a}'/\mathfrak{a}$	1.320	1.796
M_S	2.068	2.757
p'/p	4.82	9.25

(3) What is the highest flow Mach number that can be produced by a shock wave?

This problem can be answered by means of Eqs. (VI.e.1 and 2) where all constant terms can be neglected compared to M_S which tends toward infinity. One obtains thus:

$$\frac{\rho'}{\rho} = \frac{\hat{\mathfrak{u}}}{\hat{\mathfrak{u}}'} \rightarrow \frac{\gamma + 1}{\gamma - 1}$$

$$\frac{p'}{p} \rightarrow \frac{2\gamma}{\gamma + 1} M_S{}^2$$

showing that the density ratio approaches a finite limit. The ratio of these relations yields

$$\frac{p'/p}{\rho'/\rho} = \left(\frac{\mathfrak{a}'}{\mathfrak{a}}\right)^2 \rightarrow \frac{2\gamma(\gamma - 1)}{(\gamma + 1)^2} M_S{}^2$$

One obtains further

$$1 - \frac{\hat{\mathfrak{u}}'}{\hat{\mathfrak{u}}} = \frac{\hat{\mathfrak{u}} - \hat{\mathfrak{u}}'}{\mathfrak{a}} \cdot \frac{\mathfrak{a}}{\hat{\mathfrak{u}}} = \frac{|\Delta\mathfrak{u}|}{\mathfrak{a}} \cdot \frac{1}{M_S} \to \frac{2}{\gamma + 1}$$

(see the derivation of Eqs. (VI.e.10 and 11)). Since $\Delta\mathfrak{u}$ approaches infinity in the same manner as M_S, any finite initial flow velocity can be neglected, and one can write $\Delta\mathfrak{u}$ for the velocity $\mathfrak{u}'$ that is produced by the shock wave. The last two relations can then be combined to give the required answer

$$\frac{\mathfrak{u}'}{\mathfrak{a}'} = M' \to \left[\frac{2}{\gamma(\gamma - 1)}\right]^{1/2}$$

Thus, although the flow velocity, speed of sound, temperature, and pressure produced by a shock wave all tend to become infinite as the shock Mach number approaches infinity, the Mach number of the flow produced approaches only the rather low limit given above. For air and helium, this limit becomes 1.890 and 1.342, respectively. Note that this limitation applies to flow in a constant-area duct. If the duct in which a shock produces a supersonic flow is connected to a diverging channel, the flow can be accelerated to any desired Mach number by appropriate expansion of the duct [60] (see the example at the end of Chapter VII.h).

For another problem, involving a shock wave of variable strength, see Example IX.e.

VI.f. Contact Surfaces (Interfaces)

It happens quite frequently that problems must be investigated in which the gas column in the duct is composed of two (or more) sections, each containing a different gas or the same gas at different entropy levels. If the gas is at rest or flowing with constant velocity, the contact surface, or interface, becomes a slowly widening transition zone as a result of diffusion. If, however, the two gases are accelerated toward the one of higher density, a plane interface is unstable and disintegrates rapidly by forming streamers of the denser gas reaching deep into the other gas.[90] Acceleration in the opposite direction does not create instability. It has been shown, however, that impulsive acceleration of a contact surface should lead to instability regardless of the direction of the acceleration.[101] For the purpose of wave diagram construction, an interface will be assumed in the following to be a discontinuity. The errors that are introduced as a result of diffusion or interface instability have not yet been studied.

It is obvious from the nature of a contact surface that its velocity must be equal to the flow velocity on either side and also that the pressure must be the same on both sides. The wave diagrams on either side must, therefore, be matched at the interface by satisfying these conditions.

So far, all wave diagram procedures involved only one gas, and the flow variables were made nondimensional by expressing them in terms of an arbitrary reference state (Chapter III.b). In the case of two (or more) gases, the reference states could be selected completely independent of each other. It is, however, desirable for obvious reasons to use a single reference value of the speed of sound, a_0, for the entire wave diagram. Since the matching conditions involve the pressure, it is convenient to use also the same reference pressure, p_0, for both gases.

The reference states are thus defined except for the values of the specific entropy at these states. They may be taken as zero for both gases, although any other value would serve equally well. One might, for instance, prefer such a value for S_0 in one gas that the entropy becomes zero for a state that is of more importance for the problem under investigation than the reference state. Since only entropy differences are required for the wave diagram procedures, the actual choice is immaterial. Care must be exercised, however, to use consistent relations whenever pressures or densities are computed from the speed of sound and entropy. The following will be based on the assumption that the entropy is zero at the reference state for both sides of the interface.

The flow conditions on the left and right side of the interface will be characterized by subscripts L and R, respectively. The variables that are known for a point of the interface are then P_L, Q_R, S_L, and S_R; the unknowns are $\mathcal{a}_L$, $\mathcal{u}_L$, $\mathcal{a}_R$, and $\mathcal{u}_R$. Two relations are provided by the definition of the known Riemann variables and two more are provided by the matching conditions

$$\left.\begin{aligned} \mathcal{u}_L &= \mathcal{u}_R \\ p_L &= p_R \end{aligned}\right\} \tag{VI.f.1}$$

This introduces the two additional variables p_L and p_R. One obtains the required additional relations from Eq. (III.c.6) by taking state 1 in this equation as the reference state on either side of the interface. This yields

$$\left.\begin{aligned} \mathcal{a}_L &= \left(\frac{p_L}{p_0}\right)^{\frac{\gamma_L-1}{2\gamma_L}} e^{\frac{\gamma_L-1}{2} s_L} \\ \mathcal{a}_R &= \left(\frac{p_R}{p_0}\right)^{\frac{\gamma_R-1}{2\gamma_R}} e^{\frac{\gamma_R-1}{2} s_R} \end{aligned}\right] \tag{VI.f.2}$$

The best method to solve this system of equations is a trial-and-error procedure based on a guess for one of the unknowns, say, $\mathcal{a}_L$. From Eqs. (VI.f.2), one can then calculate p_L/p_0 and, since this equals p_R/p_0, one obtains the corresponding value of $\mathcal{a}_R$. Since $\mathcal{u}_L$ follows from $\mathcal{a}_L$ and P_L, one can compute Q_R from $\mathcal{u}_R(= \mathcal{u}_L)$ and $\mathcal{a}_R$; if this does not agree with the known value of Q_R, the original estimate for $\mathcal{a}_L$ must be modified until agreement is reached.

This procedure is considerably simplified if the value of γ is the same on both sides of the interface, although the gases may be different. For $\gamma_L = \gamma_R = \gamma$, the foregoing equations can be solved directly. One obtains the sum and the ratio of the speeds of sound as

$$\mathcal{a}_L + \mathcal{a}_R = \frac{\gamma - 1}{2}(P_L + Q_R) \tag{VI.f.3}$$

$$\frac{\mathcal{a}_L}{\mathcal{a}_R} = e^{\frac{\gamma-1}{2}(S_L - S_R)} \tag{VI.f.4}$$

The solution of these equations is given by

$$\left.\begin{aligned} \mathfrak{a}_L &= \frac{P_L + Q_R}{\dfrac{2}{\gamma - 1}\left[1 + e^{\frac{\gamma-1}{2}(S_R - S_L)}\right]} \\ \mathfrak{a}_R &= \frac{P_L + Q_R}{\dfrac{2}{\gamma - 1}\left[1 + e^{\frac{\gamma-1}{2}(S_L - S_R)}\right]} \end{aligned}\right\} \qquad \text{(VI.f.5)}$$

Ordinarily, one would use only one of these equations and compute the second unknown from either Eq. (VI.f.3) or (VI.f.4). The flow velocity is then obtained from P_L or Q_R.

It is seen that, as long as the flow on both sides of an interface is isentropic or multi-isentropic, the denominators of Eqs. (VI.f.5) are constant, which makes the procedure particularly simple.

The path of an interface is easily plotted in a wave diagram since its velocity is always given by $\mathfrak{u}_L(= \mathfrak{u}_R)$.

An alternative procedure can be derived by taking the difference between the known Riemann variables at the interface,

$$P_L - Q_R = \frac{2}{\gamma - 1}\,\mathfrak{a}_R\left(\frac{\mathfrak{a}_L}{\mathfrak{a}_R} - 1\right) + 2\mathfrak{u}$$

where no subscript for $\mathfrak{u}$ is needed since the velocities are the same on both sides of the interface. Substituting the second of Eqs. (VI.f.5) and Eq. (VI.f.4), one obtains

$$P_L - Q_R = (P_L + Q_R)Z + 2\mathfrak{u}$$

where

$$Z = \frac{e^{\frac{\gamma-1}{2}(S_L - S_R)} - 1}{e^{\frac{\gamma-1}{2}(S_L - S_R)} + 1} = \tanh\left[\frac{\gamma - 1}{4}(S_L - S_R)\right] \qquad \text{(VI.f.6)}$$

has been introduced as an abbreviation.

The Riemann variable of the transmitted characteristic is then determined, by using the general relations (see Eq. (IV.a.4))

$$Q_R = P_R - 2\mathfrak{u} \quad \text{or} \quad P_L = Q_L + 2\mathfrak{u}$$

which yield

$$\left.\begin{aligned} P_R &= P_L - (P_L + Q_R)Z \\ Q_L &= Q_R + (P_L + Q_R)Z \end{aligned}\right\} \qquad \text{(VI.f.7)}$$

If the flow on both sides of the interface is isentropic or multi-isentropic, Z has a constant value. There is little to choose between the two methods described. If the reflected characteristic need not be entered in the wave diagram, the second method would actually be a little faster.

For small entropy changes across the interface, the procedure reduces effectively to that for entropy gradients described in Chapter V.c.1. In order to prove this, one must calculate the increments Δ_+P or Δ_-Q across the interface under the condition that $S_L - S_R = \Delta_-S = -\Delta_+S$ is so small that higher powers of this quantity than the first can be neglected. Under this condition, one obtains from the first of Eqs. (VI.f.5)

$$\mathfrak{a}_L = \frac{(\gamma - 1)(P_L + Q_R)}{2\left(2 - \dfrac{\gamma - 1}{2}\Delta_-S\right)}$$

$$= \frac{\gamma - 1}{4}(P_L + Q_R)\left(1 + \frac{\gamma - 1}{4}\Delta_-S\right)$$

Since

$$Q_L = \frac{2}{\gamma - 1}\mathfrak{a}_L - \mathfrak{u}_L = \frac{4}{\gamma - 1}\mathfrak{a}_L - P_L$$

the increment Δ_-Q is given by

$$\Delta_-Q = Q_L - Q_R = \frac{4}{\gamma - 1}\mathfrak{a}_L - (P_L + Q_R)$$

Substitution for $\mathfrak{a}_L$ from the foregoing relation yields, finally,

$$\Delta_-Q = \frac{\gamma - 1}{4}(P_L + Q_R)\Delta_-S$$

$$= \frac{\gamma - 1}{4}(P_R + Q_R - \Delta_+P)\Delta_-S = \mathfrak{a}_R\Delta_-S$$

The product $(\Delta_+P)(\Delta_-S)$ can be neglected since the increments of the Riemann variables are small of the same order of magnitude as $\Delta_\pm S$. Similarly, one obtains

$$\Delta_+P = \mathfrak{a}_L\Delta_+S$$

These expressions for Δ_+P and Δ_-Q are identical with Eqs. (V.c.1), which proves the above statement.

Numerical Example: Suppose a pipe is filled with two adjacent columns of hot and cold air of such temperatures that

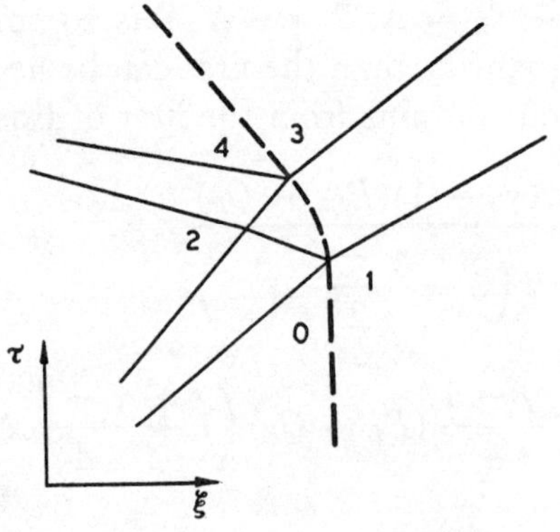

FIG. VI.f.1. Interaction of an expansion wave with a contact surface that is initially at rest

the speed of sound in the hot gas is 1.5 times higher than that in the cold gas. The air is initially at rest, and an expansion wave which reduces the initial pressure by 20% reaches the interface from the cold side. How strong are the transmitted and reflected waves, and what is the final velocity of the interface?

The corresponding wave diagram is sketched in Fig. VI.f.1 where only the first and last characteristics of each wave are indicated since the details of the transition from the initial to the final state are of no interest here. Let the initial state of the cold air be the reference state. The following conditions are, therefore, prescribed: $\mathcal{U}_0 = \mathcal{U}_1 = 0$, $\mathcal{A}_0 = 1.000$, $\mathcal{A}_1 = 1.500$, and $p_2/p_0 = 0.8$. The problem is to find $\mathcal{U}_3 (= \mathcal{U}_4)$, p_4/p_2, and p_3/p_1.

First, one must solve for the flow conditions following the arriving wave (region 2). From

$$Q_2 = 5\mathcal{A}_2 - \mathcal{U}_2 = Q_0$$

and

$$\frac{\mathcal{A}_2}{\mathcal{A}_0} = \left(\frac{p_2}{p_0}\right)^{1/7}$$

one obtains $\mathcal{A}_2 = 0.969$, $\mathcal{U}_2 = -0.155$, and, therefore, $P_2 = 4.690$.

The procedure described in this chapter can now be applied because both $P_4 = P_2$ and $Q_3 = Q_1$ are known. Since the flow on either side of the interface is isentropic, the denominators in Eqs. (VI.f.5) are constant. Instead of calculating $S_1 - S_0$ from Eq. (III.c.6) and then calculating the denominator, one can determine the latter directly by substituting in either of Eqs. (VI.f.5) the known conditions 0 and 1. One obtains thus for the denominator of the first equation:

$$\text{Denominator} = \frac{P_0 + Q_1}{\mathcal{A}_0} = \frac{5.000 + 7.500}{1.000} = 12.500$$

and, therefore,

$$\mathcal{A}_4 = \frac{P_4 + Q_3}{\text{Denominator}} = 0.975$$

Eq. (VI.f.3) yields, then, $\mathcal{A}_3 = 1.463$.

The final interface velocity is, therefore, given by

$$\mathcal{U}_3 = \mathcal{U}_4 = P_4 - 5\mathcal{A}_4 = -0.185$$

The strength of the reflected wave is then determined by

$$\frac{p_4}{p_2} = \left(\frac{\mathcal{A}_4}{\mathcal{A}_2}\right)^7 = 1.042$$

and, for the transmitted wave, one obtains

$$\frac{p_3}{p_1} = \frac{p_4}{p_0} = \frac{p_4}{p_2} \times \frac{p_2}{p_0} = 0.834$$

One sees that, in this example, the reflected wave is a compression wave and the transmitted wave is stronger than the incident wave.

The interaction of waves with a helium-air interface is illustrated in points VIII and IX of Example IX.b.3. See also the example in Chapter VI.h.2.2.

VI.g. Discontinuous Change of the Duct Cross Section as Approximation to a Gradual Change

1. *General Remarks.* It was pointed out in Chapter V.b.1 that gradual changes of the duct area may be approximated in the wave diagram by discontinuous changes. Such an approximation can, of course, only be used to evaluate the effects of area changes at some distance from the change. The details of the wave phenomena, particularly their timing, must be expected to be somewhat distorted by this simplification. The extent of this distortion depends also on where one locates the discontinuous change of cross section. Since the best location can be found only by comparison of the results with those obtained from a wave diagram based on gradual area changes, the actual choice must be made more or less arbitrarily, but in view of the simplifying assumptions that must often be made before a wave diagram can be prepared, the procedures based

on a discontinuous change of area will usually lead to sufficiently accurate results at a considerable saving of time.

One must visualize the flow phenomena that occur in a gradually varying duct and can then assume that the entire process through which the changes of the flow variables take place is concentrated at a single discontinuity section. Since these changes then take place instantaneously, they may be computed on the basis of steady-flow relations, but it must, of course, be assumed that no flow separation occurs at the discontinuity section.

Whenever the flow is subsonic on one side and supersonic on the other side of the discontinuity section, the flow pattern in the actual duct with continuously varying area becomes so complex that auxiliary stations must be introduced which are then also assumed to coincide with the discontinuity section. Thus, a change of procedures becomes necessary whenever the Mach number on one side passes through unity. It is, therefore, always advisable to include those characteristics in the wave diagram that lead exactly to $M = 1.0$. Usually, these waves must be found by interpolation.

The flow conditions on the left and right side of the discontinuity will be indicated by subscripts L and R, respectively. In addition to these, subscripts 1, 2, $\cdots$ will also be used with the convention that odd numbers refer to the smaller area irrespective of whether it is located on the left or right side of the discontinuity. Additional auxiliary stations will be introduced as needed.

In all cases, the variables which are known, in addition to the two duct areas A_L and A_R, are the Riemann variables P_L and Q_R, and the entropy for the flow entering the discontinuity section. The unknown variables are thus $\mathfrak{a}_L$, $\mathfrak{u}_L$, $\mathfrak{a}_R$, $\mathfrak{u}_R$ and the entropy of the flow leaving the discontinuity section. The definitions of the known Riemann variables supply two relations, and two more follow from the conservation of energy and mass flow across the discontinuity. If no shocks are located between A_L and A_R, one has $S_L = S_R$, and there are then enough equations to solve for the four remaining unknowns. If a shock wave is located somewhere between A_L and A_R, the entropy of the flow through the discontinuity section is modified, and the shock location represents a further unknown. The conditions of conservation of mass flow and energy between the location of the shock and either A_L or A_R yield the necessary two additional equations. The detailed procedures for the various possible cases are described in the following chapters.

2. *Subsonic Flow on Both Sides of the Change of Cross Section.* This is the case that occurs most frequently. The conservation conditions for energy and mass flow across the discontinuity may be written in the nondimensional form (see Appendix)

$$\mathcal{U}_L^2 + \frac{2}{\gamma - 1}\mathcal{A}_L^2 = \frac{2}{\gamma - 1}\mathcal{A}_L^2\left(1 + \frac{\gamma - 1}{2}M_L^2\right) \tag{VI.g.1}$$

$$= \frac{2}{\gamma - 1}\mathcal{A}_R^2\left(1 + \frac{\gamma - 1}{2}M_R^2\right)$$

$$\mathcal{A}_L^{\frac{2}{\gamma-1}}A_L|\mathcal{U}_L| = \mathcal{A}_L^{\frac{\gamma+1}{\gamma-1}}A_LM_L = \mathcal{A}_R^{\frac{\gamma+1}{\gamma-1}}A_RM_R \tag{VI.g.2}$$

where the second equation already implies that no entropy changes are involved. If $\mathcal{A}$ is eliminated from these equations, one obtains

$$A_L\frac{M_L}{\left(1 + \frac{\gamma - 1}{2}M_L^2\right)^{\frac{\gamma+1}{2(\gamma-1)}}} = A_LD_L = A_RD_R \tag{VI.g.3}$$

This equation could also have been formulated in terms of Mach number functions other than the function D defined here. Of course, all of these must be proportional to either D or $1/D$. Since the procedures of this chapter require frequent application of the continuity equation, it is essential to have a table or plot of a suitable Mach number function available. Functions that are equal or proportional to either D or $1/D$ are included in most gas dynamics tables for isentropic flow.* With the aid of such a table, one can relate the Mach numbers on both sides of the discontinuity section. The Mach number function has an extreme value at $M = 1.0$, but there is never any doubt whether the subsonic or the supersonic branch must be used, since M_L and M_R must both be either smaller or larger than 1.0.

A convenient way to determine the unknown flow variables by means of a simple iteration procedure is based on a guess for one of the unknowns. In principle, the guess may be made for any one of them, but since the variations of the flow variables in the narrower duct are larger than in the wider one, any error of the guess shows up better if one selects $\mathcal{A}$ or $\mathcal{U}$ in the wider duct.

From the preceding equations, it is seen that it is unimportant whether the larger duct is on the left or right side, and subscripts 1 and 2 will be used to refer to the smaller and larger duct area, respectively. From the estimated value of $\mathcal{A}_2$ (or $\mathcal{U}_2$) and the known value of the Riemann variable for the wider duct, one obtains $\mathcal{U}_2$ (or $\mathcal{A}_2$) and, therefore, also $M_2 = |\mathcal{U}_2|/\mathcal{A}_2$. With the aid of a suitable table or plot of a Mach number function equal or related to that defined by Eq. (VI.g.3), one can then determine M_1 for the given area ratio A_2/A_1. $\mathcal{A}_1$ follows then from Eq. (VI.g.1) and, from this and M_1, one obtains $\mathcal{U}_1$. The sign of $\mathcal{U}_1$ does not follow from this procedure, but it must, of course, be the same as that of $\mathcal{U}_2$ and the flow direction is known from the wave diagram. One can thus, finally, compute the Riemann variable for the narrower duct which must agree with the already known value of P_1 or Q_1. In case of disagreement, the estimate for $\mathcal{A}_2$ must be revised until agreement is reached.

In any wave diagram which involves a sudden change of cross section, the described procedure must be repeated every

* For example, the function D is tabulated for $\gamma = 1.4$ in Foa's table,[78] while Emmons' table [76] lists a function $A/A^* = D_{M=1}/D$ (A^* is the area where $M = 1.0$).

time a characteristic crosses the discontinuity section. If one prepares, therefore, an auxiliary chart in which the Mach number on one side and the speed of sound ratio are plotted as functions of the Mach number on the other side of the discontinuity (from Eqs. (VI.g.1 and 3)), a great deal of computing can be saved. This plot is, of course, only valid for the area ratio for which it was prepared. Experience has shown, however, that the time spent on its preparation is more than recovered in the drawing of a single wave diagram.* Chart 4a represents a sample plot prepared for $A_2/A_1 = 1.70$ and $\gamma = 1.4$ (see Chap. XI).

While M_1 can have any value between zero and 1.0, M_2 can vary only between zero and a maximum value that depends on the area ratio (for the example presented in Chart 4a, M_2 cannot exceed 0.369). It is possible that no solution can be found by means of the described procedure, namely, when $D_1 = D_2A_2/A_1$ becomes greater than the possible maximum value of D at $M = 1.0$. This indicates that the flow cannot be subsonic on both sides of the discontinuity section. In this case, one should find the last wave for which the procedures

* An alternative procedure which does not require any iteration is described in reference 113 (see also references 41 and 51). From the expression

$$\frac{P_L}{Q_R} = \frac{\mathfrak{a}_L\left(\dfrac{2}{\gamma - 1} + M_L\right)}{\mathfrak{a}_R\left(\dfrac{2}{\gamma - 1} - M_R\right)}$$

one eliminates the ratio $\mathfrak{a}_L/\mathfrak{a}_R$ with the aid of Eq. (VI.g.2) and obtains thus a relation between M_L and M_R with the ratio of the Riemann variables as parameter. Similarly, Eq. (VI.g.3) represents a relation between these Mach numbers for any value of the area ratio. One can, therefore, plot two families of curves M_L versus M_R in the same coordinate system for various values of A_L/A_R and P_L/Q_R, respectively. For any given pair of these parameters, the Mach numbers can then be read off directly at the intersection of the corresponding curves. In order to obtain a chart of general usefulness, both parameters must be varied over a considerable range. A chart that allows interpolations to be carried out with adequate accuracy must, therefore, contain many closely spaced curves, and great care would have to be exercised to avoid mistakes in reading the data. This author favors the iteration procedure described above, since it is also quite fast and involves only the preparation and use of a simple chart of two curves.

can still be carried out—the characteristic for which M_1 just becomes 1.0—and then change to the procedures described in VI.g.4 or 5.

A numerical example of these procedures is presented in Example IX.c.2.

3. *Supersonic Flow on Both Sides of the Change of Cross Section.* If the flow that approaches the section of discontinuity is supersonic, both $\mathcal{U} + \mathcal{A}$ and $\mathcal{U} - \mathcal{A}$ are of the same sign, and therefore both P- and Q-waves enter the discontinuity section from the upstream side. Thus P and Q, and therefore also $\mathcal{A}$, $\mathcal{U}$, and M, are known there. If the flow on the downstream side is also supersonic, the Mach number and the speed of sound there can be directly computed with the aid of Eqs. (VI.g.3 and 1) where the supersonic branch of the Mach number function D must be used. From M and $\mathcal{A}$, one obtains $\mathcal{U}$, and the problem is thus completely solved.

It is possible that, for certain area ratios, Eq. (VI.g.3) has no solution, and the flow then cannot be supersonic on both sides of the discontinuity section. In this case, find the last characteristic for which a solution can still be found—the Mach number becomes unity in the narrower duct—and then change to the appropriate procedure described in the following chapters.

4. *Supersonic Flow Entering and Subsonic Flow Leaving the Change of Cross Section.* This case corresponds to the flow in the diverging part of a supersonic diffuser. The flow pattern is indicated in Fig. VI.g.1 for a supersonic flow coming from the left side. A shock wave must be located somewhere in the duct—station σ—and the flow conditions on the downstream side will be indicated by a prime ($'$). In drawing the wave diagram, it is then assumed that stations 1, 2, σ, and σ' are coinciding. As a consequence of this approach, the shock Mach number is equal to the flow Mach number at the location of the shock ($M_S = M_\sigma$).

Pressure waves may approach the region of changing cross section from either side, but as long as the flow enters the divergent duct with supersonic velocity, the flow conditions at station 1 are completely determined, as in the preceding chapter.

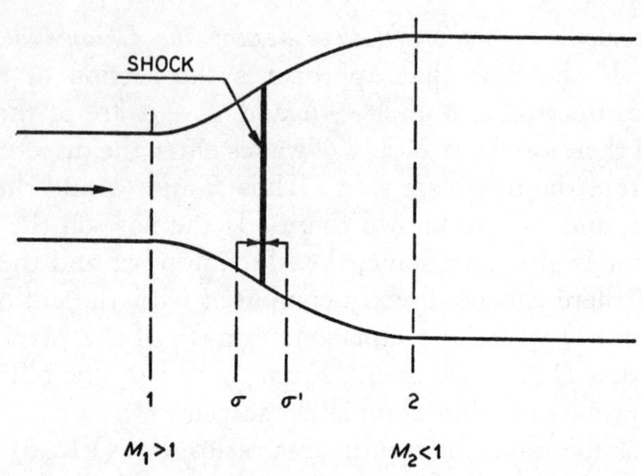

Fig. VI.g.1. Supersonic flow entering a divergent duct with a shock wave formed in the divergent section

The energy equation between stations 1 and 2 is again

$$\frac{2}{\gamma-1}\mathfrak{a}_2^{\,2}+\mathfrak{u}_2^{\,2}=\frac{2}{\gamma-1}\mathfrak{a}_1^{\,2}+\mathfrak{u}_1^{\,2}=\frac{2}{\gamma-1}\mathfrak{a}_{1,s}^{\,2}$$

where $\mathfrak{a}_{1,s}$ is the known stagnation value of the speed of sound in the supersonic flow. A second condition is provided by the known Riemann variable at station 2. The solution of these two relations presents exactly the same mathematical problem as that of isentropic flow into a duct from a reservoir (Chapter VI.d.3). The chart that was prepared to facilitate its solution may therefore be used for the problem under consideration here if subscripts e and E of Chart 1 are replaced by 2 and $1,s$, respectively.

One obtains thus a solution for $\mathfrak{a}_2$ and $\mathfrak{u}_2$ and, therefore, also for M_2, but S_2 remains to be determined since the entropy rise between the two stations depends on the location of the shock wave.

If the state of the gas in Eq. (III.c.5) is expressed in terms of stagnation pressure and temperature, one can calculate the entropy rise across a shock wave as

$$S' - S = \frac{1}{\gamma} \ln \frac{p_s}{p_s'} \tag{VI.g.4}$$

since the stagnation temperature remains constant in this case. The continuity equation for a steady adiabatic but not isentropic flow can be written in the form (see Appendix)

$$p_s AD = \text{const.} \tag{VI.g.5}$$

Since the flow is isentropic between stations 1 and σ, and also between σ' and 2 (see Fig. VI.g.1), one has the following relations

$$\left.\begin{aligned} A_1 D_1 &= A_\sigma D_\sigma \\ A_2 D_2 &= A_\sigma D_\sigma' \\ p_{s,\sigma} D_\sigma &= p_{s,\sigma}' D_\sigma' \end{aligned}\right\} \tag{VI.g.6}$$

These combine with Eq. (VI.g.4) to

$$S_2 - S_1 = S_\sigma' - S_\sigma = \frac{1}{\gamma} \ln \frac{A_2 D_2}{A_1 D_1} \tag{VI.g.7}$$

Since M_1 and M_2 are already known, Eq. (VI.g.7) allows the entropy rise to be computed which completes the calculations. The shock Mach number can be obtained with the aid of Table 1, and from the first of Eqs. (VI.g.6) one can determine the shock location (A_σ). Clearly, for the foregoing procedure to be valid, A_σ must lie between A_1 and A_2. This condition should always be verified because, if it is not satisfied, the waves are so strong that the shock is pushed away from the region of area change. The shock then either travels upstream into the supersonic flow or is swept downstream by the flow. Once the shock wave leaves the region of area change, the flow there becomes isentropic and either completely subsonic or completely supersonic. In these cases, the appropriate procedures of the preceding chapters apply.

Numerical Example: With reference to Fig. VI.g.1, let $A_2/A_1 = 1.4$ and assume that a steady supersonic flow of air enters through the narrow duct with $\mathcal{a}_1 = 1.000$ and $\mathcal{u}_1 = 1.500$ so that $M_1 = 1.500$. A shock occurs somewhere in the diverging section, and pressure waves arrive there from the downstream side. Determine the reflection of the characteristic $Q_2 = 5.300$, and the entropy rise of the gas particle that passes through the change of cross section at this instant. Find also the extreme variations of Q_2 for which the shock can remain within the diverging portion of the duct.

The stagnation speed of sound in the supersonic flow is given by $\mathcal{a}_{1,s} = (1 + 1.5^2/5)^{1/2} = 1.204$. For $Q_2/\mathcal{a}_{1,s} = 4.402$, one obtains from Chart 1a $\mathcal{a}_2/\mathcal{a}_{1,s} = 0.977$ and, therefore, $\mathcal{a}_2 = 1.176$. From this and Q_2 follow $\mathcal{u}_2 = 0.580$ and then $M_2 = 0.493$. The reflected wave is, therefore, determined by $P_2 = 6.460$. Since $D_1 = 0.492$ and $D_2 = 0.428$, one obtains from Eq. (VI.g.7), $S_2 - S_1 = 0.142$ as the entropy rise of a gas particle that passes through the shock wave at the instant of arrival of the Q-wave under consideration. Table 1a shows that the shock Mach number for this entropy rise is given by $M_\sigma = 1.785$ ($D_\sigma = 0.407$). The first of Eqs. (VI.g.6) yields for the position of the shock wave $A_\sigma/A_1 = 1.209$ so that the condition $A_1 \leq A_\sigma \leq A_2$ is satisfied.

In order to answer the second part of the question, one must determine the flow conditions at 2 if the shock occurs either at the narrowest or at the widest duct area. Only the results of these simple, steady-flow calculations will be given here. In the first case, one has, clearly, $M_\sigma = M_1 = 1.5$ which leads to $M_2 = 0.419$, $\mathcal{a}_2 = 1.184$, and $\mathcal{u}_2 = 0.496$. In the second case, the supersonic flow reaches $M_\sigma = 1.971$ before the shock occurs. This leads to $M_2 = 0.582$, $\mathfrak{a}_2 = 1.165$, and $\mathfrak{u}_2 = 0.678$. It is thus seen that, as soon as Q_2 exceeds a maximum value of 5.424, the shock travels upstream into the supersonic flow; if Q_2 drops below 5.147, the shock can no longer remain in the diverging section of the duct and is swept downstream.

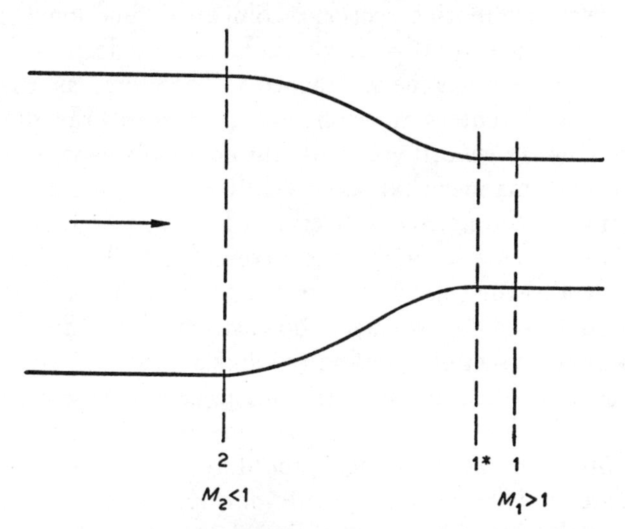

FIG. VI.g.2. Subsonic flow entering and supersonic flow leaving a region where the cross section of the duct is reduced

5. *Subsonic Flow Entering and Supersonic Flow Leaving the Change of Cross Section.* A subsonic flow becoming supersonic in a simple converging nozzle (Fig. VI.g.2) represents a flow pattern that cannot exist in steady flow. In order to show how it can be established in nonsteady flow, assume at first that the flow in the narrower duct is just sonic ($M_1 = 1.0$). At the larger section 2, the flow is then subsonic and has a Mach number that depends on the area ratio A_2/A_1. Since no characteristic can travel upstream, only waves coming from the subsonic side have to be considered.

Let a compression wave approach the contraction region. As it overtakes any gas layer, it increases its Mach number, and it would thus appear as if sonic flow could be reached somewhere in the converging part of the duct. However, this is not possible since the mass flow through such a section could never pass through the following smaller one. This is easily seen from the condition $AD =$ const. (Eq. VI.g.3) which must hold between two neighboring sections. Since the function D has its maximum value at $M = 1$, the area cannot decrease further once sonic flow has been reached. Therefore, as the compression wave enters the contracting part of the duct, the resulting wave reflections must automatically slow down the flow in such manner that sonic velocity is reached only at the beginning of the narrowest section which will be indicated by 1*. The pressure level there is increased over its value before the arrival of the compression wave, and it is thus seen that the latter continues to propagate downstream into the region of constant cross section, making the flow at station 1 supersonic.

In a wave diagram where gradual area changes are considered, the described wave phenomena emerge quite automatically. In the simplified procedure where sections 1, 1*, and 2 are made to coincide, steady-flow relations can be applied between the flow conditions at 2 and 1*, while wave diagram procedures must be used to obtain the conditions at 1.

The general case ($M_1 > 1$) can now be explained with the aid of Fig. VI.g.3, which is based on a flow direction from left to right. Assume that the problem has already been solved for points 1, 1*, and 2. The waves entering and leaving the discontinuity are labeled with the respective Riemann variables. On the left side are the arriving and reflected waves P_2 and Q_2, respectively. In going through the discontinuity, P_2 changes to P_{1*} and immediately to P_1. Since the flow at 1 is supersonic, Q_1 is traveling in the same direction as P_1. At 1* the Mach number is 1.0 and the wave Q_{1*} propagates, therefore, with zero velocity.

Let P_4 be the next arriving characteristic. The steady-flow relations between conditions 4 and 3* are given again by Eqs.

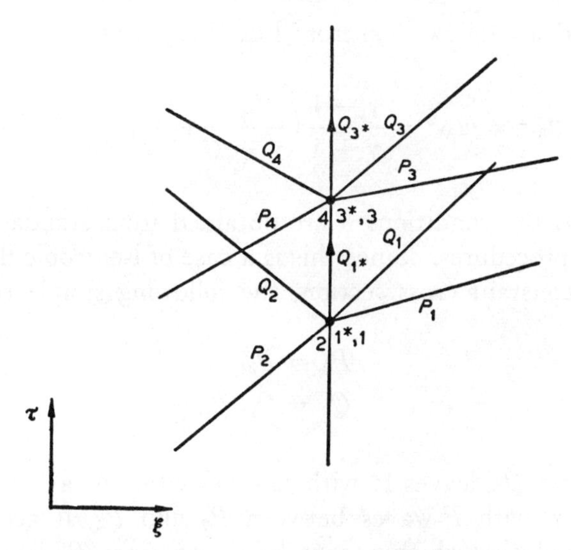

FIG. VI.g.3. Appearance of wave diagram when subsonic flow enters and supersonic flow leaves a region where the cross section of the duct is reduced

(VI.g.1 and 3) with the additional condition $M_{3*} = 1.0$. Eq. (VI.g.3) yields thus

$$D_4 = D_{3*} \frac{A_{3*}}{A_4} = D_{M=1} \frac{A_1}{A_4} \qquad \text{(VI.g.8)}$$

Once this is solved for M_4, one can compute $\mathfrak{a}_4$ and $\mathfrak{u}_4$ from the known value of the Riemann variable of the arriving wave as

$$\left.\begin{aligned} \mathfrak{a}_4 &= \frac{P_4 \text{ (or } Q_4)}{\dfrac{2}{\gamma - 1} + M_4} \\ \mathfrak{u}_4 &= \pm M_4 \mathfrak{a}_4 \end{aligned}\right\} \qquad \text{(VI.g.9)}$$

In these equations, allowance has been made for the possibility that the flow direction and contraction of the duct may be from left to right or vice versa so that either P_4 or Q_4 is known.

$\mathfrak{A}_{3*}$ and $\mathfrak{U}_{3*}$ follow then from Eq. (VI.g.1) as

$$\mathfrak{A}_{3*}^2 = \mathfrak{U}_{3*}^2 = \frac{\gamma - 1}{\gamma + 1}\left(\frac{2}{\gamma - 1} + M_4^2\right) \mathfrak{A}_4^2 \quad \text{(VI.g.10)}$$

Finally, the conditions 3 are obtained from standard wave diagram procedures. Since this is a case of isentropic flow in a duct of constant cross section, the following simple relations apply

$$\begin{aligned} P_3 &= P_{3*} \\ Q_3 &= Q_{1*} \end{aligned} \quad \text{(VI.g.11)}$$

The wave Q_{1*} leaves 1* with zero velocity. As a result of its interaction with P-waves between P_2 and P_4, it acquires a positive velocity which cannot be shown since P_4 is the next wave that is actually entered in the wave diagram. It follows, however, that Q_{1*} is the Q-wave that passes through 3 as expressed by the above equation.

With the determination of P_3 and Q_3, the problem is completely solved. Since no entropy changes are introduced at the discontinuity, each gas layer in passing through it maintains the entropy level at which it arrives there.

VI.h. Flame Fronts

1. *General Remarks.* It was pointed out in Chapter V.c.2.1 that combustion processes in nonsteady flow might be treated in a wave diagram by assuming the chemical reaction to take place instantaneously as the unburned gas passes through an advancing flame front. The main advantage of this approach lies in the ease with which the wave diagrams on either side of the flame front may be constructed since time-consuming procedures are limited to points along the flame front. Two parameters are required to characterize the combustion pro-

cess:* the amount of heat q released per unit weight of combustible mixture, given by Eq. (V.c.4), and the burning velocity v_f. (The nondimensional form of v_f is defined in the usual manner by $\mathfrak{V}_f = v_f/a_0$.) Under the usual burning conditions, the average velocity with which the flame advances relative to the unburned gas is greatly increased over the burning velocity in laminar flow since turbulence and flame front instabilities make the effective area of the flame front considerably larger than the cross section of the duct. For a one-dimensional treatment of flame propagation, only this "effective" burning velocity is of interest, and it is this quantity to which v_f refers; v_f may be many times the laminar burning velocity which is only of the order of one foot per second for most air/fuel mixtures. In the following, v_f, as a measure of the chemical reaction rate, will always be treated as a positive quantity. Variables pertaining to the burned and unburned gases on the two sides of a flame front will be indicated by subscripts b and u, respectively.

Considering the large temperature differences that appear across flame fronts, one might wish to use different values of the specific heats on either side. If this is done, one deals essentially with two different gases and, as in Chapter VI.f, the same reference values, a_0 and p_0, will be used on both sides of the discontinuity; the entropy level at these reference states will be taken as zero.

As the gas passes though the flame front, it expands and in order to maintain the continuity of mass flow, it must undergo an acceleration. The associated pressure drop across the flame front for various values of the temperature ratio T_b/T_u and the burning velocity was computed [33] by means of procedures that are essentially those to be described in Chapter VI.h.3; the specific heats were assumed to be constant, and the results are presented in the following table.

* This does not apply in the case of detonation where only one parameter is required (see Chapter VI.h.4).

PRESSURE RATIO ACROSS A FLAME FRONT, $\frac{p_u}{p_b}$

T_b/T_u \ v_f/a_u	0.10	0.15	0.20	0.25	0.30
6	1.05	1.12	1.20	1.31	1.45
7	1.06	1.13	1.22	1.34	1.48
8	1.07	1.14	1.24	1.36	1.50

It is seen that for moderate burning velocities—less than 10–15% of the speed of sound in the unburned gas—the pressure change across a flame front is so small that it can usually be neglected, particularly in view of the sweeping assumptions that must be made for the combustion process.

The value of v_f depends not only on the combustible mixture but also on the state of the unburned gas (particularly its temperature) and on the past history of the burning process. Not enough is known as yet to state how v_f must be varied in the course of wave diagram construction, and one is thus forced to make some reasonable assumptions about the magnitude and behavior of v_f. It will be seen that, if the specific heats and the pressure are taken to be the same on both sides of a flame front, a simple procedure results if one assumes that the burning velocity is directly proportional to the absolute temperature of the unburned gas. Other assumptions, including the apparently simpler one, v_f = const., lead to time-consuming trial-and-error procedures. Although q and v_f may be given arbitrary values, certain limitations are imposed on the choice of these quantities since the burned gas cannot expand to higher than sonic velocity.[69, 105]

The velocity of the flame front relative to the duct will be denoted by w_f (in nondimensional form, $\mathcal{W}_f = w_f/a_0$). One sees immediately that

$$\mathcal{W}_f = \mathcal{U}_u \pm \mathcal{V}_f \qquad \text{(VI.h.1)}$$

where the upper sign must be used if the unburned gas is on

the right side of the flame front and the lower sign refers to the opposite case.

The Riemann variables P_L and Q_R for the waves arriving at the flame front, and the entropy level of the unburned gas S_u, are known. The additional conditions required to solve the matching problem at the flame front are provided by the continuity, energy, and momentum equations for steady flow. The last of these may be replaced by the assumption of equal pressures on both sides of the flame front in the case of moderate burning velocities.

2. *Moderate Burning Velocities*

2.1. *Variable Specific Heats.* The most commonly used combustible gases are mixtures of hydrocarbons and air with air/fuel ratios of 15 or higher. Changes of the gas constant during combustion need not then be considered, but for the sake of generality, these will be included in the following equations. However, the specific heat of air changes at high temperatures, and one may wish to use different values of γ for the unburned and burned gases. A convenient choice is $\gamma_b = 4/3$ (see also the footnote to the Table of Properties of Some Important Gases in Chapter XI).

The values of γ to be used on the two sides of the flame front must first be selected. Some assumptions as to the variation of the burning velocity with the flow conditions in the unburned gas must also be made. However, the procedures derived in the following do not involve the actual assumptions.

The conservation of mass flow through the flame front can be expressed in the form

$$\frac{\rho_u}{\rho_b} = \frac{\mathcal{U}_b - \mathcal{W}_f}{\mathcal{U}_u - \mathcal{W}_f} \qquad \text{(VI.h.2)}$$

In this relation, $\mathcal{W}_f$ may be eliminated with the aid of Eq. (VI.h.1), and since it is here assumed that the pressure is the same on both sides of the flame front, one obtains

$$\frac{\rho_u}{\rho_b} = 1 \pm \frac{\mathcal{U}_u - \mathcal{U}_b}{\mathcal{V}_f} = \frac{R_b T_b}{R_u T_u} = \frac{\gamma_u}{\gamma_b}\left(\frac{\mathcal{a}_b}{\mathcal{a}_u}\right)^2 \qquad \text{(VI.h.3)}$$

In this and all following equations where the $\pm$ sign appears, the upper sign again applies if the unburned gas is on the right side of the flame front.

Since only moderate burning velocities are being considered here, the flow velocities relative to the flame front are comparatively low, and the stagnation temperature in a coordinate system in which the flame front is at rest differs only insignificantly from the static temperature at the flame front. The temperature ratio T_b/T_u can, therefore, be computed from the energy equation as

$$c_{pb} T_b = c_{pu} T_u + q \qquad \text{(VI.h.4)}$$

Since $a_u^2 = \gamma_u R_u T_u = gJ(\gamma_u - 1) c_{pu} T_u$
one obtains

$$\frac{T_b}{T_u} = \frac{c_{pu}}{c_{pb}}\left(1 + \frac{K_1}{\mathcal{a}_u^2}\right) \qquad \text{(VI.h.5)}$$

where

$$K_1 = \frac{gJ(\gamma_u - 1)}{a_0^2} q \qquad \text{(VI.h.6)}$$

The selected values of the specific heats in the burned and unburned gas may be assumed to be constant because temperature changes produced by waves in these regions are small compared with the temperature change across the flame front. The heat released by the chemical reaction q depends somewhat on the final temperature,* but only a preselected constant value is used. These simplifications seem to be quite reasonable in view of the other approximations that must be made. K_1 may

* See p. 224 of reference 83.

thus be considered as a known constant, and Eqs. (VI.h.3 and 5) can be solved for $\mathfrak{A}_b$ and $\mathfrak{U}_b$ in terms of $\mathfrak{A}_u$:

$$\mathfrak{A}_b = \mathfrak{A}_u \left[\frac{\gamma_b - 1}{\gamma_u - 1}\left(1 + \frac{K_1}{\mathfrak{A}_u^2}\right)\right]^{1/2} \qquad \text{(VI.h.7)}$$

$$\mathfrak{U}_b = \mathfrak{U}_u \mp \mathfrak{V}_f \left[\frac{\gamma_b - 1}{\gamma_b} \cdot \frac{\gamma_u}{\gamma_u - 1}\left(1 + \frac{K_1}{\mathfrak{A}_u^2}\right) - 1\right] \qquad \text{(VI.h.8)}$$

These two equations form the basis of a trial-and-error procedure. First, one estimates the value of $\mathfrak{A}_u$; then, $\mathfrak{U}_u$ follows from the known value of the Riemann variable in the unburned gas, P_u or Q_u. Since S_u is also known, all flow conditions in the unburned gas are determined, and the value of $\mathfrak{V}_f$ follows from whatever assumptions have been made for this quantity. $\mathfrak{A}_b$ and $\mathfrak{U}_b$ can then be computed from Eqs. (VI.h.7 and 8), and these results must be consistent with the known Riemann variable on the side of the burned gas; otherwise, the estimate for $\mathfrak{A}_u$ must be modified until agreement is reached.

After these calculations have been completed, it remains to establish the entropy level in the burned gas. The entropy rise across the flame front cannot be determined directly since one deals effectively with different gases on the two sides of the flame front; it must be computed from other known state variables and assumptions about the entropy level at the selected reference states. With the convention that the entropy level at the reference state is zero on both sides of the flame front, one obtains from Eq. (III.c.6)

$$S_b = \frac{2}{\gamma_b - 1} \ln \mathfrak{A}_b - \frac{1}{\gamma_b} \ln \frac{p_b}{p_0}$$

and

$$S_u = \frac{2}{\gamma_u - 1} \ln \mathfrak{A}_u - \frac{1}{\gamma_u} \ln \frac{p_u}{p_0}$$

Considering that $p_b = p_u$, we can combine these relations to give

$$S_b = \frac{\gamma_u}{\gamma_b} S_u + \frac{2}{\gamma_b - 1} \ln \mathfrak{a}_b - \frac{2\gamma_u}{\gamma_b(\gamma_u - 1)} \ln \mathfrak{a}_u \quad \text{(VI.h.9)}$$

2.2. *Constant Specific Heats and Burning Velocity Proportional to Temperature of Unburned Gas.* The procedures described in the preceding chapter may be considerably shortened if the same value of the specific heat is used on both sides of the flame front, and if one makes, furthermore, the assumption that the burning velocity varies proportionately with the temperature of the unburned gas. If k is the proportionality factor,* this assumption may be expressed in the form

$$v_f = kT_u = \frac{k}{\gamma R} a_u^2$$

or, in nondimensional form,

$$\mathfrak{v}_f = \frac{ka_0}{\gamma R} \mathfrak{a}_u^2 \quad \text{(VI.h.10)}$$

With these assumptions, Eq. (VI.h.8) simplifies to

$$\mathfrak{u}_b = \mathfrak{u}_u \mp K_2 \quad \text{(VI.h.11)}$$

where, using Eqs. (VI.h.5) and (V.c.4),

$$K_2 = \frac{ka_0}{\gamma R} K_1 = \frac{kq}{c_p a_0} = \frac{kH\eta_c}{c_p a_0(\alpha + 1)} \quad \text{(VI.h.12)}$$

It is easily verified that the sum of the known Riemann variables combined with Eq. (VI.h.11) leads to

$$\mathfrak{a}_b + \mathfrak{a}_u = \frac{\gamma - 1}{2} (P_L + Q_R + K_2) = K_3 \quad \text{(VI.h.13)}$$

* Essentially, this assumption amounts to determining k from the value of v_f that one selects for some gas temperature.

regardless of which side of the flame front contains the unburned gas. From Eq. (VI.h.7), one obtains then

$$\mathfrak{a}_b^2 = \mathfrak{a}_u^2 + K_1$$

which, together with Eq. (VI.h.13), yields the simple solutions

$$\left.\begin{aligned} \mathfrak{a}_u &= \frac{K_3}{2} - \frac{K_1}{2K_3} \\ \mathfrak{a}_b &= \frac{K_3}{2} + \frac{K_1}{2K_3} \end{aligned}\right\} \qquad \text{(VI.h.14)}$$

The value of K_3 must be determined for every point along the flame front from Eq. (VI.h.13), whereas K_1 and K_2 remain constant as defined by Eqs. (VI.h.5 and 12). The velocities $\mathfrak{U}_u$ and $\mathfrak{U}_b$ follow from the known Riemann variables, and the entropy rise across the flame front can be calculated from Eq. (VI.h.9) which now simplifies to

$$S_b = S_u + \frac{2}{\gamma - 1} \ln \frac{\mathfrak{a}_b}{\mathfrak{a}_u} \qquad \text{(VI.h.15)}$$

It is seen that all equations can be solved directly and that no time-consuming trial-and-error procedures are required. This method should, therefore, be applied whenever flame fronts must be considered in a wave diagram unless there are good reasons for using the more laborious procedures described in the preceding and following chapters.

Numerical Example: If the Riemann variables on the two sides of the flame front are known, the procedures of this chapter should not present any difficulties. The following example illustrates a case in which the Riemann variables are not directly available from the wave diagram.

Consider a closed duct of constant cross section filled with air. Gasoline vapor (H = 19,000 Btu/lb) is added to a portion of the air adjacent to a closed end until the air/fuel ratio reaches $\alpha = 20$. Let the mixture be ignited at the closed end and assume that a flame front advances into the unburned gas with a burning velocity that is taken proportional to the absolute temperature of the unburned gas. Suppose that $v_f = 100$ ft/sec at 540°R, so that $k = 100/540$, and assume a combustion efficiency $\eta_c = 90\%$. Let the initial temperature of the air be 540°R, so that the speed of sound is 1140 ft/sec. Neglect changes of pressure and specific heat across the flame front.

Prepare a wave diagram for this case and investigate the phenomena that occur when the flame front reaches the interface between air and combustible mixture.

A sketch of the wave diagram (not drawn to scale) is shown in Fig. VI.h.1. Let the initial state of the gas in the duct be taken as the reference state. The interface merely separates combustible mixture from air, but as long as it is not reached by the flame front, the state of the gas is the same on both sides. Since P_L and Q_R at the flame front are not known, the value of K_3 in Eq. (VI.h.13) cannot be found, and Eqs. (VI.h.14) therefore cannot be applied. However, since the burned gas is adjacent to a closed end of the duct, one knows that $\mathcal{U}_b = 0$, and Eq. (VI.h.11) yields then $\mathcal{U}_u = K_2 = 0.551$ for the numerical values prescribed here. The gas ahead of the flame front must be accelerated from rest to this velocity. Since the flame is assumed to reach its burning velocity instantly upon ignition, the acceleration of the gas takes place through a shock wave traveling ahead of the flame front. The shock strength is given by $|\Delta\mathcal{U}|/\mathcal{A} = \mathcal{U}_u/\mathcal{A}_0 = 0.551$, and one finds then $\mathcal{A}'/\mathcal{A} = \mathcal{A}_u = 1.116$, $M_S = \mathcal{W} = 1.383$, and $\Delta S = S_u = 0.028$ from Table 1a.* The burning velocity is given by $v_f = 100\ T_u/T_0 = 100\ \mathcal{A}_u^2$ (since the burning velocity was prescribed at the reference temperature) so that $\mathcal{V}_f = v_f/a_0 = 0.109$. From Eq. (VI.h.1), then, follows the propagation

* One finds also that $p'/p = p_u/p_0 = 2.07$. This shows that the unburned gas is precompressed by the shock wave. Both precompression and acceleration of the gases ahead of a propagating flame are phenomena of major importance in nonsteady-flow devices, such as the pulsejet, where combustion energy is utilized.

velocity of the flame front as $\mathcal{W}_f = 0.660$. Eqs. (VI.h.5 and 7) yield $\mathcal{a}_b = 2.719$ and, finally, one obtains $S_b = 4.480$ from Eq. (VI.h.15).

After a certain time, when the flame front reaches the interface, combustion stops. One must, therefore, expect pressure

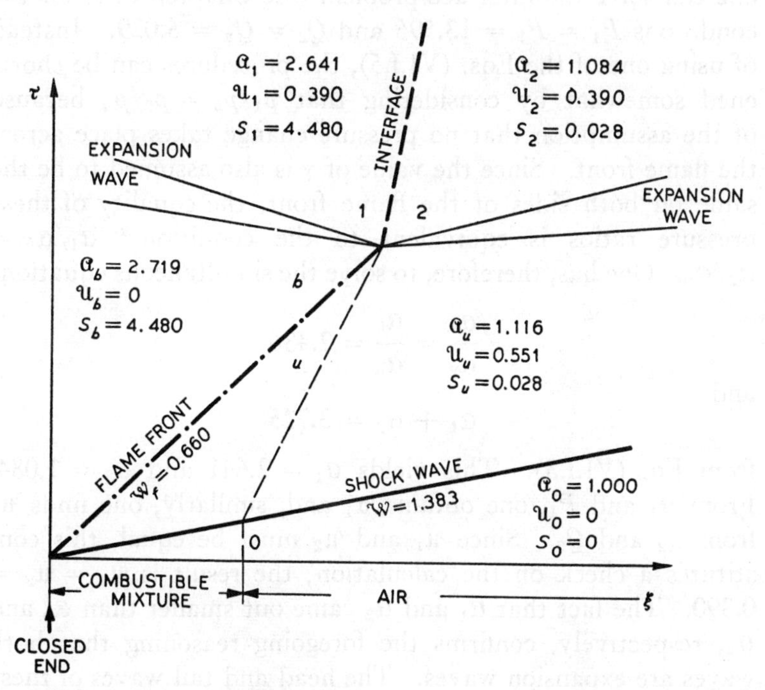

FIG. VI.h.1. Flame front advancing into a combustible mixture. Ignition at a closed end of the duct (see the numerical example)

waves traveling both upstream and downstream to originate at that point. The gas on the right side of the interface receives no further impulses from the expanding burned gas. On the contrary, its own inertia tends to carry the latter along. The resulting waves must, therefore, be of a kind that accelerates the burned gas and decelerates the air. These physical considerations show that both waves must be expansion waves.

The same conclusion is, of course, reached by carrying out the following procedures which do not require any advance information as to the nature of the waves.

Denoting the flow conditions of the burned gas and the air after passage of these waves by subscripts 1 and 2, respectively, one can solve the interface problem (see Chapter VI.f) for the conditions $P_1 = P_b = 13.595$ and $Q_2 = Q_u = 5.029$. Instead of using one of the Eqs. (VI.f.5), the procedures can be shortened somewhat by considering that $p_1/p_b = p_2/p_u$ because of the assumption that no pressure change takes place across the flame front. Since the value of γ is also assumed to be the same on both sides of the flame front, the equality of these pressure ratios is equivalent to the condition * $\mathcal{a}_1/\mathcal{a}_b = \mathcal{a}_2/\mathcal{a}_u$. One has, therefore, to solve the simultaneous equations

$$\frac{\mathcal{a}_1}{\mathcal{a}_2} = \frac{\mathcal{a}_b}{\mathcal{a}_u} = 2.436$$

and

$$\mathcal{a}_1 + \mathcal{a}_2 = 3.725$$

from Eq. (VI.f.3). This yields $\mathcal{a}_1 = 2.641$ and $\mathcal{a}_2 = 1.084$. From $\mathcal{a}_1$ and P_1, one obtains $\mathcal{u}_1$ and, similarly, one finds $\mathcal{u}_2$ from $\mathcal{a}_2$ and Q_2. Since $\mathcal{u}_1$ and $\mathcal{u}_2$ must be equal, this constitutes a check on the calculation; the result is $\mathcal{u}_1 = \mathcal{u}_2 = 0.390$. The fact that $\mathcal{a}_1$ and $\mathcal{a}_2$ came out smaller than $\mathcal{a}_b$ and $\mathcal{a}_u$, respectively, confirms the foregoing reasoning that both waves are expansion waves. The head and tail waves of these travel, therefore, with the velocities $(\mathcal{u} + \mathcal{a})_u$ and $(\mathcal{u} + \mathcal{a})_2$, respectively, on the right side of the interface and, similarly, with $(\mathcal{u} - \mathcal{a})_b$ and $(\mathcal{u} - \mathcal{a})_1$ on the left side. All numerical results obtained are collected in Fig. VI.h.1.

* This implies that the waves are either expansion waves or, if they are shock waves, that they are so weak that the changes of state which they produce can be treated as isentropic. In this particular case, it has already been shown on physical grounds that the waves are expansion waves, but even if this were not known, it would be best to guess that the waves will produce isentropic changes of state. A change to the more complicated shock relations is then postponed until the need for this has been established (see Chapter VII.a).

3. *High Burning Velocities.* If the burning velocity is so high that the pressure change across the flame front cannot be neglected, the exact energy and momentum equations must be used for the derivation of the matching procedures for the flame front. Different values for the specific heats of the burned and unburned gases may again be used.

From the momentum equation

$$p_b - p_u = \rho_u(\mathcal{U}_u - \mathcal{W}_f)(\mathcal{U}_u - \mathcal{U}_b)a_0^2$$

one can compute the pressure ratio

$$\frac{p_b}{p_u} = 1 - \frac{\gamma_u}{\mathcal{A}_u^2}(\mathcal{W}_f - \mathcal{U}_u)(\mathcal{U}_u - \mathcal{U}_b) \qquad \text{(VI.h.16)}$$

If this is combined with the continuity equation in the form of Eq. (VI.h.2), one obtains the relation

$$\left(\frac{\mathcal{A}_b}{\mathcal{A}_u}\right)^2 = \frac{\gamma_b}{\gamma_u}\left[1 - \frac{\gamma_u}{\mathcal{A}_u^2}(\mathcal{W}_f - \mathcal{U}_u)(\mathcal{U}_u - \mathcal{U}_b)\right]\frac{\mathcal{U}_b - \mathcal{W}_f}{\mathcal{U}_u - \mathcal{W}_f} \qquad \text{(VI.h.17)}$$

The energy equation must now include a term for the chemical energy to be released and, therefore, takes the form

$$(\mathcal{U}_u - \mathcal{W}_f)^2 + \frac{2}{\gamma_u - 1}\mathcal{A}_u^2 + \frac{2gJq}{a_0^2} = (\mathcal{U}_b - \mathcal{W}_f)^2 + \frac{2}{\gamma_b - 1}\mathcal{A}_b^2 \qquad \text{(VI.h.18)}$$

The following trial-and-error procedure may then be used. First, estimate the value of $\mathcal{A}_u$. This, together with S_u and the known Riemann variable, determines the flow conditions in the unburned gas. From Eq. (VI.h.1) and whatever assumptions have been made for the burning velocity, one obtains $\mathcal{W}_f$ so that the left-hand side of Eq. (VI.h.18) can be computed and a quadratic relation between $\mathcal{A}_b$ and $\mathcal{U}_b$ remains. The known Riemann variable in the burned gas provides an additional linear relation between these unknowns, and the two equations must be solved. This is, probably, best done by using the actual numerical values instead of writing a complicated general solution in terms of all the quantities involved. It must then be checked whether or not the results obtained are consistent with Eq. (VI.h.17). In case of disagreement, the estimate for $\mathfrak{a}_u$ must be modified until agreement is reached.

Finally, the entropy level S_b must be determined. Using again the convention that the entropy level of both reference states is zero, one obtains from Eq. (III.c.6)

$$S_b = \frac{\gamma_u}{\gamma_b} S_u + \frac{2}{\gamma_b - 1} \ln \mathfrak{a}_b - \frac{2\gamma_u}{\gamma_b(\gamma_u - 1)} \ln \mathfrak{a}_u + \frac{1}{\gamma_b} \ln \frac{p_u}{p_b} \tag{VI.h.19}$$

The ratio p_u/p_b in this equation must be computed from Eq. (VI.h.16).

It is seen that these procedures are quite lengthy. If high rates of burning must be treated in a wave diagram, it may be well to consider whether some other model of the combustion process would not be equally acceptable and at the same time involve less time-consuming procedures. In particular, instantaneous volume combustion (see Chapter VII.j) may offer a good compromise in such cases.

4. *Detonation Waves.* In the preceding chapters, the reaction rate in the flame front $\mathfrak{v}_f$ could be assigned any "reasonable" value. This is no longer true in the case of detonation, where only one reaction rate is possible. The latter is determined by the Chapman-Jouguet rule which requires that the reaction front, when observed from the burned gas behind it, move with the speed of sound in the burned gas (see, for instance, page 212 of reference 13). In addition to Eq. (VI.h.1) and the conservation conditions for mass flow, momentum, and energy, one has, therefore, also the relation

$$\mathfrak{w}_f - \mathfrak{u}_b = \pm \mathfrak{a}_b \tag{VI.h.20}$$

where again the upper sign applies if the unburned gas is to the right of the detonation wave.

If one substitutes Eqs. (VI.h.1 and 20) into Eq. (VI.h.17), one obtains, after some rearranging,

$$\mathcal{a}_b = \frac{\gamma_b \mathcal{v}_f}{\gamma_u(\gamma_b + 1)}\left[\gamma_u + \left(\frac{\mathcal{a}_u}{\mathcal{v}_f}\right)^2\right] \qquad \text{(VI.h.21)}$$

and elimination of $\mathcal{w}_f$ from Eqs. (VI.h.1 and 20) yields

$$\mathcal{u}_b = \mathcal{u}_u \pm \mathcal{v}_f \mp \mathcal{a}_b \qquad \text{(VI.h.22)}$$

In the wave diagram, the flow variables ahead of the detonation wave are known, and the speed of sound and the flow velocity behind the wave can then be computed from Eqs. (VI.h.21 and 22), provided the detonation velocity is known. This is the case for certain gas mixtures (see, for instance, reference 83). In general, however, one must compute $\mathcal{v}_f$ from the heat of reaction q of the mixture considered. For this purpose, one substitutes Eqs. (VI.h.1, 20 and 21) into Eq. (VI.h.18) and, after rearranging, obtains a quadratic equation for $\mathcal{v}_f^2$

$$\left(\frac{\mathcal{v}_f}{\mathcal{a}_u}\right)^4 - 2\left[\frac{(\gamma_b^2 - 1)gJq}{a_0^2\mathcal{a}_u^2} + \frac{\gamma_b^2 - 1}{\gamma_u - 1} - \frac{\gamma_b^2}{\gamma_u}\right]\left(\frac{\mathcal{v}_f}{\mathcal{a}_u}\right)^2 + \left(\frac{\gamma_b}{\gamma_u}\right)^2 = 0 \qquad \text{(VI.h.23)}$$

This equation has two positive solutions of which the larger one corresponds to the Chapman-Jouguet detonation velocity, while the other solution represents a Chapman-Jouguet deflagration (see, for instance, Chapter III.86 of reference 13) that is not considered here. Small deviations from the Chapman-Jouguet detonation velocity may be observed because of the three-dimensional structure of actual detonation waves.[118, 121]

The entropy level in the burned gas is again given by Eq. (VI.h.19) where the pressure ratio can now be expressed in the form

$$\frac{p_b}{p_u} = 1 + \frac{\gamma_u}{\mathcal{a}_u^2}\mathcal{v}_f(\mathcal{v}_f - \mathcal{a}_b) \qquad \text{(VI.h.24)}$$

which is obtained by substituting Eqs. (VI.h.1 and 20) into Eq. (VI.h.16).

VI.i. Duct Terminated by a Short Nozzle

1. *Outflow.* One is sometimes interested in flow through ducts which are terminated by a short converging or diverging section. If this part of the duct is so short that the flow conditions do not change significantly during the time required for a wave to pass through it, one may treat the gradual area change as if it were discontinuous and the nozzle had zero length. The same procedures may, therefore, also be used if the duct is terminated by an orifice plate. (Regarding the errors that may be introduced by such simplifications, see the remarks in Chapter VI.g.1.) The problems of wave reflection from the end of the duct are essentially modifications of those treated in Chapter VI.g. Instead of both P_L and Q_R being known at the duct section where the area changes discontinuously, one of these variables must now be replaced by the boundary conditions discussed in Chapters VI.d.2 and 4.

Subscript e will be used to denote the quantities relating to the exit section, whereas the flow conditions in the duct adjacent to the end section will be indicated by subscript 1. In the wave diagram, stations 1 and e are then made to coincide.

The flow in the vicinity of the exit is not one-dimensional, and it may be desirable to assume an "effective" exit area to allow for this. For example, the flow through a converging nozzle contracts further after leaving the exit section and the boundary conditions apply at the minimum section (vena contracta). Only crude estimates can be made for the effective exit area since little is known as yet about three-dimensional, nonsteady flow. In the following, A_e will be taken as the chosen, effective exit area at which the boundary conditions apply.

The flow between stations 1 and e can be treated in the same manner as described in Chapter VI.g.1. In the case of subsonic outflow, the boundary condition directly determines $\mathcal{A}_e$ at any time (Eq. VI.d.1). The corresponding values of $\mathcal{A}_1$ and $\mathcal{U}_1$ are conveniently obtained from the known Riemann variable of the arriving wave and an auxiliary chart prepared for the prescribed values of A_1/A_e and γ. This chart may be constructed in the following manner. For any value of M_e between zero and 1.0, the corresponding values of M_1 and $\mathfrak{a}_1/\mathfrak{a}_e$ are determined by Eqs. (VI.g.3 and 1), respectively; since $|\mathfrak{u}_1|/\mathfrak{a}_e = M_1 \times (\mathfrak{a}_1/\mathfrak{a}_e)$, one can also determine * P_1(or $Q_1)/\mathfrak{a}_e$. It is, therefore, possible to plot $\mathfrak{a}_1/\mathfrak{a}_e$ or $\mathfrak{u}_1/\mathfrak{a}_e$ as a function of P_1(or $Q_1)/\mathfrak{a}_e$. Since both $\mathfrak{a}_e$ and the Riemann variable are known, $\mathfrak{a}_1$ and $\mathfrak{u}_1$ are obtained with a minimum of further calculations. The value of $\mathfrak{u}_e$ could easily be determined too (the relation between $\mathfrak{u}_e/\mathfrak{a}_e$ and P_1(or $Q_1)/\mathfrak{a}_e$ could also be plotted), but this may not be required.

If the exit velocity becomes sonic, Eq. (VI.g.3) yields M_1 directly. $\mathfrak{a}_1$ and $\mathfrak{u}_1$ can then be computed from this and the known Riemann variable. If required, $\mathfrak{a}_e = \mathfrak{u}_e$ is obtained from Eq. (VI.g.1).

If the flow at station 1 is supersonic, no reflected waves can travel upstream, and the flow in the end portion of the duct can be treated by means of steady-flow relations such as discussed in Chapters VI.g.3 and 4.

2. *Inflow.* Flow into a short flare at the end of a duct has been covered in Chapter VI.d.3 where it was pointed out that the assumption of isentropic flow implies just such an inlet configuration. In the case of flow into a duct that is constricted at the end, one may still neglect the entropy rise due to flow separation at the inlet section and treat the flow as isen-

* Note that, by definition, the Mach number is always positive and one has, therefore,

$$P = \mathfrak{a}\left(\frac{2}{\gamma - 1} \pm M\right)$$

$$Q = \mathfrak{a}\left(\frac{2}{\gamma - 1} \mp M\right)$$

where the upper sign applies for positive flow velocities. Since the direction of the flow (outflow) is known, there is no doubt about the sign to be used.

tropic. The flow conditions at station 1 can then be obtained directly by means of the procedures of Chapter VI.d.3. It is necessary only to ascertain that the flow at the constriction (station e) is not choked. The continuity equation in the form of Eq. (VI.g.3) makes this check quite easy. For sonic inflow velocity, the problem becomes identical with that described in Chapter VI.g.4.

If the effects of flow separation at the inlet section are to be taken into account, the required iteration procedure is essentially that described in Chapter VI.d.5.1. The energy equation in the form of the first of Eqs. (VI.d.4) and an estimate for the Riemann variable of the wave arriving at the end of the duct represent two relations in which $\mathfrak{a}_1$ and $\mathfrak{u}_1$ are the only unknowns. The solution again can be obtained with the aid of Chart 1, and the entropy rise during inflow must then be estimated on the basis of suitable assumptions or experimental data. This leads to an improved value of the Riemann variable, and the procedure is repeated until the resulting flow conditions do not change any more.

The assumptions that one can make about the entropy rise depend greatly on the actual configuration of the inlet and may range all the way from isentropic flow to complete loss of the dynamic head in the narrowest part of the inlet nozzle. To illustrate the approach, consider the duct to be terminated by an orifice A_e that is sharp-edged on the inside and rounded on the outside. For this configuration, one may assume that the flow is isentropic from the external reservoir to the narrowest section of the stream tube. Flow separation occurs at the discontinuous widening of the duct area from A_e to A_1 and the associated entropy rise must be evaluated. As long as the flow in the orifice remains subsonic, the pressure p_e acts on the entire internal face of the orifice plate, and the momentum equation between stations e and 1 can be written as

$$p_1 A_1 = p_e A_1 + A_e \rho_e u_e (u_e - u_1)$$

If this is divided by the continuity equation in the form

$$\rho_1 = \rho_e \frac{A_e u_e}{A_1 u_1}$$

one can express p/ρ as a^2/γ and substitute $a_E{}^2 - (\gamma - 1)u^2/2$ for a^2 from the energy equation. One obtains thus a quadratic equation for u_1 in terms of u_e and known constants. Its solution is given by

$$\left.\begin{aligned} \mathcal{U}_1/\mathcal{a}_E &= \{B - [B^2 - 2(\gamma + 1)]^{1/2}\}/(\gamma + 1) \\ B &= \left[\gamma - \frac{\gamma - 1}{2}\frac{A_1}{A_e} + \frac{A_1}{A_e}\left(\frac{\mathcal{a}_E}{\mathcal{U}_e}\right)^2\right]\frac{\mathcal{U}_e}{\mathcal{a}_E} \end{aligned}\right\} \quad \text{(VI.i.1)}$$

where dimensionless variables have again been introduced. The second solution of the quadratic equation must be discarded because $\mathcal{U}_1$ must approach zero if the orifice is made small enough. For any selected value of $\mathcal{U}_e/\mathcal{a}_E$ one can now compute $\mathcal{a}_e/\mathcal{a}_E$ from the energy equation and $\mathcal{U}_1/\mathcal{a}_E$ from Eq. (VI.i.1). The entropy rise is, therefore, finally derived from Eq. (III.c.7) with the aid of the energy and continuity equations as

$$S_1 - S_E = S_1 - S_e$$

$$= \frac{1}{\gamma(\gamma - 1)} \ln \frac{\dfrac{2}{\gamma - 1} - \left(\dfrac{\mathcal{U}_1}{\mathcal{a}_E}\right)^2}{\dfrac{2}{\gamma - 1} - \left(\dfrac{\mathcal{U}_e}{\mathcal{a}_E}\right)^2} - \frac{1}{\gamma} \ln \frac{A_e \dfrac{\mathcal{U}_e}{\mathcal{a}_E}}{A_1 \dfrac{\mathcal{U}_1}{\mathcal{a}_E}} \quad \text{(VI.i.2)}$$

If the pressure in the duct drops sufficiently low to produce sonic flow through the orifice, p_e need no longer be equal to the pressure that acts on the orifice plate, and the above momentum equation does not apply. The continuity equation can then be expressed in the form of Eq. (VI.g.5) as

$$p_{e,s} A_e D_e = p_{1,s} A_1 D_1$$

and since the stagnation temperature is constant, the entropy rise is given by

$$S_1 - S_E = \frac{1}{\gamma} \ln \frac{p_{e,s}}{p_{1,s}} = \frac{1}{\gamma} \ln \frac{A_1 D_1}{A_e D_{M=1}} \qquad \text{(VI.i.3)}$$

For any value of $\mathcal{A}_1/\mathcal{A}_E$ one can then obtain $|\mathcal{U}_1|/\mathcal{A}_E$ from the energy equation; this yields M_1, and the corresponding entropy rise follows from Eq. (VI.i.3).

As pointed out in Chapter VI.d.5.1, it is best to plot $S_1 - S_E$ as function of $\mathcal{A}_1/\mathcal{A}_E$ for the area ratio required. Since $\mathcal{A}_1$ and $\mathcal{U}_1$ are known for each step of the iteration procedure, the corresponding entropy rise is then easily obtained which, in turn, leads to an improved value of the Riemann variable P_1 or Q_1. The iteration procedure is thus readily carried out.

VII

INTERACTION OF DISCONTINUITIES

VII.a. General Remarks

The procedures of Chapter VI deal with the interaction of characteristics and the various types of discontinuities that may be encountered in the course of the construction of a wave diagram. If any discontinuity occurs, the corresponding procedures must, therefore, be frequently applied. The speed with which the calculations can be carried out is, thus, of major importance, and the time expended on the preparation of auxiliary charts is easily recovered. In addition to those cases, there are a number of other important boundary and matching problems which occur comparatively infrequently. Typical examples are the intersection of two shock waves, or the crossing of a contact surface by a shock wave. In many of these cases, a direct solution is not practicable even if possible, and the problems are best solved by trial-and-error methods. The interaction of shock waves with flame fronts is discussed in some detail in reference 105.

Although the number of possible interaction problems is quite large, the approach to their solution is always the same so that the unusual cases not included in the following chapters should not present undue difficulties (see, for instance, the numerical example at the end of Chapter VI.h.2.2). The discontinuity fronts before and after interaction plus any reflected waves or newly created contact surfaces divide the wave diagram in the vicinity of the interaction point into a number of steady-flow regions. Fortunately, the nature of the waves or discontinuities that separate these regions can, in general, be anticipated so that the conservation laws for mass flow, energy, and momentum between adjacent regions need not be written in their most general form. Instead, the appropriate, simpler transition equations—the Rankine-Hugoniot equation in the case of a shock wave, or constant pressure and flow velocity for a contact surface, and so forth—can be immediately utilized. In the case of pressure waves, it may sometimes be doubtful whether they are shocks or centered expansion waves. As long as a shock is weak enough to treat changes of state across it as isentropic, there is no difference in the procedure—conservation of the Riemann variable of the crossing characteristic. It is thus best to assume first that changes of state are isentropic and to change to the more complicated Rankine-Hugoniot relations only after the presence of a sufficiently strong shock wave has been established.

VII.b. Collision of Two Shock Waves

1. *Weak Shocks.* The wave diagram for this interaction problem appears as in Fig. VII.b.1. The flow conditions in

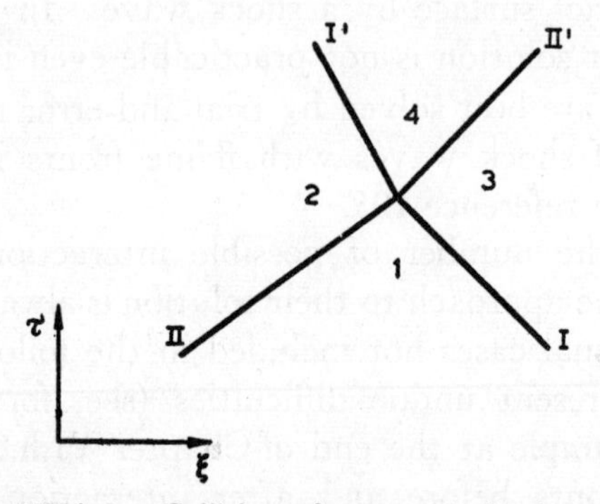

Fig. VII.b.1. Collision of two weak shock waves

region 1 and the known strengths of the colliding shock waves I and II completely determine the conditions in regions 2 and 3. Region 4 between the two shock waves I′ and II′ that emerge from the interaction point must have uniform pressure and flow velocity. If this were not so, additional pressure waves through this region would have to originate at the interaction point which cannot exist, however, since they would immediately combine with the shock waves.

Since weak shocks do not introduce entropy changes, the value of S is the same in regions 1 to 4. It was pointed out in Chapter VI.e.2 that, under these conditions, the Riemann variable of a characteristic that crosses a shock wave remains unchanged. From $Q_2 = Q_1$, the relations

$$\mathcal{U}_2 = \mathcal{U}_1 + \frac{2}{\gamma - 1}(\mathcal{a}_2 - \mathcal{a}_1)$$

and, therefore,

$$P_4 = P_2 = \frac{2}{\gamma - 1}(2\mathcal{a}_2 - \mathcal{a}_1) + \mathcal{U}_1$$

follow. Similarly, one obtains

$$Q_4 = Q_3 = \frac{2}{\gamma - 1}(2\mathcal{a}_3 - \mathcal{a}_1) - \mathcal{U}_1$$

The speed of sound in region 4 is thus given by

$$\mathcal{a}_4 = \frac{\gamma - 1}{4}(P_2 + Q_3) = \mathcal{a}_2 + \mathcal{a}_3 - \mathcal{a}_1$$

It is seen from this relation that

$$\left.\begin{aligned} \Delta\mathcal{a}_{\mathrm{I}}' &= \mathcal{a}_4 - \mathcal{a}_2 = \mathcal{a}_3 - \mathcal{a}_1 = \Delta\mathcal{a}_{\mathrm{I}} \\ \Delta\mathcal{a}_{\mathrm{II}}' &= \mathcal{a}_4 - \mathcal{a}_3 = \mathcal{a}_2 - \mathcal{a}_1 = \Delta\mathcal{a}_{\mathrm{II}} \end{aligned}\right\} \qquad \text{(VII.b.1)}$$

Thus, the increment $\Delta\mathcal{a}$ of weak shock waves is not affected by the crossing of another weak shock. This property of weak shock waves makes the handling of the intersection problem extremely simple. It must be remembered, however, that $\Delta\mathcal{a}$ is not an absolute measure of the shock strength which is characterized by the speed of sound *ratio* or one of the other quantities discussed in Chapter VI.e.3. Indeed, $\mathfrak{a}_4/\mathfrak{a}_2$ is not equal to $\mathfrak{a}_3/\mathfrak{a}_1$ which is seen if the first of Eqs. (VII.b.1) is written in the form

$$\frac{\mathfrak{a}_4 - \mathfrak{a}_2}{\mathfrak{a}_2} = \frac{\mathfrak{a}_3 - \mathfrak{a}_1}{\mathfrak{a}_1} \cdot \frac{\mathfrak{a}_1}{\mathfrak{a}_2} \tag{VII.b.2}$$

The left side of this equation is an absolute measure of the strength of shock I′ while the right side involves the strength of both shocks before collision. Clearly, for constant strength of shock I, the strength of shock I′ is reduced as shock II becomes stronger. One arrives thus at the conclusion that colliding shock waves are always weakened by the interaction.

2. *Strong Shocks.* It was pointed out in the preceding chapter that the wave diagram region between the shocks after collision must have uniform velocity and pressure and it was also shown that the shocks are weakened by the interaction. If the colliding shocks are so strong that entropy changes must be taken into account, the problem becomes not only complicated because the exact Rankine-Hugoniot relations must be used but also because the entropy rise of a fluid element depends on which of the shocks is crossed first. Consequently, the particle path through the intersection point of the shocks becomes a contact surface that separates fluid elements of different entropy levels (Fig. VII.b.2). The need for the existence of the contact surface can be shown by proving that the flows in regions 4 and 5, in general, cannot be identical. If they were identical, the following conditions would have to be satisfied

$$p_4 = p_5 \quad \text{or} \quad \left(\frac{p_4}{p_2}\right)\left(\frac{p_2}{p_1}\right) = \left(\frac{p_5}{p_3}\right)\left(\frac{p_3}{p_1}\right)$$

$$S_4 = S_5 \quad \text{or}$$

$$(S_4 - S_2) + (S_2 - S_1) = (S_5 - S_3) + (S_3 - S_1)$$

Here, the strength of each shock is expressed once by pressure ratio and once by entropy rise. These two equations could only be compatible for arbitrary strength of the colliding shocks if the entropy rise across a shock were proportional to the logarithm of the corresponding shock pressure ratio. Inspection of Eq. (VI.e.5) shows that this is not the case, since

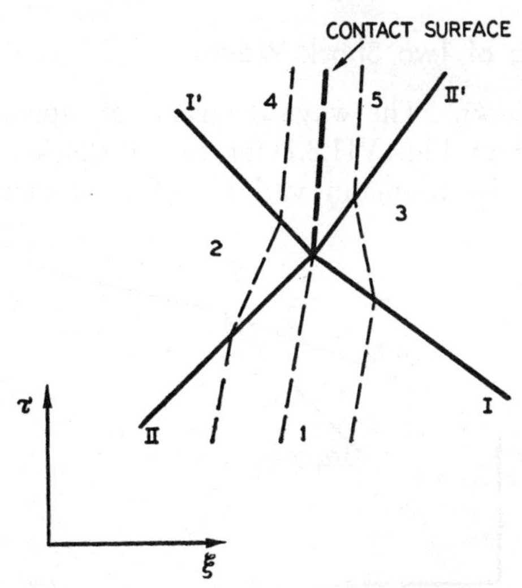

FIG. VII.b.2. Collision of two shock waves (general case)

the first factor after the ln-sign represents p'/p from Eq. (VI.e.2), whereas the second factor is a more complicated function of the pressure ratio. Thus, a contact surface in the region between the shocks after collision must be included. In the special case, when the colliding shocks are of equal strength, no interface appears for reasons of symmetry.

A direct solution of all equations that relate the neighboring regions of the wave diagram at the interaction point is not practicable. It is best to guess the value of one of the unknowns, say, $\mathfrak{u}_4$. This determines the strength of shock I′ and the other flow variables in region 4. The strength of shock II′ follows from the pressure ratio p_5/p_3 which must be equal to p_4/p_3. One obtains then $\mathfrak{U}_5$, and this must come out equal to $\mathfrak{U}_4$. In case of disagreement, the estimate of $\mathfrak{U}_4$ must be modified until agreement is reached.

A numerical example for the collision of two shock waves is given at the end of Chapter VII.e.2.

VII.c. Merging of Two Shock Waves

1. *Weak Shocks.* The wave diagram corresponding to this case is shown in Fig. VII.c.1 for two *P*-shocks. From the constancy of the Riemann variable of a characteristic that

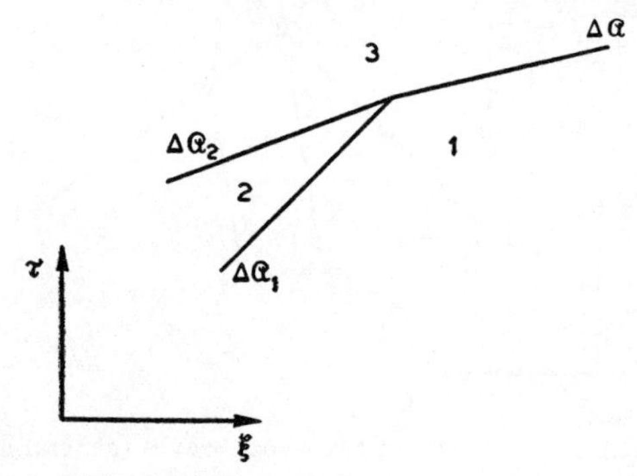

FIG. VII.c.1. Merging of two weak shock waves

crosses a weak shock wave, it follows that the flow conditions in region 3 are uniquely determined. The strength of the shock that is formed by the merging process is thus obtained from the extremely simple condition

$$\Delta\mathfrak{A} = \Delta\mathfrak{A}_1 + \Delta\mathfrak{A}_2 \qquad \text{(VII.c.1)}$$

Once the resultant shock is known, the problem is completely solved. It may happen, however, that the shock strength after merging is considered too large to neglect entropy changes. In this case, the procedures of the following chapter must be used.

2. *Strong Shocks.* When one shock wave overtakes another to form a single resultant shock, and entropy changes have to be taken into account, the wave diagram in the vicinity of the interaction point appears as shown in Fig. VII.c.2 for two merging *P*-shocks. As in the case of two colliding shocks, the particle path through the interaction point becomes an interface separating the gases that have been compressed by the

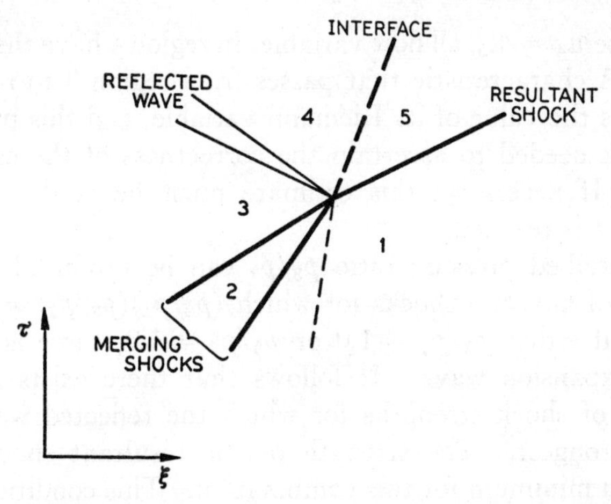

FIG. VII.c.2. Merging of two shock waves (general case)

resultant shock from those that have been affected by the two separate shocks. In addition, a reflected wave is created, and von Neumann [29] could show that the latter must be an expansion wave for all gases.* (Solutions for which the reflected wave is a shock are possible only for $\gamma > \frac{5}{3}$, which condition cannot be satisfied by any real gas.) This reflected wave is usually quite weak.

The interaction problem is best solved by iteration. Region 1 and the strengths of the merging shocks are either prescribed or known from the wave diagram, so that the flow conditions in regions 2 and 3 can be computed. One makes then an es-

* See also reference 43 and page 178 of reference 13.

timate for one of the unknowns, say, $\mathfrak{u}_5$. This determines the strength of the resultant shock and, therefore, also the remaining flow conditions in region 5. As pointed out above, the reflected wave is a (centered) expansion wave, so that $S_4 = S_3$. Since also $p_4 = p_5$, one can calculate $\mathfrak{a}_4$ from the condition (see Eq. III.c.6).

$$\mathfrak{a}_4 = \mathfrak{a}_5 e^{\frac{\gamma-1}{2}(S_4 - S_5)} \tag{VII.c.2}$$

Because $\mathfrak{u}_4 = \mathfrak{u}_5$, all flow variables in region 4 have then been found. A characteristic that passes from region 3 to region 4 maintains the value of its Riemann variable, and this provides the check needed to ascertain the correctness of the estimate for $\mathfrak{u}_5$. If necessary, this estimate must be modified until agreement is reached.

A prescribed pressure ratio p_3/p_1 can be produced by the merging of any two shocks for which $(p_2/p_1)(p_3/p_2) = p_3/p_1$. Clearly, if either $p_2/p_1 = 1.0$ or $p_3/p_2 = 1.0$, there is no reflected expansion wave. It follows that there exists a combination of shock strengths for which the reflected wave becomes strongest. The strength of the resultant shock goes through a minimum for this combination. This condition does not correspond to the merging of two equal shocks as one might expect; instead, the first shock must be somewhat stronger than the second.[43]

Numerical Example: With reference to Fig. VII.c.2, let the flow conditions in region 1 be prescribed by $\mathfrak{a}_1 = 1.000$, $\mathfrak{u}_1 = 0$ and $S_1 = 0$ (reference state). Consider the merging of two shock waves with shock Mach numbers of 1.5 and 3.0, respectively. Determine the interaction for the case in which the stronger shock overtakes the weaker one, and repeat the calculation for the reversed sequence. Assume the gas to be air.

Since the state in region 1 is the reference state, Table 1a yields directly $\mathfrak{a}_2 = 1.149$, $\mathfrak{u}_2 = 0.694$, and $S_2 = 0.052$. For the overtaking shock, one obtains $\mathfrak{a}_3/\mathfrak{a}_2 = 1.636$, $(\mathfrak{u}_3 -$

$\mathfrak{u}_2)/\mathfrak{a}_2 = 2.222$, and $S_3 - S_2 = 0.795$ from which the conditions in region 3 follow as $\mathfrak{a}_3 = 1.880$, $\mathfrak{u}_3 = 3.247$, and $S_3 = 0.847$. These determine also $P_3 = 12.647$, which is the Riemann variable of the wave that crosses the reflected expansion wave from region 3 to region 4.

According to the described procedure, one must now make a guess for $\mathfrak{u}_5 = \mathfrak{u}_4$. Nothing is known about this velocity except that it must be greater than $\mathfrak{u}_3$ (because the reflected wave is an expansion wave). Suppose the guess made is $\mathfrak{u}_5 = 3.280$. Since under the conditions of this example $\mathfrak{u}_5 = |\Delta\mathfrak{u}|/\mathfrak{a}_1$, one obtains from Table 1a directly the values $\mathfrak{a}_5 = 2.080$ and $S_5 = 1.516$. Eq. (VII.c.2) yields then $\mathfrak{a}_4 = 1.820$, which leads to $P_4 = 12.380$. This is not equal to P_3, indicating that the guess for $\mathfrak{u}_4$ is in error.

The next guess might be $\mathfrak{u}_5 = 3.502$. (In order to avoid unnecessary interpolation, one should, of course, use values that are actually listed in the table.) Repetition of the foregoing calculation yields $P_4 = 12.757$. It is now seen that the correct value of $\mathfrak{u}_5$ must lie between the two guesses made. Linear interpolation indicates that the next guess should be $\mathfrak{u}_5 = 3.435$, leading to $P_4 = 12.640$. Finally, one obtains $\mathfrak{u}_5 = 3.439$; the resulting value $P_4 = 12.649$ gives the best agreement with P_3 that can be achieved within the accuracy used. For these final values, one obtains from Table 1a also the Mach number and pressure ratio of the resultant shock.

The calculations for the second case where the weaker shock overtakes the stronger one are entirely similar. The important results of both cases are collected in the table on page 134.

The shocks in this example are so strong that the flow velocity becomes supersonic in regions 3, 4, and 5 and, in the second case, also in region 2. Both $\mathfrak{u}_3 - \mathfrak{a}_3$ and $\mathfrak{u}_4 - \mathfrak{a}_4$ are, therefore, positive, so that the reflected expansion wave does not travel to the left (as indicated in Fig. VII.c.2) but is swept downstream. The results also show that the state in region 3 is

RESULTS OF NUMERICAL EXAMPLE

VARIABLE	OVERTAKING SHOCK	
	Strong	Weak
$\mathfrak{a}_2$	1.149	1.636
$\mathfrak{u}_2$	0.694	2.222
S_2	0.052	0.795
p_2/p_1	2.46	10.33
$\mathfrak{a}_3$	1.880	1.880
$\mathfrak{u}_3$	3.247	3.357
$S_3 = S_4$	0.847	0.847
p_3/p_1	25.41	25.41
$\mathfrak{a}_4$	1.842	1.851
p_4/p_3	0.896	0.894
$\mathfrak{a}_5$	2.151	2.180
$\mathfrak{u}_5 = \mathfrak{u}_4$	3.439	3.502
S_5	1.623	1.665
p_5/p_1	21.97	22.71

not affected if the sequence of the two shocks is reversed, but the flow velocity is higher if the strong shock comes first.

See also Example (2) at the end of Chapter VII.e.2.

VII.d. Crossing of a Contact Surface by a Shock Wave

1. *General Case.* This type of interaction appears in a wave diagram as shown in Fig. VII.d.1 for a P-shock crossing an interface. The reflected wave can be either a (centered) expansion wave, as shown in the figure, or a shock wave. Since the interface conditions $p_1 = p_2$ and $p_4 = p_5$ must be satisfied, it is seen that the pressure ratio of the transmitted shock is smaller or greater than that of the incident shock, depending

on whether the reflected wave is an expansion or a shock wave.

Impulsive interface acceleration by the incident shock wave should be expected to lead to instability (see the remarks on page 88). Since the interaction process takes place practically instantaneously, interface instability should have little effect on the interaction phenomena. The breaking up of the interface may, however, affect the later flow conditions. The errors

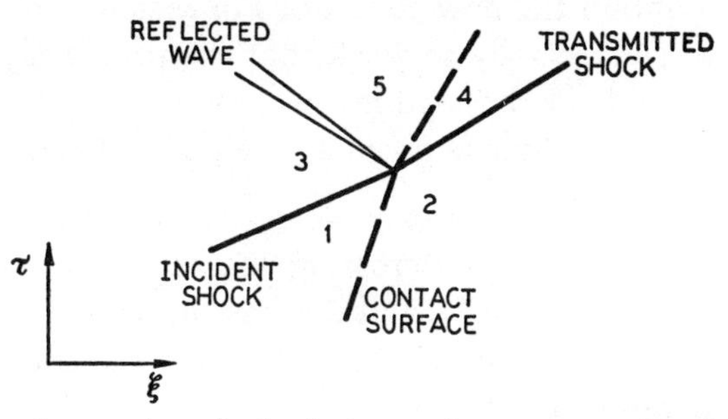

FIG. VII.d.1. Interaction of a shock wave and a contact surface (general case)

that are introduced by treating the contact surface as a well-defined discontinuity, not only before but also after interactions, have not yet been investigated.

It is not practicable to solve the interaction problem analytically except in special cases (see Chapters VII.d.2 and 3), and it is best to obtain the results by a trial-and-error procedure. Let the known flow conditions on both sides of the interface before arrival of the shock wave be indicated by subscripts 1 and 2 and let the incident shock travel in region 1. From the known strength of the shock, one obtains directly the flow conditions in region 3. Regions 4 and 5 represent the final conditions after passage of the transmitted shock and the reflected wave, respectively. It is convenient to guess for $\mathfrak{a}_4$ which determines the strength of the transmitted shock and, therefore, all flow conditions in region 4. Since $\mathfrak{u}_5 = \mathfrak{u}_4$, one knows the velocity change across the reflected wave, $\mathfrak{u}_5 - \mathfrak{u}_3$. If this change corresponds to an expansion wave ($\mathfrak{U}_5 > \mathfrak{U}_3$ for a primary P-shock as shown in Fig. VII.d.1), $\mathfrak{a}_5$ follows from the constant Riemann variable of the characteristics that connect regions 3 and 5. Since S_5 then equals S_3, one can compute p_5 which should come out equal to p_4 in order to satisfy the matching conditions for the interface. If necessary, the estimate for $\mathfrak{a}_4$ must be modified until agreement is reached. If the reflected wave is a shock wave, it is most convenient to use Table 1 to obtain the flow conditions in region 5, although the reflected wave is usually so weak that the previously described procedure could also be used in this case.

A numerical example is given at the end of Chapter VII.e.2.

2. *Weak Shock Waves and* $\gamma_L = \gamma_R$. For weak shock waves, the Riemann variable of a crossing characteristic remains constant (see Chapter VI.e.2), and if γ has the same value on both sides of the interface, a direct solution for the shock-interface interaction can be obtained.

The conditions that the waves are weak can be set down immediately. They are (for the case of an incident P-shock as shown in Fig. VII.d.1)

$$Q_3 = Q_1, \quad Q_4 = Q_2, \quad P_5 = P_3$$

(The last of these is rigorously satisfied if the reflected wave is an expansion wave.) These relations can be combined with Eqs. (VI.f.7) which express the change of a Riemann variable in crossing the interface. One obtains thus

$$\left.\begin{aligned} P_4 &= P_5 - (P_5 + Q_4)Z = P_3 - (P_3 + Q_2)Z \\ Q_5 &= Q_4 + (P_5 + Q_4)Z = Q_2 + (P_3 + Q_2)Z \end{aligned}\right\} \quad \text{(VII.d.1)}$$

This already determines the flow conditions in regions 4 and 5 and, therefore, also the transmitted and reflected waves. One may apply Eqs. (VI.f.7) also to the characteristics that cross the interface before the arrival of the incident shock wave.

This yields the relations

$$P_2 = P_1 - (P_1 + Q_2)Z$$

$$Q_3 = Q_1 = Q_2 + (P_1 + Q_2)Z$$

If these are combined with Eqs. (VII.d.1), one obtains

$$\left.\begin{aligned} P_4 - P_2 &= (P_3 - P_1)(1 - Z) \\ Q_5 - Q_3 &= (P_3 - P_1)Z \end{aligned}\right\} \qquad \text{(VII.d.2)}$$

These equations give an indication about the strength of the transmitted and reflected waves with respect to that of the arriving wave. The differences of the Riemann variables are, however, not true measures of shock strength (see Chapter VI.e.3). If the quantities

$$Q_4 - Q_2 = (Q_3 - Q_1)(1 - Z) = 0$$

$$P_5 - P_3 = (Q_3 - Q_1)Z = 0$$

are added to the first and second of the Eq. (VII.d.2), respectively, one obtains

$$\left.\begin{aligned} \mathfrak{a}_4 - \mathfrak{a}_2 &= (\mathfrak{a}_3 - \mathfrak{a}_1)(1 - Z) \\ \mathfrak{a}_5 - \mathfrak{a}_3 &= (\mathfrak{a}_3 - \mathfrak{a}_1)Z \end{aligned}\right\} \qquad \text{(VII.d.3)}$$

These relations express the change of the speed of sound across the transmitted and reflected wave, respectively, in terms of the corresponding increment of the arriving wave. Since $|Z|$ cannot exceed 1.0, the increment of the transmitted wave is always positive, indicating that the latter is always a shock wave. The type of the reflected wave depends on whether Z is positive or negative.

For $Z = 0$, no reflected wave exists. This case is discussed in the following chapter.

3. *Conditions for No Reflection from the Contact Surface.* Since the reflected wave can be either a compression or an expansion wave, conditions must be possible under which a shock

passes through an interface without producing a reflected wave. This condition is occasionally of interest and the matching equations can be solved analytically even in the general case of strong shocks and different values of γ on both sides of the interface. The wave diagram for this interaction appears as Fig. VII.d.2. The conditions for the absence of a reflected wave

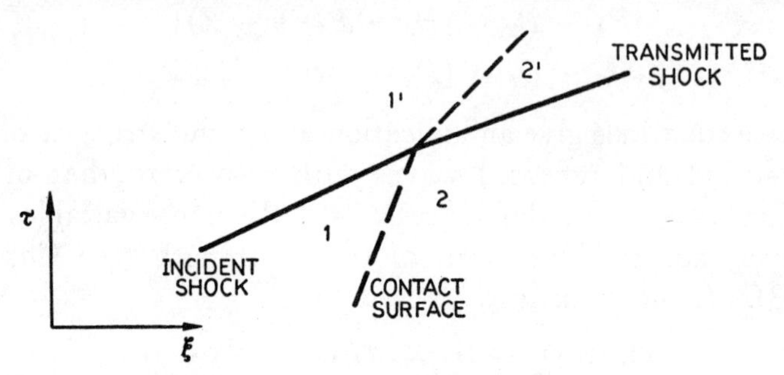

FIG. VII.d.2. Interaction of a shock wave and a contact surface (case of no reflection)

follow immediately from the matching conditions at the interface

$$\left.\begin{aligned} \frac{p_1'}{p_1} = \frac{p_2'}{p_2} = \frac{p'}{p} \\ \frac{\mathfrak{u}_1' - \mathfrak{u}_1}{\mathfrak{a}_1} \cdot \mathfrak{a}_1 = \frac{\mathfrak{u}_2' - \mathfrak{u}_2}{\mathfrak{a}_2} \cdot \mathfrak{a}_2 \end{aligned}\right\} \tag{VII.d.4}$$

The first equation indicates that the pressure ratio of the shock wave is not affected by the interaction. In the second equation, only the first terms are measures of shock strength. Although these terms are related to the corresponding pressure ratios, these relations are still functions of γ (see Chapter VI.e.1). In other words, the strengths of the incident and transmitted shocks are equal only if expressed by their pressure ratio but not in terms of other measures of shock strength.

With the aid of Eqs. (VI.e.1 and 2), one can express $|\Delta\mathfrak{u}|/\mathfrak{a}$ in terms of p'/p (the procedure is similar to that used to derive

Eqs. (VI.e.10 and 11)). The relation thus obtained,

$$\frac{|\Delta \mathfrak{u}|}{\mathfrak{a}} = \left(\frac{2}{\gamma(\gamma - 1)}\right)^{1/2} \frac{\left(\frac{p'}{p} - 1\right)}{\left(1 + \frac{\gamma + 1}{\gamma - 1}\frac{p'}{p}\right)^{1/2}} \tag{VII.d.5}$$

combines with Eqs. (VII.d.4) to

$$\frac{\gamma_1(\gamma_1 - 1)\left(1 + \frac{\gamma_1 + 1}{\gamma_1 - 1}\frac{p'}{p}\right)}{\mathfrak{a}_1^{\,2}} = \frac{\gamma_2(\gamma_2 - 1)\left(1 + \frac{\gamma_2 + 1}{\gamma_2 - 1}\frac{p'}{p}\right)}{\mathfrak{a}_2^{\,2}} \tag{VII.d.6}$$

Eq. (VII.d.6) must be satisfied whenever a shock wave passes through an interface without creating a reflected wave. It is immediately obvious that, for equal values of γ on both sides of the interface, the condition for no reflection becomes $\mathfrak{a}_1 = \mathfrak{a}_2$ regardless of the strength of the shock.

Eq. (VII.d.6) can be solved for the strength of the shock wave that produces no reflected wave

$$\frac{p'}{p} = \frac{\gamma_2(\gamma_2 - 1)\mathfrak{a}_2^{\,-2} - \gamma_1(\gamma_1 - 1)\mathfrak{a}_1^{\,-2}}{\gamma_1(\gamma_1 + 1)\mathfrak{a}_1^{\,-2} - \gamma_2(\gamma_2 + 1)\mathfrak{a}_2^{\,-2}} \tag{VII.d.7}$$

In order for this shock to be physically possible, it is necessary that the calculated pressure ratio be greater than one. This implies that the values of γ_1, γ_2, $\mathfrak{a}_1$, and $\mathfrak{a}_2$ cannot be entirely arbitrary, but must satisfy certain restrictive conditions. It is clear from the symmetry of Eq. (VII.d.7) that it is immaterial from which side the incident shock wave approaches the interface. One can, therefore, assume without loss of generality that the numerator of Eq. (VII.d.7) is positive. A physically significant solution for p'/p can then exist only if the denominator is also positive. These conditions require that

$$\frac{\gamma_2(\gamma_2 - 1)}{\gamma_1(\gamma_1 - 1)} \geq \left(\frac{\alpha_2}{\alpha_1}\right)^2 \geq \frac{\gamma_2(\gamma_2 + 1)}{\gamma_1(\gamma_1 + 1)}$$

This inequality can be satisfied only if the upper limit is greater than the lower limit, or if γ_2 is greater than γ_1. A further restriction for the upper limit of α_2/α_1 follows from the requirement $p'/p \geq 1.0$. For a positive numerator in Eq. (VII.d.7), this is satisfied if $\alpha_2/\alpha_1 \leq \gamma_2/\gamma_1$. Since $(\gamma_2/\gamma_1)^2$ is smaller than the upper limit in the foregoing inequality, the speed of sound ratio must satisfy the condition

$$\frac{\gamma_2}{\gamma_1} \geq \frac{\alpha_2}{\alpha_1} \geq \left[\frac{\gamma_2(\gamma_2 + 1)}{\gamma_1(\gamma_1 + 1)}\right]^{1/2} \quad \text{with} \quad \gamma_2 > \gamma_1 \qquad \text{(VII.d.8)}$$

One sees, therefore, that Eq. (VII.d.7) yields a physically significant solution only if the speed of sound ratio lies between rather narrow limits. At the upper limit, the value of γ/α is the same on both sides of the interface. This is equivalent to $\rho_2 a_2 = \rho_1 a_1$ since $\gamma/a = \rho a/p$; the quantity ρa is known as the characteristic acoustic impedance of a material (see any textbook on acoustics). Substituted into Eq. (VII.d.7), the upper limit leads to $p'/p = 1.0$, and one obtains thus the well-known condition that matching of the characteristic acoustic impedances at a contact surface prevents the reflection of sound waves. If, on the other hand, α_2/α_1 becomes equal to the lower limit, the denominator in Eq. (VII.d.7) vanishes, and only an infinitely strong shock wave would produce no reflection.

If one substitutes $\alpha = (\gamma p/\rho)^{1/2}$ into relation (VII.d.8), the inequalities take the form

$$\left.\begin{aligned} \gamma_2\rho_2 &\geq \gamma_1\rho_1 \\ (\gamma_2+1)\rho_2 &\leq (\gamma_1+1)\rho_1 \end{aligned}\right\} \qquad \text{(VII.d.9)}$$

The conditions (VII.d.8 or 9) can be satisfied for any pair of gases, provided that the temperatures on both sides of the interface can be properly adjusted. If the same temperature prevails on both sides, they are, in general, not satisfied.

Numerical Example: The case of shock-interface interaction for constant temperatures on both sides of the interface was considered by Paterson.[30] This author listed neon ($\gamma = \frac{5}{3}$, $\rho/\rho_{\text{air}} = 0.696$) and acetylene ($\gamma = 1.26$, $\rho/\rho_{\text{air}} = 0.905$) as a combination of gases for which the reflected wave may be absent. At what shock strength does this occur?

Making the same substitution for the speed of sound as above but this time into Eq. (VII.d.7), one obtains

$$\frac{p'}{p} = \frac{(\gamma_2 - 1)\rho_2 - (\gamma_1 - 1)\rho_1}{(\gamma_1 + 1)\rho_1 - (\gamma_2 + 1)\rho_2}$$

For the special combination of gases under consideration here, the shock strength for which there is no reflected wave is, therefore, given by $p'/p = 1.21$.

The term *taylored interface* has been used to indicate the absence of shock reflection. This condition has become important for the operation of shock tunnels designed to produce extremely high flow velocities.[111, 120, 131]

VII.e. Shock Reflection from an End of the Duct

1. *General Case.* Let subscripts 1, 2, and 3 refer to the flow regions ahead and behind the incident shock wave, and behind the reflected wave, respectively (Fig. VII.e.1). From the given

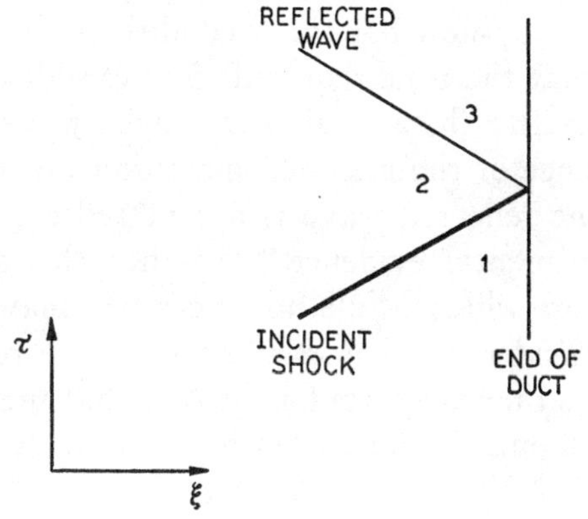

Fig. VII.e.1. Shock reflection from an open end

flow conditions in region 1 and the known strength of the incident shock wave, one obtains directly the conditions in region 2. The type and strength of the reflected wave are determined by the boundary conditions in region 3 which are usually, but not always, the same as in region 1. If the pressure in region 3 is greater than that in region 2, the reflected wave is also a shock wave; in the opposite case, it is an expansion wave. If inflow is taking place, region 3 is split into two regions by a contact surface which separates the gases that were in the tube before shock reflection and those entering later (see Chapter VII.e.3.2). The various boundary conditions that may apply at the end of a duct are discussed in Chapter VI.d.

In general, one can always solve the reflection problem by varying an estimate for the type and strength of the reflected wave until the boundary conditions are satisfied. However, in some important cases, solutions can be obtained directly.

It was pointed out in Chapter VI.d.1 that the reflection of pressure waves from an open end may be determined by using the same boundary conditions that would apply if the flow were steady. The reasoning to justify this approximation—short travel time of waves across the duct compared to the time in which the flow changes appreciably—cannot be applied when a shock wave reaches the open end since the flow changes then practically instantaneously. Nevertheless, it is general practice to apply the steady-flow boundary conditions also in this case. This implies that the time required to re-establish steady-flow exit conditions after the arrival of the shock wave is neglected. As a consequence of the assumed instantaneous readjustment of the flow, the reflected wave is a centered expansion wave. There is experimental evidence [89, 92, 95] that this simplification introduces errors which, while not excessive, may occasionally have to be considered. The results of a more refined theory which considers the lag effects for the case that the gas is at rest both inside and outside the duct before the arrival of the shock wave are included in Chapter VII.e.3.1. Otherwise, the procedures in the following chapters are all based on the steady-flow boundary conditions which should be adequate in most practical applications.

2. *Closed End or Moving Piston.* Fig. VII.e.2 illustrates this case for the right end of a duct. The boundary condition requires that $\mathfrak{U}_3$ and the piston velocity $\mathfrak{V}$ be equal. The change of velocity $\Delta\mathfrak{U}$ across the incident and reflected waves must,

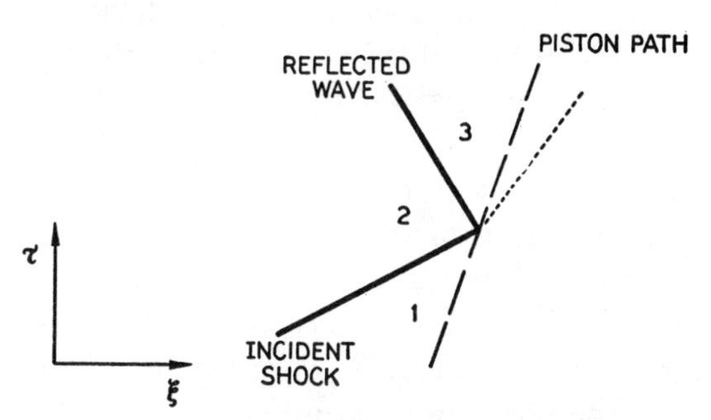

FIG. VII.e.2. Shock reflection from a moving piston

therefore, be equal and of opposite sign. The reflected wave is thus a shock wave whose strength is given by $|\mathfrak{U}_3 - \mathfrak{U}_2|/\mathfrak{A}_2$. The remaining variables are then determined by means of the Rankine-Hugoniot relations (Table 1).

For weak shock waves, the condition $|\mathfrak{U}_3 - \mathfrak{U}_2| = |\mathfrak{U}_2 - \mathfrak{U}_1|$ becomes equivalent to $\mathfrak{A}_3 - \mathfrak{A}_2 = \mathfrak{A}_2 - \mathfrak{A}_1$: The value of $\Delta\mathfrak{A}$ is the same for both incident and reflected shocks.

On rare occasions, the piston velocity may suddenly be changed at the instant of shock reflection (dotted line in Fig VII.e.2). As long as the piston does not recede from the gas with a velocity greater than $\mathfrak{U}_2$, the reflected wave remains a shock wave; for $\mathfrak{V} = \mathfrak{U}_2$ no reflected wave exists, and for still greater piston velocities an expansion wave is reflected. In the latter case, the constancy of the appropriate Riemann variable must be used instead of the Rankine-Hugoniot shock relations to determine the flow conditions in region 3.

Numerical Example: (1) Let a strong shock wave (M_s = 4.600) travel in a constant-area duct filled with air of room temperature (T_1 = 540°R). What temperature will be reached if this shock is reflected from a closed end of the duct? What is the ratio of final to initial pressures?

With reference to Fig. VII.e.2, let the conditions in region 1 serve as reference conditions ($\mathcal{A}_1 = 1.000$). Following the arriving shock wave, one has then $\mathcal{A}_2 = 2.248$, $\mathcal{U}_2 = 3.652$, and $p_2/p_1 = 24.52$ from Table 1a. The strength of the reflected shock wave is, therefore, given by $|\Delta\mathcal{U}|/\mathcal{A} = \mathcal{U}_2/\mathcal{A}_2 = 1.624$, for which value Table 1a yields $\mathcal{A}_3/\mathcal{A}_2 = 1.419$ and p_3/p_2 = 6.39; the Mach number of the reflected shock is 2.371. This shows that the strength of the reflected shock is considerably below that of the arriving shock. For the final pressure ratio, one obtains, therefore, $p_3/p_1 = (p_3/p_2)(p_2/p_1) = 156.7$. The corresponding temperature ratio is given by $T_3/T_1 = \mathcal{A}_3{}^2$ = 10.175 so that the final temperature is found as T_3 = 5490°R. This example shows that one can heat gases rapidly to extremely high temperatures by means of shock waves. This method has found important applications in studies of the properties of high-temperature gases [72] and of the kinetics of chemical reactions.[73,74,91] At temperatures such as involved in this example, the errors introduced by assuming γ and R to be constants are beginning to become noticeable, and results such as obtained here must be considered as approximations.

(2) Let a constant-area duct be filled with air at rest and take the state of this gas as the reference state. Suppose that a piston is moving into the gas from the left side traveling with a velocity $\mathcal{V} = 0.500$, while a Q-shock (M_S = 2.000) advances from the right side. Determine the resulting interaction phenomena and compare the pressure at the piston after shock reflection with that obtained if the piston had remained stationary.

The wave diagram for this case is shown in Fig. VII.e.3. The piston motion produces a P-shock of such strength that

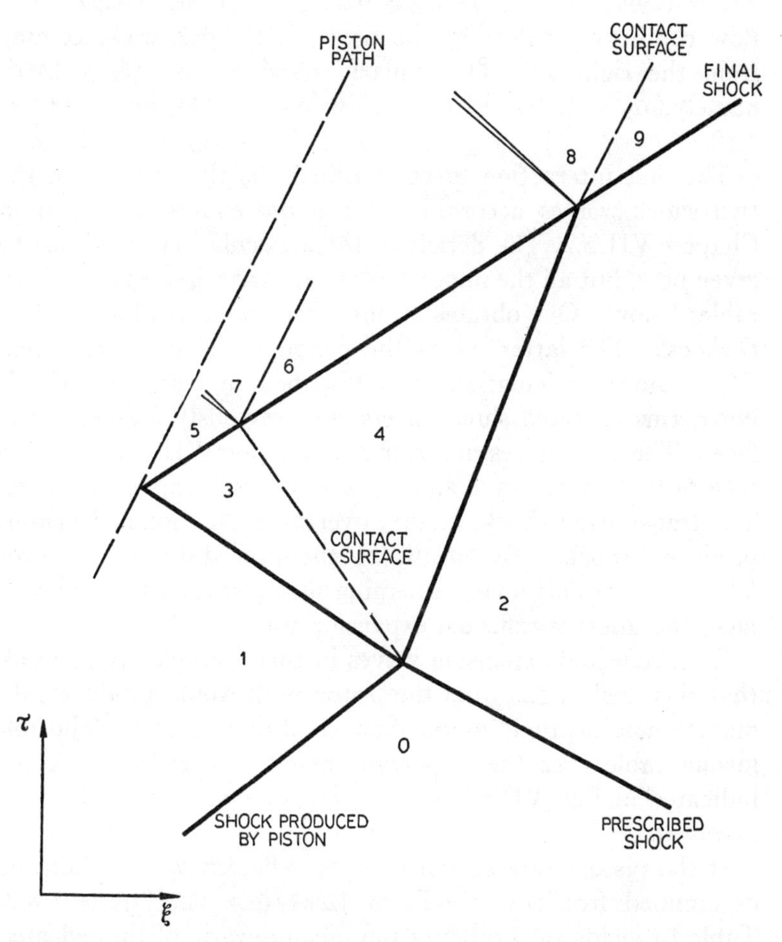

FIG. VII.e.3. Piston, set into motion, colliding with a shock wave (see example (2)) (the units for the ξ and τ scales are equal)

$\mathfrak{u}_1 = \mathfrak{v}$. (The formation of this shock depends on the manner in which the piston is accelerated to its final velocity. These details are of no concern as long as only interactions are to be studied and the exact timing of the events is not required.) For $|\Delta \mathfrak{u}|/\mathfrak{a} = \mathfrak{u}_1/\mathfrak{a}_1 = 0.500$, one obtains, from Table 1a, $M_S = 1.344$, $\mathfrak{a}_1 = 1.104$, $S_1 = 0.021$, and $p_1/p_0 = 1.94$. The flow conditions following the prescribed shock wave coming from the right side are similarly obtained for $M_S = 2.000$, namely, $\mathfrak{a}_2 = 1.299$, $\mathfrak{u}_2 = -1.250$, $S_2 = 0.234$, and $p_2/p_0 = 4.50$.

The first interaction to be considered is the collision of the two shock waves according to the procedures described in Chapter VII.b.2. The details of these calculations need not be given here, but all the important results are collected in the two tables below. One obtains an interface and a modified *P*- and *Q*-shock. The latter meets the piston path and is reflected. The required calculations are described in Chapter VII.e.2. Next, this reflected shock meets the previously created interface. The procedures according to Chapter VII.d.1 yield the transmitted shock wave and a weak reflected expansion wave. The transmitted shock, finally, overtakes the modified piston-produced shock. By means of the procedures of Chapter VII.c.2 one finds a finally emerging shock wave, a second interface, and another reflected expansion wave.

Both reflected expansion waves in this example are so weak that their reflections from the piston path would produce only minute modifications of the flow conditions found. The following tables list the important results for the ten regions indicated in Fig. VII.e.3 and the pressure ratios for all waves involved.

If the piston were stationary, the reflected shock would be determined from the condition $|\Delta \mathfrak{u}|/\mathfrak{a} = |\mathfrak{u}_2|/\mathfrak{a}_2 = 0.962$. Table 1a yields then directly the pressure ratio of the reflected shock as 3.34. The resulting pressure at the piston is, therefore, given by $3.34 \times (p_2/p_0) = 15.03$, which is considerably lower than the pressure produced when the piston is moving ($p_9/p_0 = 23.04$).

See also the example at the end of Chapter VII.f.

RESULTS OF THE NUMERICAL EXAMPLE

Region	$\mathcal{A}$	$\mathcal{U}$	S	p/p_0
0	1.000	0	0	1.00
1	1.104	0.500	0.021	1.94
2	1.299	−1.250	0.234	4.50
3	1.393	−0.738	0.201	7.65
4	1.404	−0.738	0.245	7.65
5	1.669	0.500	0.301	23.56
6	1.681	0.506	0.334	23.41
7	1.668	0.506	0.301	23.41
8	1.673	0.544	0.334	23.04
9	1.739	0.544	0.528	23.04

WAVE STRENGTHS FOR THE NUMERICAL EXAMPLE

Wave Strength	*Remark*
$p_1/p_0 = 1.94$	Shock wave produced by piston
$p_2/p_0 = 4.50$	Prescribed *Q*-shock
$p_4/p_2 = 1.70$	
$p_5/p_3 = 3.08$	
$p_6/p_4 = 3.06$	
$p_7/p_5 = 0.993$	Expansion wave
$p_8/p_6 = 0.984$	Expansion wave
$p_9/p_2 = 5.12$	Final shock wave

3. *Open Ends*

3.1. *Outflow.* The three possible cases of outflow from a duct are illustrated in Fig. VII.e.4 where the flow conditions in region 1 are known and in region 2 can be directly determined from the strength of the arriving shock wave. The reflected wave is an expansion wave which reduces the exit pressure to the level required by the boundary conditions (discussed in Chapter VI.d.2). If the flow in region 3 is subsonic, the entire centered fan of the expansion wave travels back into the duct. If the exit velocity becomes sonic before complete expansion to the external pressure has taken place, no further reduction of the pressure is possible since the waves can no longer travel upstream. Final expansion to the exterior pressure must then take place outside the duct and is of no concern here. If the flow is already supersonic in region 2, no reflected expansion wave can appear.

In the first case and, for a weak shock wave, $\mathfrak{a}_1$ and $\mathfrak{a}_3$ are equal (see Eq. (VI.d.1)) and the increments of $\mathfrak{a}$ across the

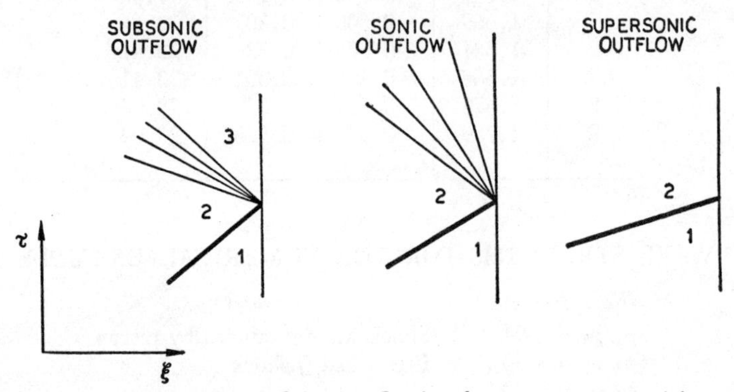

FIG. VII.e.4. Various cases of shock reflection from an open end with outflow

incident shock and the reflected expansion wave are, therefore, equal and of opposite sign. Because of the constancy of the appropriate Riemann variables, the corresponding increments of the flow velocity are then also equal and of equal sign.

It was mentioned in Chapter VII.e.1 that the lag in establishing the steady-flow boundary conditions may, occasionally, be significant. Using acoustic theory, this author [95] could derive a relation for the effective exit pressure * as function of the time after arrival of the shock wave at the exit. The results of this theory are shown in Fig. VII.e.5 where the

* The flow phenomena at the exit following the arrival of a shock wave are highly complex and certainly not one-dimensional (see, for instance, reference 93). For use in wave diagram procedures, an effective mean pressure is needed which leads to the correct wave phenomena inside the duct where the flow is again one-dimensional. This effective pressure is obtained by the above-mentioned theory.

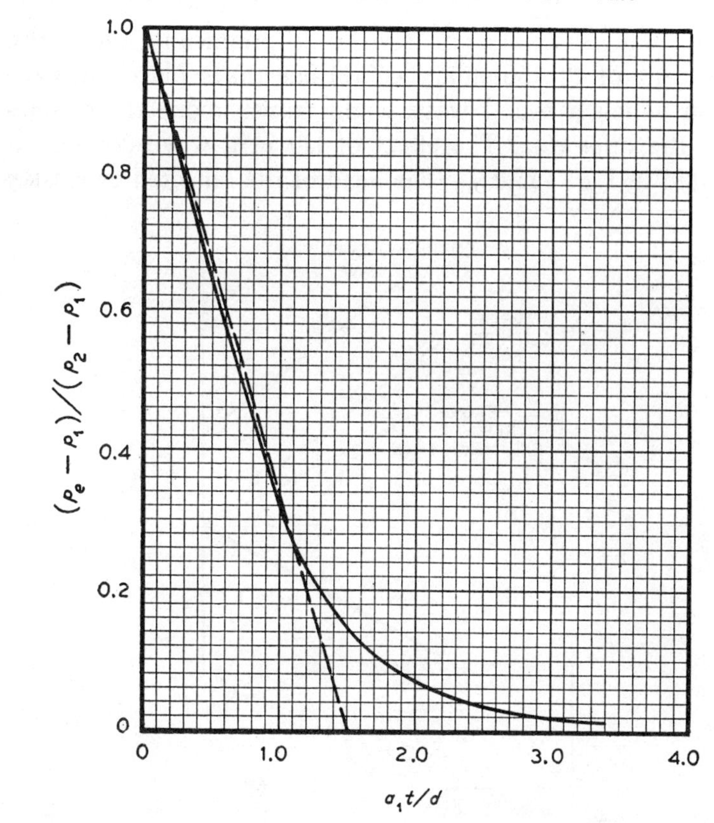

FIG. VII.e.5. Pressure transient at an open end following the arrival of a shock wave

Solid line—Acoustic theory
Dashed line—Possible approximation

pressure ratio $I(\tau) = (p_e - p_1)/(p_2 - p_1)$ is plotted against a nondimensional time based on the duct diameter d and the speed of sound before arrival of the shock as reference variables. This curve does not depend on the gas involved. The influence of the gas properties is included in the speed of sound used to obtain a nondimensional time for the abscissa scale. Values of the function $I(\tau)$ are also given in the table in Appendix II.A.1.

From Fig. VII.e.5, it is seen that practically the entire adjustment of the boundary conditions takes place during a time in which a sound wave could travel about three times across the duct exit. The effect of lag in a wave diagram is indicated in Fig. VII.e.6. The head of the reflected expansion

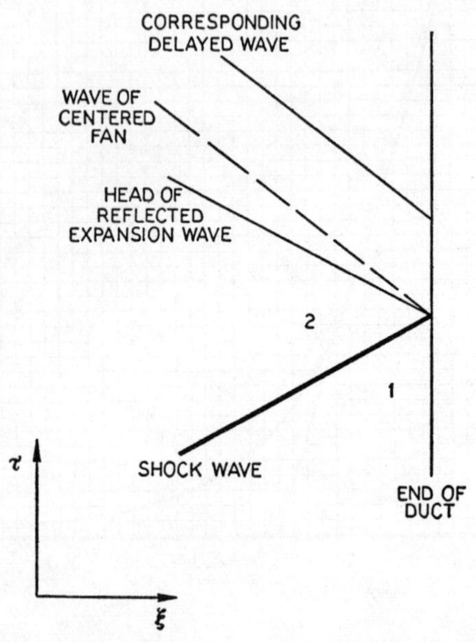

Fig. VII.e.6. Shock reflection from an open end, allowing for lag in adjustment of the exit pressure to its steady-flow value

wave originates at the instant when the shock reaches the end of the duct. Each later characteristic corresponds to exit conditions that are determined by the velocity (slope) of the particular wave, $\mathfrak{u} - \mathfrak{a}$, and by $P = P_2$ (for the right end of the duct). Since the expansion wave produces only isentropic changes of state, these conditions determine also the corresponding pressure at the exit. If one uses steady-flow boundary conditions, the entire change of the exit pressure from p_2 to its final value is assumed to take place instantaneously and one obtains a centered expansion wave. Taking lag into account, any particular exit conditions are delayed for a time that can now be obtained from Fig. VII.e.5. Note that the slope of the delayed wave is the same as that of the corresponding characteristic of the centered expansion wave, provided that the steady flow in region 2 is not disturbed by further arriving waves.

Since shock waves involved in problems of this kind are rather weak (for a shock pressure ratio of about 1.94 in air, the reflected expansion wave produces already sonic outflow), one may consider the pressure ratio plotted in Fig. VII.e.5 as equal to the corresponding speed of sound ratio. This linearization, which is often not too satisfactory, would not significantly affect the decay curve in this case. A further simplification that may be permissible is to represent the decay curve as a straight line. The decay process may then be approximated by the relation

$$\frac{\mathcal{A}_e - \mathcal{A}_1}{\mathcal{A}_2 - \mathcal{A}_1} = 1 - \beta \frac{a_1 t}{d} = 1 - \beta \mathcal{A}_1 \frac{L_0}{d} \tau \qquad \text{(VII.e.1)}$$

where β is a suitably chosen constant; the dotted line in Fig. VII.e.5 corresponds to $\beta = \frac{2}{3}$. The nondimensional time τ (measured from the instant of arrival of the shock wave at the open end) is here again stated in terms of the same unit that is used in the rest of the wave diagram (that is, based on the reference variables L_0 and a_0).

Since L_0 always represents a significant dimension of the physical system under consideration, Eq. (VII.e.1) may be used to estimate the importance of lag. Suppose that the lag of the characteristic corresponding to two thirds of the entire pressure drop through the expansion wave ($(\mathcal{A}_e - \mathcal{A}_1)/(\mathcal{A}_2 - \mathcal{A}_1) = \frac{1}{3}$) should not exceed 10% of the time in which a sound wave travels through the reference length ($\tau = 0.1$). For $\beta = \frac{2}{3}$ and $\mathcal{A}_1 = 1.0$, this condition is satisfied for $L_0/d > 10$.

The theoretical results shown in Fig. VII.e.5 were found to be in good agreement with experimental data for shock pressure ratios up to at least 1.9. This is also the range in which any lag correction would be of importance since the significance of this phenomenon decreases with increasing shock strength. The lag time as indicated in Fig. VII.e.5 is independent of shock strength. On the other hand, any characteristic of the reflected wave must advance upstream into a flow that is first produced by the shock and then further accelerated by the preceding portion of the expansion wave; the velocities with which these waves travel decrease, therefore, with increasing shock strength, and the rate at which pressure changes take place in the expansion wave also decreases. The fixed lag times become thus relatively less important with increasing shock strength.

Whether or not lag effects are considered significant depends entirely on the problem under investigation and on the particular information that one wishes to obtain from the wave diagram. It should be stressed, however, that using such refined procedures may not be warranted in view of the errors introduced into a wave diagram by other approximations. The results represented by Fig. VII.e.5 may be used to derive computing procedures for other problems of wave reflection. These are discussed in Appendix II.A.

Numerical Example: With reference to Fig. VII.e.4, suppose that steady outflow of air from a duct is prescribed by $\mathcal{U}_1 = 0.300$ and $\mathcal{A}_1 = 1.000$ $(p_1/p_0 = 1.00)$. Let a shock wave be reflected from the end of the duct and determine the strength of the reflected wave if the strength of the arriving shock is given by (1) $p_2/p_1 = 1.50$ and (2) $p_2/p_1 = 2.00$, respectively. Neglect entropy changes across these shocks.

(1) The speed of sound following the arriving shock wave is obtained from Eq. (III.c.6) as $\mathcal{A}_2/\mathcal{A}_1 = \mathcal{A}_2 = (p_2/p_1)^{1/7} = 1.060$, and the condition $Q_2 = Q_1 = 4.700$ yields $\mathcal{U}_2 = 0.600$. The exit conditions are determined by $P_3 = P_2 = 5.900$ and by the boundary condition $p_3/p_0 = p_1/p_0 = 1.00$ (provided the re-

sulting exit velocity is subsonic). The boundary condition alone determines $\mathcal{A}_3 = 1.000$, and the exit velocity follows then as $\mathcal{U}_3 = 0.900$. The pressure ratio across the reflected expansion wave is the reciprocal of that of the arriving shock wave, or $p_3/p_2 = 1/1.50$.

(2) Similarly to the first case, the stronger shock wave leads to $\mathcal{A}_2 = 1.104$ and $\mathcal{U}_2 = 0.820$. The condition $P_3 = P_2 = 6.340$ again applies, but a boundary condition $\mathcal{A}_3 = 1.000$ would not lead to subsonic outflow. Therefore, the boundary condition must now be changed to $\mathcal{U}_3 = \mathcal{A}_3$ which results in $\mathcal{U}_3 = \mathcal{A}_3 = 1.057$ and, therefore, $p_3/p_0 = 1.057^7 = 1.48$. The reflected expansion wave is now weaker than the arriving shock being given by $p_3/p_2 = 1/1.36$. As discussed above, the final expansion to p_0 takes place outside the duct.

3.2. *Inflow.* If inflow takes place before the shock wave arrives at the end of the duct, then, depending on the flow velocity and the shock strength, the flow direction may be reversed after the passage of either the shock wave or the reflected expansion wave. As long as outflow is finally established, the procedures of the preceding chapter apply. If inflow persists after wave reflection, the corresponding, more complicated boundary conditions (Chapters VI.d.3 and 5) must be used and the problem can, in general, be solved only by a trial-and-error method.

The wave diagram appears as in Fig. VII.e.7. If one estimates the value of $\mathcal{A}_3$, say, one obtains $\mathcal{U}_3$ from the energy equation (VI.d.4), and S_3 from whatever assumption is made for the entropy rise at the duct inlet. One can thus calculate the pressure at the interface and then $\mathcal{A}_4$ from Eqs. VI.f.2 since $S_4 = S_2$ is known. If there is no change of γ across the interface, $\mathcal{A}_4$ can be directly obtained from Eq. (VI.f.4). $\mathcal{U}_4$ follows then from this and the appropriate Riemann variable ($P_4 = P_2$ in Fig. VII.e.7). If it does not come out equal to $\mathcal{U}_3$, the estimate for $\mathcal{A}_3$ must be varied until agreement is reached.

A direct solution of this problem is obtainable in the case of a weak shock wave and isentropic flow. The conditions in regions 3 and 4 become then identical.

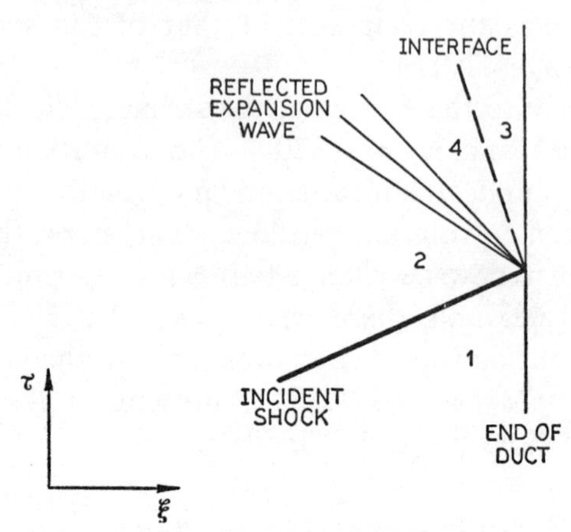

Fig. VII.e.7. Shock reflection from an open end with inflow

From the constancy of the Riemann variables, one obtains

$$\mathfrak{a}_3 = \mathfrak{a}_1 + \Delta\mathfrak{a} + \Delta\mathfrak{a}'$$

$$\mathfrak{u}_3 = \mathfrak{u}_1 \pm \frac{2}{\gamma - 1}(\Delta\mathfrak{a} - \Delta\mathfrak{a}')$$

where the upper and lower signs refer to the right and left end of a duct, respectively; $\Delta\mathfrak{a}$ is the increment across the incident shock wave, and $\Delta\mathfrak{a}'$ that across the entire reflected expansion wave ($\Delta\mathfrak{a}' = \mathfrak{a}_3 - \mathfrak{a}_2$). If the foregoing relations are substituted into the energy equation and higher order terms of the increments are neglected, one obtains the strength of the reflected wave as

$$\Delta\mathfrak{a}' = -\Delta\mathfrak{a}\frac{\mathfrak{a}_1 \pm \mathfrak{u}_1}{\mathfrak{a}_1 \mp \mathfrak{u}_1} \qquad \text{(VII.e.2)}$$

In order that the reflected wave can travel upstream, the flow in regions 2 must be subsonic. Consequently, since $\Delta\mathfrak{a}$

is positive, $\Delta\mathfrak{a}'$ is always negative, indicating again that the reflected wave is an expansion wave. It is also seen that the magnitude of $\Delta\mathfrak{a}'$ is always less than that of $\Delta\mathfrak{a}$ since Eq. (VII.e.2) applies only if the sign of $\mathfrak{u}_1$ corresponds to inflow.

VII.f. Sudden Closing of a Duct; the "Hammer" Wave

Occasionally, one is interested in the wave phenomena associated with the rapid closing of a valve in the duct. If the

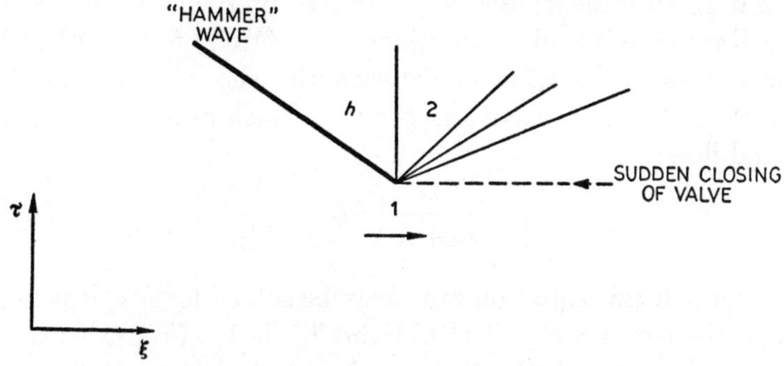

FIG. VII.f.1. Sudden closing of a valve in the duct

closing time is sufficiently short, one may treat this problem as if the flow were interrupted instantaneously. Let subscript 1 (Fig. VII.f.1) be used to denote the initial flow conditions. On one side of the valve (region 2), the inertia of the flowing gases produces a pressure drop so that the initial flow there is brought to rest by an expansion wave. Constancy of the Riemann variable of the wave that connects regions 1 and 2, together with the boundary condition $\mathfrak{u}_2 = 0$, yields the simple solution

$$\mathfrak{a}_2 = \mathfrak{a}_1 - \frac{\gamma - 1}{2}|\mathfrak{u}_1| = \mathfrak{a}_1\left(1 - \frac{\gamma - 1}{2}M_1\right) \qquad \text{(VII.f.1)}$$

where M_1 is the Mach number of the initial flow. In this form,

Eq. (VII.f.1) is valid, regardless of whether region 2 is on the left or the right side of the closing valve.

On the other side of the valve the gas "piles up" and the pressure is increased. A shock wave is formed of such strength that

$$\frac{|\Delta \mathfrak{u}|}{\mathfrak{a}_1} = \frac{|\mathfrak{u}_1|}{\mathfrak{a}_1} = M_1 \qquad \text{(VII.f.2)}$$

Because of the analogy with the phenomenon of a water hammer, this wave is called a "hammer" wave, and subscript h is used to indicate the conditions behind this wave. From the Rankine-Hugoniot equations (Eq. (VII.d.5 and VI.e.2)), one can derive a relation between the shock Mach number M_S of the hammer wave and the flow Mach number M_1 of the initial flow:

$$M_1 = \frac{2}{\gamma + 1}\left(M_S - \frac{1}{M_S}\right) \qquad \text{(VII.f.3)}$$

Although this equation can easily be solved for M_S, it is more convenient to use Eq. (VII.f.2) and Table 1. From this table, one obtains not only M_S but also the state of the gas in region h in terms of that in region 1 ($\mathfrak{a}_h/\mathfrak{a}_1$, etc.) without further calculations.

If the initial flow velocity is small enough, the hammer wave is a weak shock, and one obtains a solution similar to that expressed by Eq. (VII.f.1)

$$\mathfrak{a}_h = \mathfrak{a}_1 + \frac{\gamma - 1}{2}|\mathfrak{u}_1| = \mathfrak{a}_1\left(1 + \frac{\gamma - 1}{2} M_1\right) \quad \text{for} \quad M_1 \ll 1.0 \qquad \text{(VII.f.4)}$$

It is interesting to note that $\mathfrak{a}_h$ is higher than the stagnation value of the initial flow which is given by

$$\mathfrak{a}_{1,s} = \mathfrak{a}_1\left(1 + \frac{\gamma - 1}{2} M_1^2\right)^{1/2}$$

Consequently, the hammer pressure p_h is also higher than the stagnation pressure $p_{1,s}$. This applies, however, only for values of M_1 that are not too high. If M_1 is increased, the exact Rankine-Hugoniot relations must be used instead of Eq. (VII.f.4). It is then found that the ratio $\mathfrak{a}_h/\mathfrak{a}_{1,s}$ remains always greater than unity and approaches the value $\gamma^{1/2}$, while the entropy rise across the hammer wave causes p_h to become lower than $p_{1,s}$ at some higher Mach number. The ratio $p_h/p_{1,s}$ reaches a maximum value close to, but not exactly at, $M_1 = 1.0$ and approaches zero as M_1 approaches infinity. The following table illustrates this for the case of $\gamma = 1.4$.

COMPARISON OF HAMMER AND STAGNATION CONDITIONS

M_1	$\frac{\mathfrak{a}_h}{\mathfrak{a}_1}$	$\frac{\mathfrak{a}_{1,s}}{\mathfrak{a}_1}$	$\frac{p_h}{p_1}$	$\frac{p_{1,s}}{p_1}$	$\frac{\mathfrak{a}_h}{\mathfrak{a}_{1,s}}$	$\frac{p_h}{p_{1,s}}$
0	1.00	1.00	1.00	1.00	1.00	1.00
0.50	1.10	1.02	1.94	1.19	1.08	1.63
0.95	1.21	1.09	3.29	1.79	1.12	1.84_3 (max.)
1.00	1.23	1.10	3.48	1.89	1.12	1.84_0
1.50	1.38	1.20	5.72	3.67	1.14	1.56
2.00	1.55	1.34	8.73	7.82	1.16	1.12
2.14	1.60	1.38	9.73	9.73	1.16	1.00
2.50	1.75	1.50	12.57	17.09	1.16	0.74
∞	∞	∞	∞	∞	1.18	0

Numerical Example: The problem of the hammer wave is closely related to that of the reflection of a shock wave from a closed end of the duct. If a shock wave approaches a closed end and the flow behind the shock is considered to be the flow in region 1 of Fig. VII.f.1, then the reflected shock is, obviously, identical with the hammer wave. On the basis of this consideration, show that there is an upper limit for the strength of a shock wave that can be reflected from a closed end.

In the third example of Chapter VI.e.3, it was shown that the Mach number behind a shock wave is limited. Thus, by substituting this limit $M_1 = [2/\gamma(\gamma - 1)]^{1/2}$ into Eq. (VII.f.3), one obtains a quadratic equation for the highest possible Mach number of the reflected shock wave. The solution is given by

$$M_S = \left(\frac{2\gamma}{\gamma - 1}\right)^{1/2}$$

Thus, in air or helium, the Mach number of a shock wave that is produced by reflection from a closed end of a duct can never exceed the values 2.646 and 2.236, respectively.

See also Example IX.e for a problem involving a hammer wave.

VII.g. Sudden Opening of a Duct

1. *Opening into a Low-pressure Reservoir.* If an end of the duct is suddenly opened into a low-pressure reservoir (for instance, by breaking of a diaphragm), the gas is accelerated by an expansion wave that travels into the duct, and the wave diagram appears as in Fig. VII.g.1. Constancy of the appropriate Riemann variable, together with the exit boundary

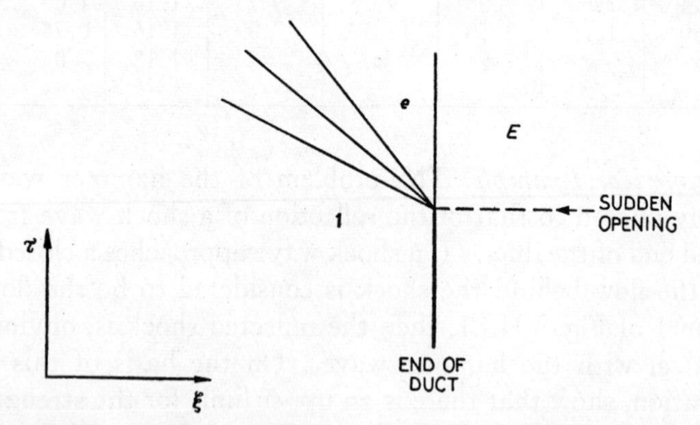

FIG. VII.g.1. Sudden opening of a duct into a low-pressure reservoir

conditions (see Chapter VI.d), determines the extent of the expansion fan. The exit velocity is either subsonic or sonic and, in these cases, the value of $\mathcal{A}_e$ is given by Eqs. (VI.d.1) or (VI.d.2).

This procedure is based on the use of the quasi-steady-flow boundary conditions. Since this case is closely related to that of a shock wave reflected from an open end, the considerations regarding the accuracy of these boundary conditions are the same as those discussed in Chapters VII.e.1 and 3.

A wave diagram for this case is presented as Example IX.b.1.

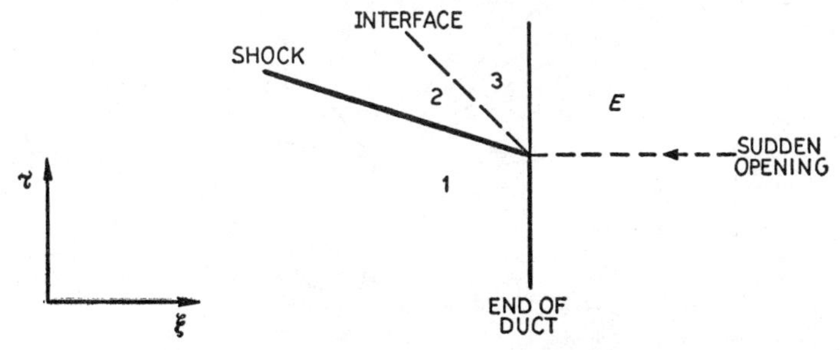

FIG. VII.g.2. Sudden opening of a duct into a high-pressure reservoir

2. *Opening into a High-pressure Reservoir.* As soon as the duct is opened, a shock wave travels into the duct followed by the interface between the gas initially in the duct and the fresh gas from the reservoir. These two gases may be different. Two conditions for the gas in region 3 (Fig. VII.g.2) are provided by the energy equation (VI.d.4) and by the relation between pressure, speed of sound, and entropy (Eq. (III.c.6)).

Depending on the assumptions that are made for the entropy rise at the inlet (see Chapter VI.d.3 and 5), the value of S_3 is either equal to the reservoir value S_E or greater. In the latter case, S_3 is a known function of $\mathcal{U}_3$ and $\mathcal{A}_3$ (for example, Eq. (VI.d.6)).

The flow variables in regions 2 and 3 are best obtained by guessing at one of the unknowns, say, $\mathcal{A}_3$. Eqs. (VI.d.4 and

III.c.6) can then be solved for $\mathfrak{U}_3$ and p_3/p_0. Since $\mathfrak{U}_2 = \mathfrak{U}_3$, the shock strength follows as $|\mathfrak{U}_2|/\mathfrak{a}_1$, and this must be compatible with the corresponding pressure ratio $\dfrac{p_2}{p_1} = \dfrac{p_3}{p_1}$. In case of disagreement, the estimate for $\mathfrak{a}_3$ must be modified until agreement is reached. The calculations can be carried out quite rapidly with the aid of Table 1.

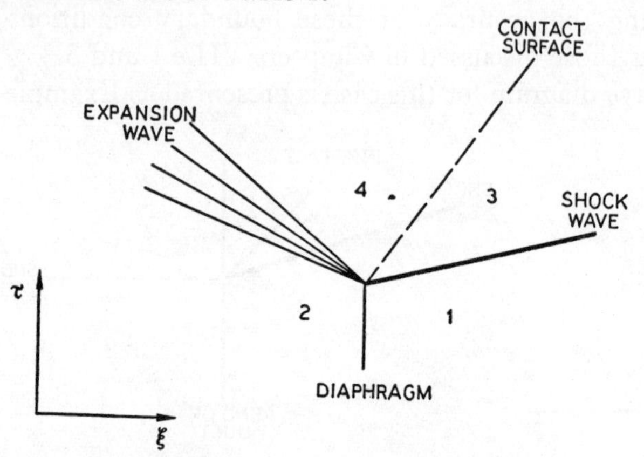

FIG. VII.g.3. Sudden removal of a partition between high- and low-pressure gases. Initial wave phenomena in a shock tube

3. *Sudden Removal of a Partition Inside the Duct; Shock Tube.* If a duct is divided by a partition into two sections in which the pressure is different, then the wave phenomena that follow a sudden removal of this partition are a combination of those described in the two preceding chapters. Experimental setups in which these phenomena can be produced are known as shock tubes. These are of great importance for a variety of investigations (see bibliography).

When the partition is suddenly removed (breaking of a diaphragm), the high-pressure gas expands and compresses the gas in the low-pressure region. A shock wave and an expansion wave are thus created (Fig. VII.g.3). In addition, the paths of the gas particles that were initially in contact with the diaphragm form, in general, a contact surface since the initial temperature on both sides of the diaphragm is arbitrary, and one may also employ two different gases.

A relation between the flow velocity in region 3 and the strength of the shock wave can be found with the aid of Eq. (VI.e.1) rearranged to the form

$$1 - \frac{\hat{\mathfrak{u}}'}{\hat{\mathfrak{u}}} = \frac{\hat{\mathfrak{u}} - \hat{\mathfrak{u}}'}{\mathfrak{a}} \cdot \frac{\mathfrak{a}}{\hat{\mathfrak{u}}} = \frac{2}{\gamma - 1} \cdot \frac{\dfrac{p'}{p} - 1}{1 + \dfrac{\gamma + 1}{\gamma - 1}\dfrac{p'}{p}}$$

In this relation, one can set $\hat{\mathfrak{u}} - \hat{\mathfrak{u}}' = |\Delta\mathfrak{u}| = |\mathfrak{u}_3|$ (the high-pressure region may be on either side of the diaphragm), $\mathfrak{a} = \mathfrak{a}_1$, $\gamma = \gamma_1$, and $p'/p = p_3/p_1$. Since $\hat{\mathfrak{u}}/\mathfrak{a} = M_S$ from Eq. (VI.e.3), one obtains thus

$$|\mathfrak{u}_3| = \frac{2\mathfrak{a}_1 M_S \left(\dfrac{p_3}{p_1} - 1\right)}{(\gamma_1 - 1)\left(1 + \dfrac{\gamma_1 + 1}{\gamma_1 - 1}\dfrac{p_3}{p_1}\right)} \tag{VII.g.1}$$

Across the expansion wave, one finds, from the constancy of the appropriate Riemann variable ($P_4 = P_2$ in Fig. VII.g.3) and the fact that changes of state take place isentropically,

$$|\mathfrak{u}_4| = \frac{2\mathfrak{a}_2}{\gamma_2 - 1}\left[1 - \left(\frac{p_4}{p_2}\right)^{\frac{\gamma_2 - 1}{2\gamma_2}}\right] \tag{VII.g.2}$$

Since $\mathfrak{u}_4 = \mathfrak{u}_3$ and $p_4 = p_3$, one can combine Eqs. (VII.g.1 and 2) to

$$\frac{p_2}{p_1} = \frac{p_3}{p_1}\left[1 - \frac{\gamma_2 - 1}{\gamma_1 - 1} \cdot \frac{\mathfrak{a}_1}{\mathfrak{a}_2} \cdot \frac{\left(\dfrac{p_3}{p_1} - 1\right) M_S}{1 + \dfrac{\gamma_1 + 1}{\gamma_1 - 1}\dfrac{p_3}{p_1}}\right]^{-\frac{2\gamma_2}{\gamma_2 - 1}} \tag{VII.g.3}$$

In this equation, M_S is related to p_3/p_1 through Eq. (VI.e.2). This relation enables one to plot or tabulate p_2/p_1 as function of p_3/p_1 for various combinations of gases and initial temperatures. Such data can be found in many publications on the shock tube (see, for instance, references 53, 55, 62, 89, 104, and 131).

If one has to start a wave diagram without the aid of special charts or tables, Eq. (VII.g.3) is not convenient if the state of the gas in regions 1 and 2 is prescribed. The flow conditions in regions 3 and 4 are then best obtained by a trial-and-error procedure. For instance, an estimate of $\mathfrak{a}_4$ yields $\mathfrak{u}_4$ from the constancy of the appropriate Riemann variable, and p_4/p_2 from the relation between pressure and the speed of sound for isentropic flow (Eq. III.c.6). Therefore, $\mathfrak{u}_3$ and p_3 have also been obtained and, if the resulting shock strengths given by $|\mathfrak{u}_3|/\mathfrak{a}_1$ and p_3/p_1 are not compatible (Table 1), the estimate for $\mathfrak{a}_4$ must be modified. In applying this procedure, care must be exercised to use the correct values of γ if two different gases are involved.

If the strength of the shock is prescribed, the diaphragm pressure ratio can be determined from Eq. (VII.g.3). Alternatively, one can find $\mathfrak{u}_3 = \mathfrak{u}_4$ and $p_3 = p_4$ with the aid of Table 1, and proceed then to compute

$$\frac{p_2}{p_1} = \frac{p_3}{p_1}\left(1 - \frac{\gamma_2 - 1}{2}\frac{|\mathfrak{u}_4|}{\mathfrak{a}_2}\right)^{-\frac{2\gamma_2}{\gamma_2 - 1}} \qquad \text{(VII.g.4)}$$

from Eq. (VII.g.2).

Numerical Example: Show that there is an upper limit of the shock strength that can be produced in a shock tube.

Clearly, the greatest shock strength is produced when the diaphragm pressure ratio becomes infinite. This requires that the bracketed term in Eq. (VII.g.3) becomes zero. One obtains thus an equation for p_3/p_1 which can be solved for various combinations of gases. If the initial temperature is the same on both sides of the diaphragm, one finds the following results [55, 62] which show the major importance of the speed of sound ratio.

MAXIMUM SHOCK STRENGTHS OBTAINABLE IN A SHOCK TUBE

Gas Combination	$\frac{\mathfrak{a}_2}{\mathfrak{a}_1}$	$\left(\frac{p_3}{p_1}\right)_{max}$
Air–Air	1.00	44
Helium–Air	2.93	132
Hydrogen–Nitrogen	3.71	574

A complete shock tube problem is presented as Example IX.d.

VII.h. Shock Wave Passing Through a Discontinuous Change of Cross Section

The case that a duct of gradually varying cross section is approximated by a discontinuous change is treated in Chapter VI.g. As a shock wave passes through such a discontinuity,

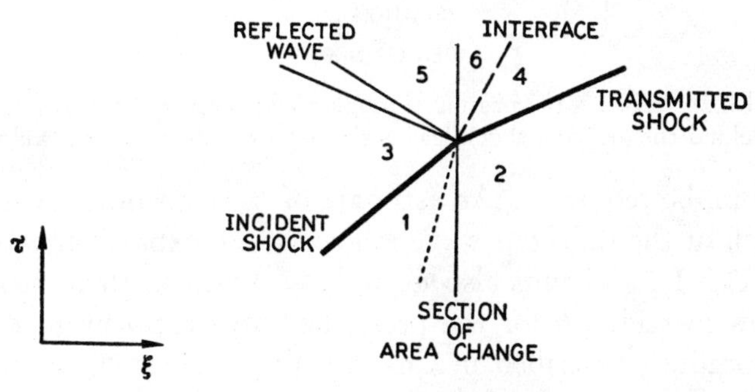

Fig. VII.h.1. Shock wave passing through a change of cross section of the duct. Subsonic flow in all regions

its strength is modified and a reflected wave is created (Fig. VII.h.1). Since a particle that passes through the transmitted shock undergoes a different entropy rise from that of a particle that crosses the incident shock and the reflected wave, an interface is formed by the particle path through the interaction point.

The flow conditions before the arrival of the incident shock wave and the strength of the latter determine regions 1, 2, and 3. Regions 4, 5, and 6 can be found by a trial-and-error procedure where the initial guess should be made for one of the variables in region 5, following the reflected wave. A guess for one of the variables in region 4 or 6 would not determine the flow in region 6 completely since S_6 depends on the entropy change across the reflected wave and an additional guess would,

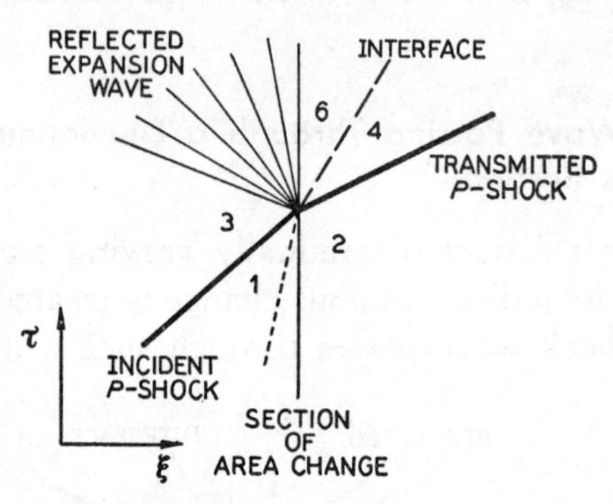

FIG. VII.h.2. Shock wave passing through a widening of the duct when the flow behind the arriving shock leaves the narrower duct at sonic velocity

therefore, be required. An estimate of $\mathfrak{a}_5$ determines type and strength of the reflected wave (shown as an expansion wave in Fig. VII.h.1), and thus also $\mathfrak{u}_5$ and S_5. One can then compute the flow in region 6 for the prescribed area ratio by means of the procedure described in Chapter VI.g. Since $\mathfrak{u}_4 = \mathfrak{u}_6$ and $p_4 = p_6$, the strength of the transmitted shock is obtained by either $|\mathfrak{u}_4 - \mathfrak{u}_2|/\mathfrak{a}_2$ or by p_4/p_2. If these two values are not compatible, the estimate for $\mathfrak{a}_5$ must be modified.

The described procedure always yields a solution for the flow conditions in the various regions of Fig. VII.h.1 as long as the flow velocity is everywhere subsonic. However, different wave patterns may be required if the flow is sonic or supersonic in one or more regions. These comparatively rare cases correspond, essentially, to those discussed in Chapter VI.g.

As an illustrative example of how the wave pattern may be modified, consider a P-shock that is moving into a widening of the duct and assume that the gas before arrival of the wave is at rest or flowing with subsonic velocity. For weak shock waves, the wave pattern is then that shown in Fig. VII.h.1. If one carries out the calculations for stronger and stronger shocks, one reaches, eventually, the condition that the velocity

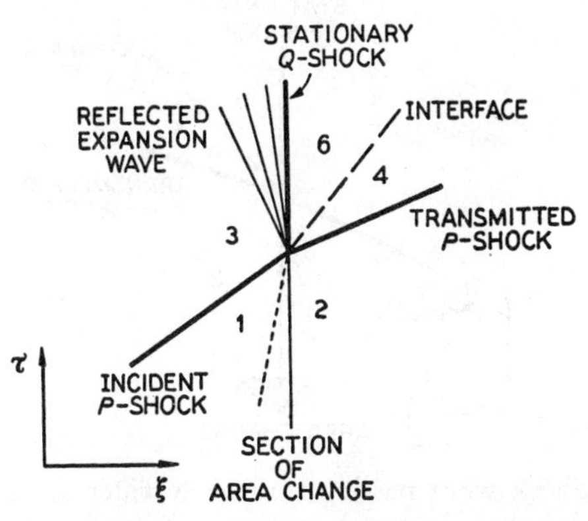

Fig. VII.h.3. Shock wave passing through a widening of the duct. Flow behind the arriving shock leaves the narrower duct with sonic velocity and a secondary shock is formed in the diverging section

with which the flow enters the discontinuity section becomes just sonic and region 5 disappears (Fig. VII.h.2). Up to this point, isentropic deceleration to the conditions in region 6 is possible. If the arriving P-shock becomes still stronger, the flow in the diverging duct section becomes supersonic at first and then a stationary Q-shock forms at such location between A_L and A_R (see Chapter VI.g.4) that both the resulting pressure and flow velocity in region 6 lead to the same strength of the transmitted shock wave (Fig. VII.h.3). The reflected expansion wave disappears when the initial shock wave is strong enough to produce sonic or supersonic flow in region 3 (Fig. VII.h.4). With further increasing strength of the P-shock, the location of the stationary shock is shifted toward the larger cross section of the duct. Once the latter is passed, a stationary Q-shock is no longer possible, since the flow velocity is then greater than the propagation velocity of the shock relative to the gas, and the shock is swept downstream (Fig. VII.h.5). A new region 7 appears between this shock and the section of the discontinuous area change. The flow between regions 3 and

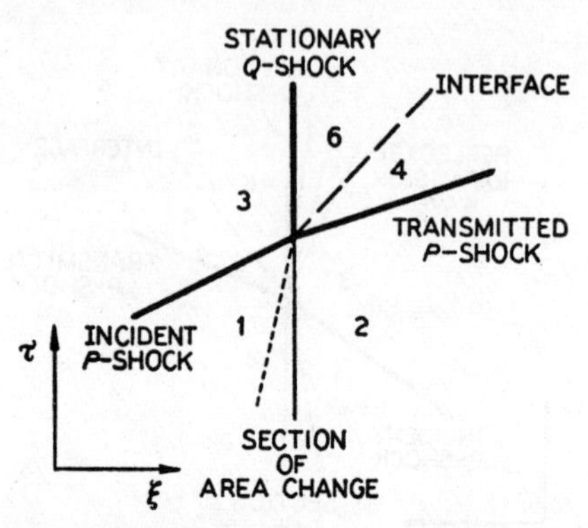

Fig. VII.h.4. Shock wave passing through a widening of the duct. Flow behind the arriving shock leaves the narrower duct at supersonic velocity and a secondary shock is formed in the diverging section

7 is steady and isentropic. The Mach number in region 7 depends only on the Mach number in region 3 and the area change and may reach very high values (see the example on p. 167). Such devices represent a form of "shock tunnels" and have become of great importance for the generation of high supersonic flow velocities (see, for instance, references 60, 120, and 131).

The shock strengths at which the various changes of the flow pattern take place depend on the area change and on the direction and magnitude of the initial flow. The correct wave pattern in any particular case is, therefore, not always known beforehand, and one must try various possibilities until a pattern is found that satisfies all conditions. This uncertainty about the wave pattern is a consequence of treating phenomena in a region of gradual area changes as if they occurred at a discontinuity. If the area change were treated as gradual, there would be no uncertainty, and the wave pattern would follow automatically from the wave diagram. In general, only one wave pattern satisfies all conditions, but if a shock wave enters a contraction of the duct and is strong enough to produce supersonic flow, there exists a range of conditions for which more than one wave pattern satisfies all requirements (see Appendix II.B).

Numerical Example: Consider a duct that includes a diverging section where the area increases by a factor of 10. This

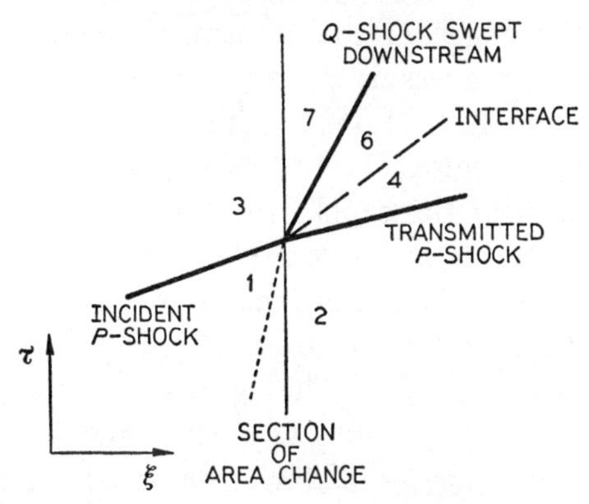

FIG. VII.h.5. Shock wave passing through a widening of the duct. Flow behind the arriving shock is supersonic in the entire region of changing cross section and a secondary shock is swept downstream

duct is filled with air at rest, and a shock wave of $M_S = 3.000$ approaches the diverging section. Compute the resulting flow pattern when the gradual area change is approximated by a discontinuous change.

In this case, the flow conditions in regions 1 and 2 are identical and may be used as reference conditions, so that $\mathcal{A}_1 = \mathcal{A}_2 = 1.000$, $\mathcal{U}_1 = \mathcal{U}_2 = 0$, and $S_1 = S_2 = 0$. For the conditions following the shock wave, one obtains $\mathcal{A}_3 = 1.637$, $\mathcal{U}_3 = 2.222$, $S_3 = 0.795$, and $p_3/p_1 = 10.33$ from Table 1a. The flow in region 3 is, therefore, supersonic ($M_3 = 1.357$), and one would expect the wave pattern to look like either Fig. VII.h.4 or 5. It is not at all clear beforehand which of the two should be selected; whether or not all transition relations can be satisfied can be determined only by trying one or the other. Suppose the calculations are based on the pattern shown in Fig. VII.h.5. In this case, the supersonic flow of region 3 is expanding as a steady, isentropic flow into the larger duct. With the aid of suitable steady-flow tables (based essentially on Eq. (VI.g.3)), one finds $M_7 = 4.021$. Eq. (VI.g.1) yields then $\mathfrak{a}_7 = 0.931$ so that $\mathfrak{u}_7 = 3.744$. The pressure in region 7 is determined by $p_7/p_1 = (p_3/p_1)(\mathfrak{a}_7/\mathfrak{a}_3)^7 = 0.199$.

The remainder of the problem is solved by iteration. For instance, the strength of the transmitted shock is estimated. Suppose the first guess is $p_4/p_1 = p_6/p_1 = 5.00$. Table 1a yields then $\mathfrak{u}_4 = \mathfrak{u}_6 = 1.359$. Thus, the change of the flow velocity between regions 7 and 6 becomes $|\mathfrak{u}_6 - \mathfrak{u}_7| = 2.385$, and the strength of the shock is obtained by dividing this by $\mathfrak{a}_7$. (Not by $\mathfrak{a}_6$ since this is a Q-shock that is swept downstream!) For this shock strength, the pressure ratio follows as $p_6/p_7 = 13.09$, whereas the original guess leads to $p_6/p_7 = (p_6/p_1)/(p_7/p_1) = 25.13$. The guess must, therefore, be modified until the two values of p_6/p_7 coincide. One obtains, finally, $p_4/p_1 = p_6/p_1 = 3.38$ as the strength of the transmitted shock for which agreement is reached. Before assuming, therefore, that Fig. VII.h.5 represents the correct wave pattern for this case, one must verify that the newly formed Q-shock is unable to travel upstream in the flow of region 7. From the pressure ratio $p_6/p_7 = 16.95$, the Mach number of the Q-shock is found to be 3.830 and, since this is less than M_7, the shock is actually swept downstream. Once the strength of both shocks has been established, the remaining flow variables can all be found with the aid of Table 1a. It is interesting to note that the Mach number of the incident shock is reduced to 1.744 in passing through the diverging duct section, whereas the newly formed Q-shock is even stronger than the incident shock. The flow is subsonic in regions 4 and 6 with flow Mach numbers of 0.799 and 0.538, respectively.

VII.i. Contact Surface Reaching an Open End of the Duct or a Discontinuous Change of Cross Section

Ordinarily, the arrival of a contact surface at an open end of the duct does not give rise to any wave phenomena. All that happens is that the gas leaving the duct changes suddenly. This is always the case if the flow before arrival of the interface is subsonic. The boundary conditions for the exit pressure (see

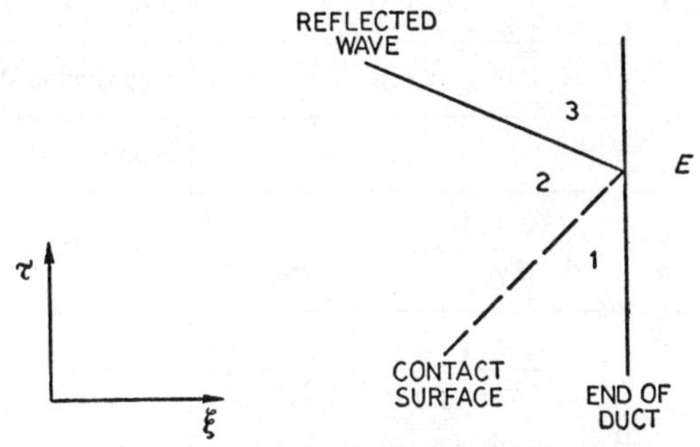

FIG. VII.i.1. Contact surface reaching an open end of the duct (general case)

Chapter VI.d.2 and 4) can then be satisfied by the flow behind the interface regardless of its Mach number.

If the initial flow is supersonic, the flow behind the interface may not be able to maintain the level of the exit pressure, and a wave travels back into the duct of such type and strength that the boundary conditions are again satisfied. The possible cases are listed in the following table for the condition that the gas surrounding the exit is at rest (Chapter VI.d.2). For subsonic flow, the exit pressure must then be equal to that of the external region, whereas for supersonic flow p_e/p_E cannot be less than the limit prescribed by Eq. (VI.d.3). In the table, subscripts 1 and 2 refer to the flow conditions on the two sides of the interface, respectively. If a reflected wave exists, the exit conditions following this wave are indicated by subscript 3 (Fig. VII.i.1). The flow behind a reflected wave is always subsonic, and p_3 is therefore equal to p_E.

WAVES CREATED WHEN AN INTERFACE REACHES AN OPEN END

(Gas in the external region at rest)

M_1	$\frac{p_1}{p_E} = \frac{p_2}{p_E}$	M_2	Reflected Wave	$\frac{p_3}{p_E}$
<1.0	1.0	$\gtrless 1.0$	None	—
>1.0	>1.0	<1.0	Expansion Wave	1.0
		>1.0	None	—
	<1.0	$< \left(\frac{\gamma_2 + 1}{2\gamma_2} \cdot \frac{p_E}{p_1} + \frac{\gamma_2 - 1}{2\gamma_2}\right)^{1/2}$	Shock Wave	1.0
		$> \left(\frac{\gamma_2 + 1}{2\gamma_2} \cdot \frac{p_E}{p_1} + \frac{\gamma_2 - 1}{2\gamma_2}\right)^{1/2}$	None	—

When a contact surface passes through a region of changing duct cross section that is approximated by a discontinuous change, transmitted and reflected waves are, in general, created even if the flow is entirely subsonic. The wave diagram for this case appears as in Fig. VII.i.2. The flow conditions in regions 1, 2, and 3 are known from preceding work on the wave diagram. One must then estimate one of the unknown flow variables, say, $\mathcal{A}_5$. This describes the type and strength of the reflected wave and, therefore, all variables in region 5. The flow from there to region 6 is steady and isentropic so that one obtains M_6 and $\mathcal{A}_6$ and, therefore, also $\mathcal{U}_6$ with the aid of Eqs. (VI.g.3 and 1). This determines both pressure and flow velocity in region 4 since these must be the same as in region 6. One of these is enough to characterize the transmitted wave, and the other serves then to check the correctness of the estimate made for $\mathcal{A}_5$.

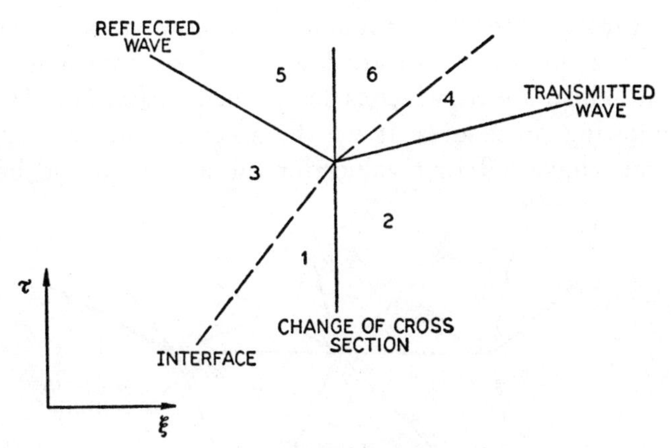

FIG. VII.i.2. Contact surface passing through a change of cross section (general case)

VII.j. Instantaneous Volume Combustion

A number of wave diagram procedures that deal with combustion phenomena were treated in Chapters V.c.2.1 and VI.h where it was pointed out that, in problems involving combustion at high burning velocities, the assumption of instantaneous volume combustion may be a good approximation. This combustion model, which is discussed in the following, requires the occurrence of a discontinuity at a certain value of the time coordinate which may be predetermined or may be assumed to depend on the wave phenomena themselves. Also the length and location of the combustion zone may be either prescribed or it may extend between the paths of two defined gas elements regardless of where they might be located at the instant of combustion (Fig. VII.j.1).

Let subscripts u and b, respectively, refer to the gas before and after it crosses the discontinuity created by the combustion process. In general, the flow variables will not be uniform along the combustion zone, and the following procedures must be carried out for any wave or particle path that passes through the discontinuity. The flow variables on the side of the unburned gas are known. Also known is q, the amount of heat added, which is determined by Eq. (V.c.4). For the following procedures, it will be assumed that the specific heats may have different values for the unburned and burned gases.

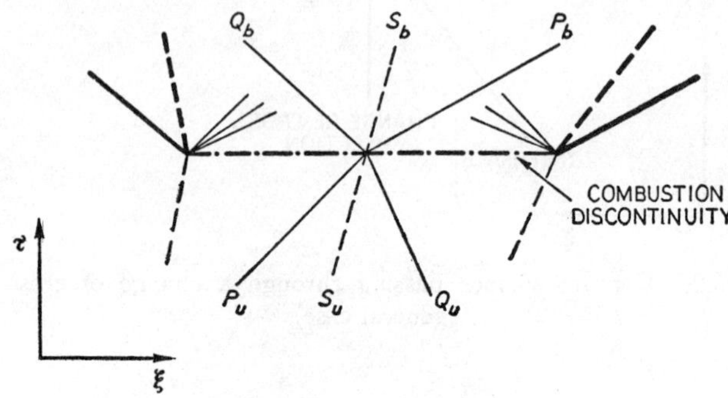

FIG. VII.j.1. Instantaneous volume combustion

The matching conditions for the wave diagrams on the two sides of the discontinuity can be derived from the consideration that instantaneous combustion of the gas contained in some region of the duct produces the same change of state as constant-volume combustion so that the gas densities before and after burning are equal. Since no expansion takes place during combustion, the flow velocities are unaffected, and the condition

$$\mathfrak{U}_b = \mathfrak{U}_u \tag{VII.j.1}$$

must be satisfied. The pressure and temperature ratios for constant-volume combustion are given by

$$\frac{p_b}{p_u} = \frac{T_b}{T_u} = \frac{c_{vu}}{c_{vb}}\left(1 + \frac{q}{c_{vu}T_u}\right) = \frac{c_{vu}}{c_{vb}}\left(1 + \frac{\gamma_u K_1}{\mathfrak{A}_u^2}\right) \tag{VII.j.2}$$

where K_1 is defined in Eq. (VI.h.6). Substitution of $(\mathfrak{A}a_0)^2/\gamma R$ for T yields a relation for the speed of sound in the burned gas

$$\mathfrak{a}_b^2 = \frac{\gamma_b(\gamma_b - 1)}{\gamma_u(\gamma_u - 1)}(\mathfrak{a}_u^2 + \gamma_u K_1) \qquad \text{(VII.j.3)}$$

Finally, the entropy rise must be obtained. If one uses the same reference values a_0 and p_0 for the unburned and burned gas, as was done before, S_b is given by Eq. (VI.h.19) where now p_b/p_u and $\mathfrak{a}_b$ must be computed from Eqs. (VII.j.2 and 3). All flow variables in the burned gas can thus be directly computed.

If one makes the assumption that the specific heats remain constant during combustion, the relations simplify to

$$\mathfrak{a}_b^2 = \mathfrak{a}_u^2 + \gamma K_1 \qquad \text{(VII.j.4)}$$

and

$$S_b = S_u + \frac{2}{\gamma(\gamma - 1)} \ln \frac{\mathfrak{a}_b}{\mathfrak{a}_u} \qquad \text{(VII.j.5)}$$

Instantaneous volume combustion creates a high-pressure region in some portion of the duct. Subsequently, the burned gas expands, and the resulting wave phenomena are the same as if a high-pressure region were suddenly released by the removal of constraining partitions. This is described in Chapter VII.g.3, and the resulting waves—shock waves and expansion waves originating at the ends of the combustion region and propagating outward and inward, respectively—are indicated in Fig. VII.j.1. (See also part of Example IX.f.)

VII.k. Shock Wave or Contact Surface Reaching a Short Nozzle at an End of the Duct

The reflection of characteristics that reach a short nozzle at an end of the duct is treated in Chapter VI.i. The wave phenomena that are caused by finite discontinuities, such as shock waves or contact surfaces, can be handled by a slight modification of the procedures described in Chapters VII.e and VII.i, respectively. The boundary conditions must now be applied to the effective exit section (see Chapter VI.i) instead of regions 1 and 3 in Figs. VII.e.1 and VII.i.1. The flow in region 1 before arrival of the discontinuity front is known, and the properties of the latter determine conditions in region 2. Type and strength of the reflected wave must then be such that the flows in region 3 and in the effective end section satisfy not only the relations for steady, isentropic flow but also the boundary conditions. If a shock wave impinges on an orifice plate mounted at the end of the duct, the lag in establishing the steady-flow boundary conditions may lead to significant changes in the form of the reflected wave. This case is discussed in Appendix II.A.2.

An estimate for one of the flow variables in region 3 determines the type and strength of the reflected wave and, therefore, all flow variables in this region. From the relations for steady isentropic flow, as described in Chapter VI.g, one derives then the flow conditions at the exit section and, if these do not also satisfy the boundary conditions, the original estimate must be modified.

If a shock wave arrives at the nozzle while inflow is taking place, an additional interface is created as described in Chapter VII.e.3.2, and the boundary conditions must again be applied to the effective end section. The relations between the flow in this section and in region 3 (see Fig. VII.e.7) are then the same as those discussed in Chapter VI.i.2.

Numerical Example: When a shock reaches an open end, it is reflected as an expansion wave, while the reflected wave is a shock when the end is closed. There should, therefore, exist an exit nozzle of such contraction ratio that the reflected wave is eliminated altogether. Assume the gas in the duct to be air at rest and suppose that a shock wave arrives at the end, specified by $M_S = 1.500$. What is the required contraction at the exit to eliminate the reflected wave?

Let the initial conditions of the air serve as reference conditions. The flow conditions following the shock wave are then given by $\mathcal{A}_2 = 1.149$, $\mathcal{U}_2 = 0.694$, and $S_2 = 0.052$. In the absence of a reflected wave, the flow from region 2 to the exit (subscript e) must be steady and at the same time must satisfy the boundary conditions at the exit.

For subsonic outflow, the boundary condition $p_e/p_1 = 1.0$ must be satisfied. Expressing pressure by speed of sound and entropy, and considering that $S_e = S_2$, one obtains

$$\mathcal{A}_e = e^{S_e(\gamma-1)/2} = 1.010$$

One knows thus $\mathcal{A}_2$, $\mathcal{U}_2$, and $\mathcal{A}_e$, and the energy equation (VI.g.1) yields $\mathcal{U}_e = 1.406$. Since this is greater than $\mathcal{A}_e$, the boundary condition used does not apply, and instead the calculations must be based on sonic outflow. In this case, the energy equation yields $\mathcal{A}_e = \mathcal{U}_e = 1.086$. Since $M_2 = 0.604$, one can find the required contraction ratio from Eq. (VI.g.3) as $A_e/A_1 = 0.845$.

VIII

TECHNIQUES OF WAVE DIAGRAM CONSTRUCTION

VIII.a. Practical Suggestions

As the first step in the preparation of a wave diagram, the initially given data must be expressed in nondimensional form, which requires also the selection of suitable reference values.

The number of characteristics that are entered in a wave diagram is arbitrary. The more lines are drawn, the more accurate the diagram becomes in principle. In practice, however, it is also well to keep in mind that extreme accuracy may become meaningless in view of the simplifying assumptions that generally must be made. (See also the remarks in Chapter IV.a.) A feeling for this problem is best acquired by working out a number of wave diagrams. The examples presented in Chapter IX may serve as a guide. They also illustrate convenient forms of recording the progress of the calculations.

Every wave in the diagram must be drawn with a slope that indicates its velocity. The time required for plotting can be greatly reduced with the aid of a triangle which carries scales on the two sides that form a right angle. This is demonstrated in Fig. VIII.a.1 where a line corresponding to a given velocity $\mathfrak{v}$ is to be drawn through point 1. The triangle *ABC* is placed so that one side passes through point 1 and is parallel to the time axis. The distance from *A* to 1 is made equal to kn/m where m and n are the lengths of the nondimensional units of

ξ and τ, respectively, and k is an arbitrary constant. On the other side of the triangle, one can then mark point 2 at the distance $k\upsilon$ from A. The value of k should be selected to allow

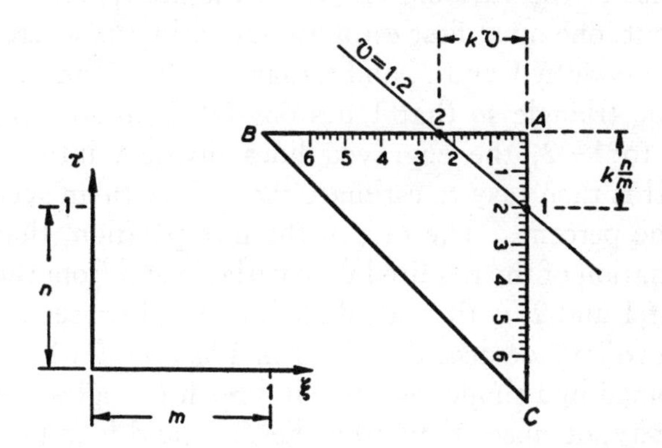

FIG. VIII.a.1. Plotting of a characteristic with the aid of a special triangle

convenient reading of the scales. Fig. VIII.a.1 illustrates the case $\upsilon = -1.2$, $m = n$, and $k = 2$. With the aid of this device, slopes can be drawn quite rapidly with adequate accuracy.

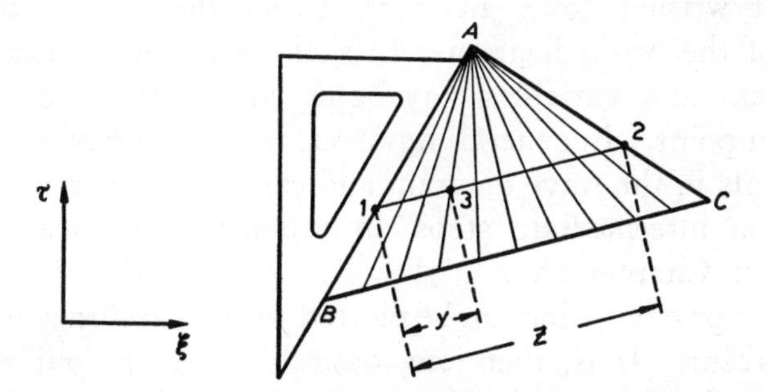

FIG. VIII.a.2. Use of a special triangle for interpolation

Another procedure that is frequently required is the insertion of waves or particle paths by interpolation. This can also be speeded up by a simple device: One side of a transparent triangle ABC (Fig. VIII.a.2) is divided into ten equal parts by engraved lines that originate at the opposite corner. Let 1 and 2 be two points at which the flow conditions are known, and 3 the point through which the interpolated wave is to be drawn. Regardless of the variable for which the interpolation is to be carried out, one must first estimate the ratio y/z, where y is the distance between 1 and 3, and z that between 1 and 2. If one places the triangle so that 1 lies on AB, 2 on AC, and BC is parallel to 1—2, the engraved lines divide z into ten equal parts. It is then easy to estimate the ratio with an accuracy of about one percent. The rest of the interpolation, that is, the determination of the required flow variable at 3 from the known values at 1 and 2, is then a simple numerical process.

The auxiliary devices described in Figs. VIII.a.1 and 2 can be combined in a single instrument. Such triangles were made with the hypotenuse about 10 inches long and found extremely convenient to use.

Other auxiliary equipment includes a good desk calculator and such charts and tables as may be necessary. Wave diagrams are best drawn on graph paper of sufficient size. For simple problems it is possible to carry out all computing steps without writing down anything except the results for each point of the wave diagram. In such cases, the values of the important flow variables may be noted on the wave diagram for each point. In general, however, it is far better to number the points in the wave diagram and carry all numerical values, including intermediate steps, in tabular form (see the examples in Chapter IX).

The slope of any line can be plotted with an accuracy of about one percent. It is, therefore, desirable to carry out the numerical calculation with an accuracy that is about one order of magnitude better.

The preparation and interpretation of wave diagrams are greatly facilitated if lines of different meanings are drawn in a distinct manner as illustrated throughout this text.

VIII.b. Checking of Wave Diagrams

The accuracy of a wave diagram, aside from that of the underlying simplifying assumptions made, depends greatly on the care with which the procedures are carried out. The slope of the characteristics changes only slightly from step to step, provided a sufficient number of waves are being drawn. Any sudden change of direction should, therefore, be suspected as a possible error unless it is associated with the crossing of a discontinuity. If a check does not reveal any error, the number of waves drawn is too small for this particular region of the wave diagram. One may then enter additional waves by interpolation or, at least, should use mean slopes even if this is not done in the rest of the diagram (see Chapter IV.a).

The detection of minor errors by inspection of a wave diagram is almost impossible. One may, however, desire to verify that the accumulated errors do not assume excessive proportions. Such a check may be based on the conservation of mass. Double integration of the continuity equation (III.c.1) between arbitrary limits of the time and distance coordinates yields

$$\int_{x_1}^{x_2} (\rho_f A_f - \rho_i A_i)\, dx + \int_{t_i}^{t_f} (\rho_2 u_2 A_2 - \rho_1 u_1 A_1)\, dt + \int_{x_1}^{x_2} \int_{t_i}^{t_f} \psi\, dx\, dt = 0 \quad \text{(VIII.b.1)}$$

where subscripts i and f have been used to indicate the flow variables of the selected initial and final values of the time coordinate, respectively, and the distance coordinate ranges between stations 1 and 2. In Eq. (VIII.b.1), the first term represents the change of mass contained in a given duct during the time interval $t_f - t_i$; the second and third terms indicate the net mass flow through the end sections and the walls of the duct, respectively.

Depending on the selection of the limits of integration, one may use this method to check either an entire wave diagram or any part of it. The integrals must be evaluated by standard graphical or numerical procedures, and the required plots of the integrands are probably best prepared in terms of the nondimensional variables that are already available from the wave diagram. The density may be expressed by the speed of sound and either the specific entropy or the pressure. The latter may be particularly convenient at open ends of a duct where it frequently has a constant value.

An alternative method of checking may be based on the fact that the mass of gas enclosed between any two selected gas layers remains constant (unless allowance has to be made for flow through the walls of the duct). Therefore, in the absence of flow through the walls of the duct, the condition

$$\int \mathcal{A}^{\frac{2}{\gamma-1}} e^{-\gamma S} \frac{A}{A_0} d\xi = \text{const.} \qquad \text{(VIII.b.2)}$$

must be satisfied at any instant if the integration is always carried out between the paths of two selected particles. The nondimensional variables have been used in this equation, and the duct area has been made nondimensional by expressing it in terms of some reference area A_0.

The two described methods of checking wave diagrams may be combined by setting up the mass balance for any region that is bounded by particle paths and lines of constant τ and ξ. By a judicial selection of this region, the computational labor may be considerably reduced. With reasonable care in the preparation of a wave diagram, it is possible to satisfy the mass balance within a few percent of the amount of gas that is present in the duct (see Example IX.f).

Wave diagrams could also be checked by setting up an energy instead of a mass balance. However, since the functions that must be integrated become then considerably more complicated, it seems that, in general, no advantage is to be derived from this procedure.

VIII.c. Periodic Phenomena

The problem of analyzing periodic phenomena frequently arises. The flow conditions at the beginning of a cycle are then part of the required solution and not known. Since a wave diagram must be started from known conditions, one can study the approach to periodic operation only through several cycles, starting from reasonably chosen initial conditions which are usually steady flow or rest.* Since the flow conditions at the end of one cycle serve as the starting conditions for the next cycle, this procedure is essentially an iteration of the conditions at the beginning of a cycle. The time required to extend the wave diagram over many cycles becomes easily prohibitive, but as few as two or three cycles often provide valuable information. More cycles may be needed to determine the periodic conditions accurately, and the number may depend critically on the choice of the starting conditions.[100] Problems of this nature might require so large an effort that only use of a digital computer would make an attempt at their solution feasible (see Appendix II.C).

The practical importance of periodic phenomena is indicated by the fact that nonsteady-flow engines, such as pulsejets or the "Comprex," are operating in a cyclic manner. Wave diagrams have been of great help to provide insight into the operating mechanism of such engines and to estimate their performance (see bibliography).

VIII.d. Alternative Methods of Wave Diagram Construction

Although it is felt that the methods for the construction of wave diagrams described in the preceding chapters are particularly convenient, they are by no means unique. Only a brief discussion of alternative procedures, however, can be given here.

* An attempt to obtain a direct solution to problems of periodic flows was presented by Schultz-Grunow.[34] However, his method could be applied only to cases of extreme simplicity and, even then, it was necessary to neglect wave interactions.

One may distinguish between two major lines of approach to the construction of wave diagrams, depending on whether the changes of the flow variables along the characteristics are evaluated numerically or graphically. The first method is represented, for instance, by references 7, 8, 11, 14, 15, 27, 45 and by this text. In order to illustrate briefly the second method, consider an isentropic flow in a duct of constant cross section so that the Riemann variables are constant along their respective characteristics. The condition of constant P or Q can then be plotted in a $\mathcal{U},\mathcal{A}$-coordinate system as a straight line. The characteristic network in this "state plane" consists, therefore, of two families of parallel straight lines that are easily drawn. To every point in the wave diagram corresponds one point in the state plane, although the reverse correspondence is not unique. The coordinates of a point in the state plane indicate the slope of the characteristics in the wave diagram, thus enabling one to proceed to the intersection with the next characteristic. This point must again be identified in the state plane—as the intersection of the two appropriate characteristics—and the construction of the wave diagram can then be carried on. By continuously referring from one diagram to the other and back, both can be constructed simultaneously. The boundary conditions are also easily indicated in this diagram as a predetermined line such as $\mathcal{U} = 0$ for a closed end, $\mathcal{A} =$ const. for subsonic and $\mathcal{A} = \mathcal{U}$ for sonic outflow from an open end; the condition of isentropic inflow (Eqs. VI.d.4) would appear as an ellipse in this representation, and so forth.

A variation of this scheme is presented in references 9, 10 and 25, where a quadratic relation between $\mathcal{U}$ and $\mathcal{A}$ (expressing essentially $\mathcal{A}_s{}^2$) is used as the ordinate of the state plane instead of just $\mathcal{A}$. In such a diagram, the characteristics become two families of congruent and parallel parabolas whose tangents at any intersection point can be shown to be orthogonal to the direction of the characteristics through the corresponding point in the wave diagram.

Numerous procedures have appeared in the literature (see, for instance, references 10, 12, 17, 19, 22, 47, and 49) by means of which the use of the state plane is modified and extended to cover also nonisentropic flows in ducts of variable cross section. These involve the construction of auxiliary plots and frequent graphical trial-and-error procedures. Because of the need to prepare two accurate diagrams, this writer prefers a graphical-numerical method that requires only a single diagram.

Other, though less important, variations of the methods of wave diagram construction result from two possible interpretations of the step-by-step method of calculation. The families of characteristics may be considered either as forming a "net" that determines discrete points or as boundaries of small "meshes." In the first case, one determines the flow variables at distinct points; in the latter, each mesh is considered as a region of steady flow separated from its neighbors by small discontinuities. The number of procedures required is somewhat reduced in the latter case compared to the former since all characteristics are treated as weak compression or "expansion" shocks having a suitably chosen strength ($\Delta \mathfrak{a} = \pm 0.02$, say), and no further instructions about the handling of characteristics are required. This is probably the reason why the "mesh" type wave diagram is sometimes preferred (see, for instance, references 7–12, 15, 45, and 51). The representation of a continuous function by a step function makes it difficult to interpolate, and since this is frequently required, particularly in more complicated wave diagrams, the methods described in this text are based essentially on the characteristic net. However, in the case of multi-isentropic flows a simplification of the procedures becomes possible by treating selected particle paths in a field of gradual entropy changes as entropy discontinuities while still following the gradual changes of the Riemann variables along the *P*- and *Q*-waves (see Chapter V.c.1).

The various techniques of wave diagram construction with

or without the use of a state plane may be carried out either in "net" or in "mesh" type diagrams. Further variations result when the procedures are expressed in terms of flow variables other than $\mathfrak{A}$, $\mathfrak{U}$, and S. In particular, the use of p instead of $\mathfrak{A}$ has been favored by some authors (see, for instance, references 7, 18, 23, and 47). This short survey indicates, therefore, the great variety of alternative procedures that could be prepared. Indeed, among the published schemes, there are hardly two that are alike in detail. Further variations of the procedures may become advantageous if the computations are to be performed by a digital computer (see Appendix II.C).

IX

EXAMPLES

IX.a. General Remarks

The great variety of procedures that may have to be applied in the construction of wave diagrams is, at first, quite bewildering, and only extensive practice can make one thoroughly familiar with the methods. Numerical examples to illustrate many of the individual procedures have already been given throughout the preceding chapters. In the following, a number of complete wave diagrams are presented, each of which involves a variety of procedures.

Progress of the work is best recorded in tabular form. Such tables may become quite extensive, but it does not seem necessary to reproduce them here in full. Only those portions are presented that refer to points of particular interest. Some additional points are listed to permit occasional checking of attempts to reconstruct the wave diagrams. In order to facilitate interpretation of the diagrams, the P- and Q-waves are always shown as thin solid lines; particle paths as thin dashed lines; and shock waves and contact surfaces as heavy solid lines and heavy dashed lines, respectively. The choice of the reference conditions used to make all variables nondimensional is stated for each example. With the exception of Example IX.f, their numerical values are not required, and the results are expressed in terms of the reference quantities.

It is pointed out in Chapter IV.a that, provided the change in slope is small, it is permissible to plot characteristics with a slope corresponding to the velocity at an end point of a characteristic segment rather than to use mean velocities. In the wave diagram for the first example, mean velocities are used throughout. The results are then compared with those obtained with the simplified procedure and it is shown that the errors would be insignificant for most practical purposes. In all other diagrams, mean velocities are used only when either the wave velocities change rather fast or there is some other compelling reason. The use of the mean velocity may be indicated when the velocities at the end of a wave segment have different signs or when one of them is equal to zero.

It is quite impracticable to include a sufficient number of examples that will include every one of the procedures described in the preceding chapters. However, after working through the given examples, the application of any of the procedures not illustrated should not be too difficult.

IX.b. Nonsteady Discharge of Compressed Air from a Duct of Constant Cross Section

1. *Isentropic Back Flow.* A duct of constant cross section is filled with air ($\gamma = 1.4$) that is isentropically compressed from atmospheric pressure, and the right end of the duct is suddenly opened to the atmosphere. What fraction of the initial mass in the duct is discharged, and how does the pressure at the closed end vary with time?

The atmospheric conditions represent convenient reference variables p_0 and a_0, and the logical choice for L_0 is the length of the duct. Let the initial conditions in the duct be prescribed by $\mathfrak{a}_1 = 1.160$ $(p_1/p_0 = \mathfrak{a}_1{}^7 = 2.83)$; in view of the problem statement, one has also $\mathfrak{u}_1 = 0$ and $S_1 = 0$.

The wave diagram for this case is shown in Fig. IX.b.1. Since the opening of the duct is assumed to take place instantaneously, a centered expansion wave originates at the exit (see Chapter IV.b) which accelerates the air instantly to a

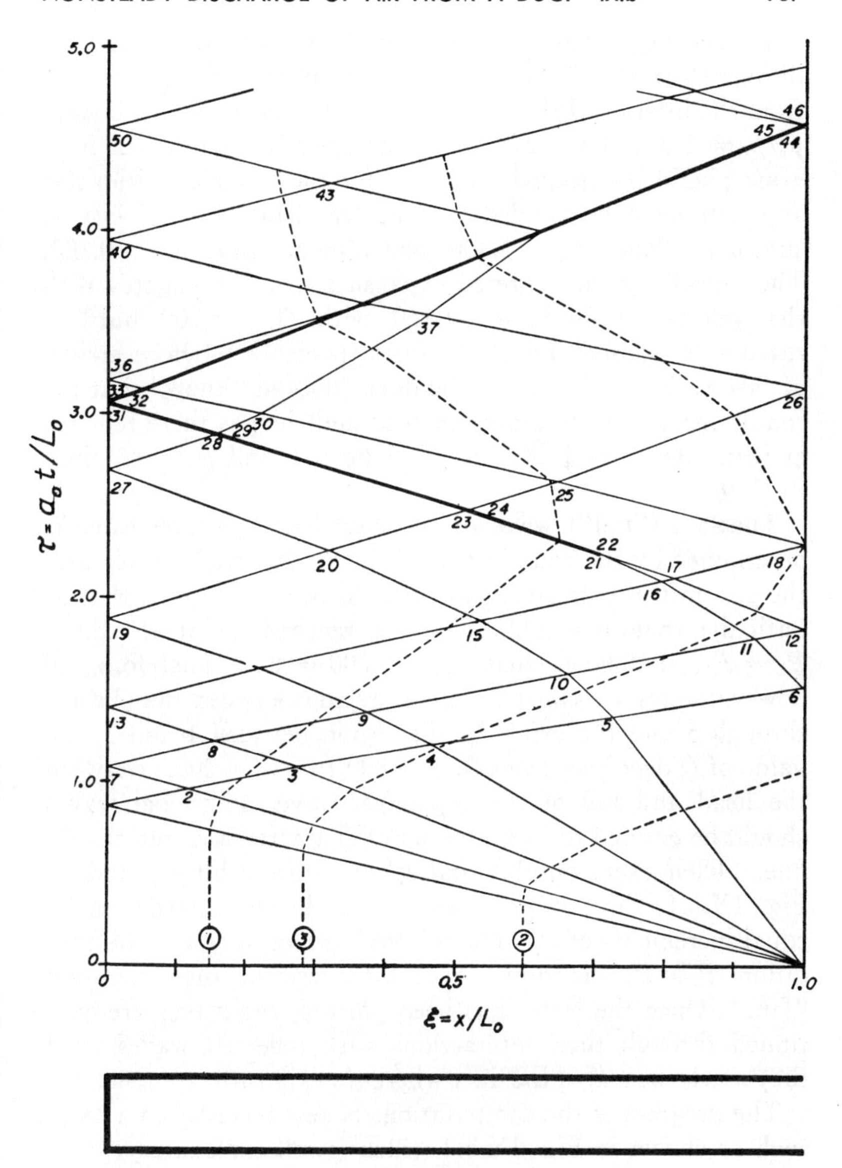

FIG. IX.b.1. Example IX.b.1. Discharge of compressed air from a duct of constant cross section (isentropic back flow)

finite velocity. The latter can be determined from the conditions $P_e = P_1$ (P is constant for a characteristic that travels from the interior of the duct to the exit), and $\mathfrak{a}_e = 1.000$ since $p_e/p_0 = 1.0$ and the gas was isentropically compressed from atmospheric conditions. In order for the latter conditions to apply, it must be verified that the resulting exit velocity is subsonic. Since $P_1 = 5.800$, one obtains, thus, $\mathfrak{u}_e = 0.800$. The "head" of the centered expansion wave propagates with the velocity $\mathfrak{u}_1 - \mathfrak{a}_1 = -1.160$ with $Q_1 = 5.800$ until it reaches the closed end of the duct (point 1). Before arrival of this wave, a point inside the duct "does not know" that the conditions at the end have changed and the gas there remains at rest. At point 1 the wave is reflected, and $P_1 = Q_1$ since $\mathfrak{u}_1 = 0$.

The last ("tail") wave of the centered expansion wave is determined by the boundary conditions. The wave accelerates the gas until the pressure has dropped to atmospheric, that is, until the value $\mathfrak{a} = 1.000$ has been reached (point 5). Since $P_5 = P_1$, it follows that $\mathfrak{u}_5 = 0.800 = \mathfrak{u}_e$. Therefore, all flow variables are constant in the region between the Q-wave through 5 and the exit. In the region between 1 and 5, the value of Q decreases from $Q_1 = 5.800$ to $Q_5 = 4.200$. Between the head and tail of the expansion wave, additional waves should be entered to keep the net of characteristics sufficiently fine. Their exact number and selection are arbitrary and, in Fig. IX.b.1, three more Q-waves are shown, determined by equal increments of Q. The selected values of Q and the condition $P = P_1$ determine everything within the expansion "fan." Once the latter has been plotted, the waves are continued through their interactions with reflected waves until they reach the end of the duct where they, in turn, are reflected.

The progress of the computations is best recorded in a table such as shown in Fig. IX.b.2 which is self-explanatory. For the sake of brevity, only the table entries for selected points are reproduced. The first 13 points should make the computing

POINT	P	Q	$\mathcal{U}$	$\mathcal{A}$	$\mathcal{U}+\mathcal{A}$	$\mathcal{U}-\mathcal{A}$	$\mathcal{W}$	p/p_0
1	5.800	5.800	0	1.160	1.160	−1.160		2.83
2	5.800	5.400	0.200	1.120	1.320	−0.920		
3	5.800	5.000	0.400	1.080	1.480	−0.680		
4	5.800	4.600	0.600	1.040	1.640	−0.440		
5, 6	5.800	4.200	0.800	1.000	1.800	−0.200		
7	5.400	5.400	0	1.080	1.080	−1.080		1.71
8	5.400	5.000	0.200	1.040	1.240	−0.840		
9	5.400	4.600	0.400	1.000	1.400	−0.600		
10, 11	5.400	4.200	0.600	0.960	1.560	−0.360		
12	5.400	4.600	0.400	1.000	1.400	−0.600		
13	5.000	5.000	0	1.000	1.000	−1.000		1.00
15	5.000	4.200	0.400	0.920	1.320	−0.520		
16	5.000	4.200	0.400	0.920	1.320	−0.520		
17	5.000	4.600	0.200	0.960	1.160	−0.760		
18	5.000	5.000	0	1.000	1.000	−1.000		
19	4.600	4.600	0	0.920	0.920	−0.920		0.56
20	4.600	4.200	0.200	0.880	1.080	−0.680		
21	4.856	4.200	0.328	0.906				
22	4.856	4.600	0.128	0.946		−0.818	−0.705	
23	4.600	4.200	0.200	0.880	1.080			
24	4.600	4.640	−0.020	0.924	0.904	−0.944	−0.822	
25	4.600	5.000	−0.200	0.960	0.760	−1.160		
26	4.600	5.280	−0.340	0.988	0.648	−1.328		
27	4.200	4.200	0	0.840	0.840	−0.840		0.30
30	4.200	5.000	−0.400	0.920	0.520	−1.320		
31	4.200	4.200	0	0.840				
32	4.200	4.720	−0.260	0.892	0.632	1.152	−1.009	0.30
33	4..720	4.720	0	0.944	0.944	−0.944	0.802	0.66
36	5.000	5.000	0	1.000	1.000	−1.000		1.00
37	4.200	5.280	−0.540	0.948	0.408	−1.488		
40	5.290	5.290	0	1.058	1.058	−1.058		1.47
44	4.260	5.410	−0.575	0.967	0.392	−1.542		
45	5.120	5.420	−0.150	1.054	0.904	−1.204	0.679	
46	5.120	4.880	0.120	1.000	1.120	−0.880		
50	5.325	5.325	0	1.065	1.065	−1.065		1.54

FIG. IX.b.2. Data sheet for Example IX.b.1 (abridged)

process quite clear. For example, point 9 is determined by $P_9 = P_8$ and $Q_9 = Q_4$; then, $\mathcal{U}_9 = \frac{1}{2}(P_9 - Q_9)$ and $\mathcal{A}_9 = \frac{1}{10}(P_9 + Q_9)$. The slopes of the characteristics at 9 are also entered in the table. Throughout this wave diagram, any line is plotted with a slope that corresponds to the mean of the velocities at its end points. Thus, since $(\mathcal{U} + \mathcal{A})_9 = 1.400$ and $(\mathcal{U} + \mathcal{A})_{10} = 1.560$, the segment of the P-wave between points 9 and 10 is drawn corresponding to a velocity $(\mathcal{U} + \mathcal{A})_{9,10} = 1.480$. These mean values could be recorded in separate columns of the table, but this is not really necessary.

The exit velocity remains constant until the first reflected wave reaches the open end (point 6). Then, it gradually decreases until it passes through zero at 18 and the flow direction is reversed. Inflow is here treated as isentropic (see Examples IX.b.2 and 3 for other assumptions). Thus, the conditions at point 26 are obtained with the aid of Chart 1a. For $P_{26} = P_{25} = 4.600$ and $\mathcal{A}_E = \mathcal{A}_0 = 1.000$, the chart yields $\mathcal{A}_{26} = 0.988$, and $\mathcal{U}_{26}$ follows from this and P_{26}.

The initial expansion wave changes to a compression wave after reflection from the open end, and the Q-waves through 16 and 17 merge to indicate the formation of a shock wave at points 21 and 22. The value of P_{21} must be found by estimating which P-wave between 15 and 20 (not entered in the wave diagram) would pass through 21. The value of $\Delta Q/\mathcal{A} = (Q_{22} - Q_{21})/\mathcal{A}_{21} = 0.442$ or of $\dfrac{\mathcal{A}_{22}}{\mathcal{A}_{21}} = 1.044$ determines the shock Mach number as $M_S = 1.140$ (Table 1a), and the shock velocity is, therefore, given by $\mathcal{W}_{22} = \mathcal{U}_{21} - M_S\mathcal{A}_{21} = -0.705$, according to Eq. (VI.e.6). (It is denoted by $\mathcal{W}_{22}$ instead of $\mathcal{W}_{21}$ to indicate that the conditions at 22 must be found in addition to those at 21 in order to obtain the strength of the shock.) The same procedure must be applied at all points along the path of the shock wave. Since the speed of sound ratio never exceeds 1.1 for any point of the shock, the entropy changes produced by the latter ($\Delta S < 0.02$) may be neglected.

The conditions at the closed end of the duct just after the reflection of the shock wave (point 33) are determined simply by $\mathcal{Q}_{33} = \mathcal{Q}_{32}$ and $\mathfrak{U}_{33} = 0$. When the shock finally reaches the open end, a centered expansion wave is reflected. Its head is given by the wave $\mathcal{Q}_{45}$ and its tail by $\mathcal{Q}_{46}$ where $P_{46} = P_{45}$ and $\mathfrak{A}_{46} = 1.000$.

In this manner, wave diagram construction could be continued as long as required. With the aid of the completed diagram, all questions about the flow pattern can be answered. To illustrate the gas motion in the duct, the paths of three particles are plotted in the wave diagram. The velocity at any point must be obtained by interpolation between the nearest points of the characteristic net. Particles ① and ② are initially located at $\xi = 0.15$ and $\xi = 0.60$, respectively. The latter particle is accelerated through the expansion wave and is immediately discharged, while the former carries out an oscillatory motion. To answer the question of what fraction of the gas initially in the duct is discharged, the path of the particle is required that passes through point 18 where the exit velocity reverses direction. The path of this particle ③ must be plotted backward from 18, and it is seen that about 72 percent of the initial mass is discharged. Since $\rho_1/\rho_0 = \mathfrak{A}_1{}^5 = 2.10$, nearly one-half of the mass (1/2.10) would have remained in the duct if the outflow had been isentropic and quasi-steady and only about 52 percent would have been discharged.*

Since $S = 0$ for the entire wave diagram, the pressure at any point is given by $p/p_0 = \mathfrak{A}^7$. In Fig. IX.b.3, the pressure at the closed end is plotted as function of the nondimensional time. The numbers of the corresponding points in the wave diagram are indicated. The pressure remains constant until the first wave from the open end arrives; then, it drops to almost one-third of atmospheric. The arrival of the shock wave, followed by further compression, raises the pressure again to a

* In this case, outflow would take place slowly through a small nozzle so that the state of the gas would be uniform throughout the duct at all times.

value above atmospheric. Such oscillations would then continue with decreasing amplitude. Note that the slope of the curve changes discontinuously at points 1 and 27 which correspond to the arrival of the head and tail of the expansion wave,

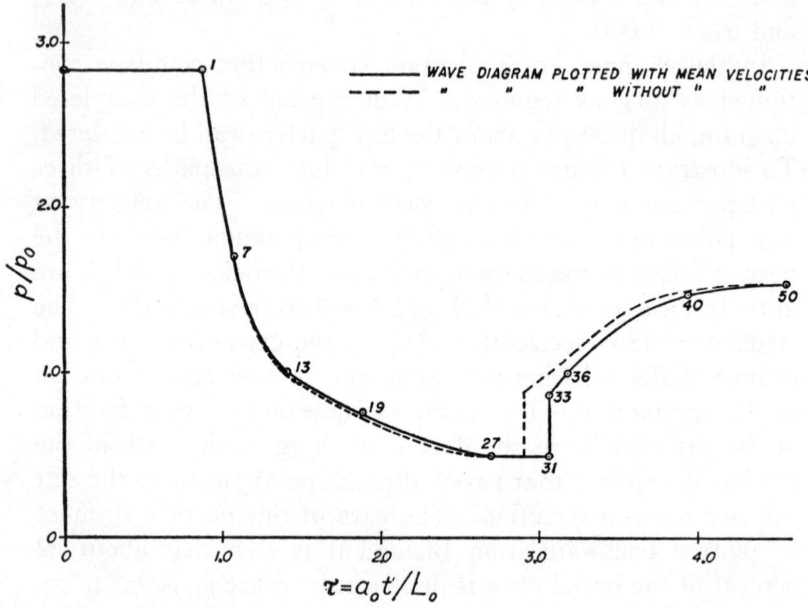

FIG. IX.b.3. Pressure history at closed end of duct (Example IX.b.1)

respectively. The reason for this is that $\frac{\partial Q}{\partial \tau} = 0$ in the regions adjacent to the expansion wave, while it has a finite value within the latter. The possibility that discontinuities of the derivatives of flow variables may occur was discussed at the end of Chapter IV.b.

As pointed out previously, all lines in Fig. IX.b.1 are drawn with slopes that represent mean velocities. The pressure history at the closed end, obtained from a wave diagram in which the mean velocities were not used, is shown as a dashed line in Fig. IX.b.3. As one would expect, the difference between the two curves indicates primarily an error in timing which is so small that it would be unimportant for most practical purposes.

The excess discharge under nonsteady-flow conditions compared to a quasi-steady discharge and the over-expansion and recompression are phenomena of major importance for all engines in which wave phenomena are utilized.

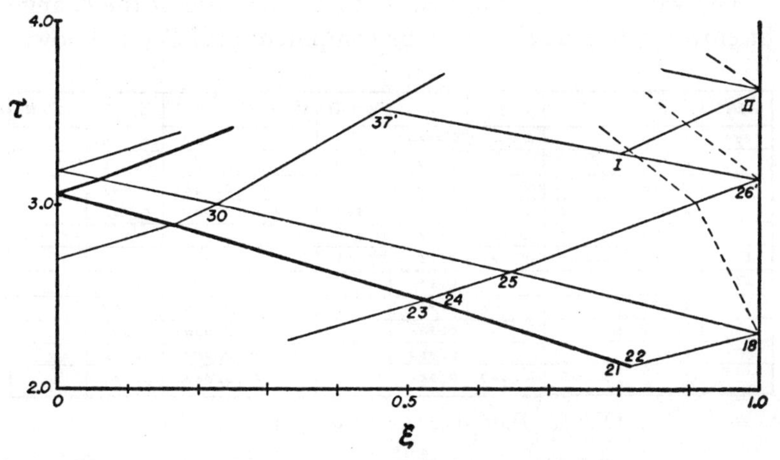

FIG. IX.b.4. Example IX.b.2. Discharge of compressed air from a duct of constant cross section (nonisentropic back flow—Borda nozzle). The earlier part of the wave diagram is identical with the corresponding portion of Fig. IX.b.1

2. *Nonisentropic Back Flow* (*Borda Nozzle*). Consider the same conditions as before, but assume that inflow is to be treated as flow into a Borda nozzle.

As long as outflow takes place, the wave diagram of this case is, of course, identical with that of Fig. IX.b.1 up to the Q-wave that originates at 18. The particle that enters at 18 starts with zero velocity and, therefore, between its path and the closed end of the duct, the condition $S = 0$ still holds. As the inflow velocity increases, the entropy of the entering particles also increases so that a multi-isentropic flow field with gradually varying entropy is created. Fig. IX.b.4 shows a section of the modified wave diagram. At the points indicated by primed numbers, the flow conditions are different from those of the preceding example, and the points that have no counterpart in Fig. IX.b.1 are marked I and II. In the table in Fig. IX.b.5, only the new and the modified points are listed; for the other points, the data can be found in Fig. IX.b.2.

The value of $P_{26'}$ is not equal to P_{25} because of the change of entropy, but $\Delta_+ P$ cannot be computed until $S_{26'}$ is known.

POINT	P	Q	$\mathfrak{U}$	$\mathfrak{A}$	$\mathfrak{U}+\mathfrak{A}$	$\mathfrak{U}-\mathfrak{A}$	S	$\Delta_+ P$	$\Delta_- Q$
26′	4.600			0.988			0.052	0.050	
	4.650			0.991			0.040	0.038	
	4.638			0.990			0.044	0.042	
	4.642	5.258	−0.308	0.990	0.682	−1.298	0.044	0.042	
37′	4.200	5.214	−0.507	0.941	0.434	−1.448	0	•	−0.044
I	4.452	5.214	−0.381	0.967	0.586		0		
II	4.452			0.980			0.081	0.079	
	4.531			0.985			0.065	0.063	
	4.515			0.984			0.068	0.066	
	4.518			0.984			0.067	0.065	
	4.517	5.323	−0.403	0.984	0.581	−1.387	0.067	0.065	

FIG. IX.b.5. Data sheet for Example IX.b.2 (abridged)

The iteration is indicated in the table. One may start by assuming that $P_{26'}$ is the same as P_{25}, which is known. The corresponding value of $\mathfrak{A}_{26'}$ is obtained from Chart 1a, and from this and Chart 2a one obtains $S_{26'}$. One can then calculate the first approximation for $\Delta_+ P$ by means of Eq. (V.c.1). This leads to an improved value for $P_{26'}$, and this procedure is repeated until no further change of the variable takes place. From the final values of $P_{26'}$ and $\mathfrak{A}_{26'}$, one obtains all remaining quantities for this point.

Point 37′ can be evaluated without iteration since the change of Q from its value at 26′ can be calculated directly and $P_{37'} = P_{30}$. The next point at the exit would be obtained when the shock wave arrives there, but it may be desirable to evaluate intermediate points. First, one selects a point I on the Q-wave between 26′ and 37′. P_{I} is found by interpolation between P_{25} and P_{30}, and $Q_{\mathrm{I}} = Q_{37'}$ because there is no change of entropy between these points. One can then evaluate point II by means of the same iteration procedure that was used for 26′.

Comparison between the flow variables at 26 and 26′ indicates the considerable effect of the entropy rise in a Borda nozzle, for instance, $\mathfrak{U}_{26} = -0.340$ while $\mathfrak{U}_{26'} = -0.308$.

3. *Discharge into a Helium Atmosphere with Isentropic Back Flow.* Consider again the same initial conditions as before, but assume that the air in the duct is now discharged into a helium reservoir ($\gamma = 5/3$) in which both pressure and temperature are atmospheric. Treat inflow into the duct as isentropic.

Since now two different gases are involved, it is most convenient to select the same reference state (defined by p_0 and a_0) for both. Since the temperature in the helium reservoir is atmospheric, the speed of sound there is given by $\mathfrak{A}_E = 2.93$ (see the table on gas properties in Chapter XI). From this and the condition of atmospheric pressure, one obtains $S_E = \dfrac{2}{\gamma - 1} \ln \mathfrak{A}_E = 3.225.$

Again, the wave diagram is identical with that of Fig. IX.b.1 up to the Q-wave through point 18 where inflow begins. The path of the particle through 18 represents the contact surface between air and helium. Since the flow remains isentropic according to the above-stated assumptions, the part of the wave diagram that applies to the flow of helium can be plotted in exactly the same manner as in Example IX.b.1, except that the value of γ is different.

A section of the modified wave diagram is reproduced in Fig. IX.b.6. The points where the flow conditions of the original example are modified are indicated by double primes, and points without counterpart in Fig. IX.b.1 are marked III, IV, V,$\cdots$

Solutions for the points along the contact surface are obtained by means of Eqs. (VI.f.2). The procedures are straightforward once the contact surface has moved some distance into the duct, but, while it is still close to the inlet, a more complicated iteration becomes necessary. The flow conditions at 18″ are immediately found since $\mathfrak{U}_{18''} = 0$ and, therefore,

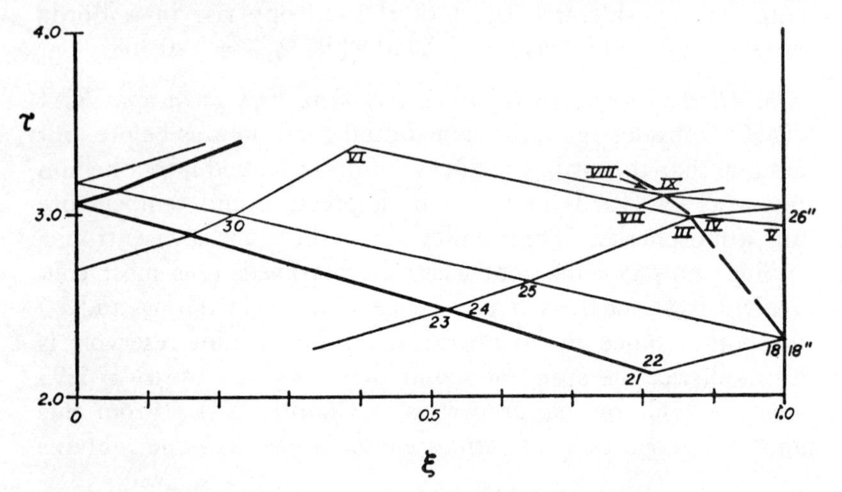

Fig. IX.b.6. Example IX.b.3. Discharge of compressed air from a duct of constant cross section into a helium atmosphere (isentropic back flow). The earlier part of the wave diagram is identical with the corresponding portion of Fig. IX.b.1

$\mathfrak{A}_{18''} = \mathfrak{A}_E$. The difficulty arises at the next step, points III and IV. On the left side, one has $P_{III} = P_{25}$, but Q_{IV} which is equal to Q_V is not known. Q_V must be found by interpolation between 18″ and 26″, and in order to solve for the conditions at the latter point, one must first obtain the solution for IV. Therefore, points III, IV, V, and 26″ must be treated simultaneously.

It is, probably, most convenient to make a guess for $\mathfrak{U}_{III} = \mathfrak{U}_{IV}$. A first approximation for the position of III can then be found as the intersection of the P-wave through 25 and the interface. This wave may be plotted with the slope $(\mathfrak{U} + \mathfrak{A})_{25}$

or $(\mathfrak{U} + \mathfrak{A})_{25,\text{III}}$ as desired, but the interface *must* be plotted with the mean velocity $\mathfrak{U}_{18,\text{III}} = \frac{1}{2}\mathfrak{U}_{\text{III}}$ since $\mathfrak{U}_{18} = 0$. The progress of the calculation is indicated in the second table of Fig. IX.b.7. The first of Eqs. (VI.f.2) yields the ratio p_{III}/p_0

POINT	P	Q	$\mathfrak{U}$	$\mathfrak{A}$	$\mathfrak{U} + \mathfrak{A}$	$\mathfrak{U} - \mathfrak{A}$
18"	8.790	8.790	0	2.930		−2.930
III	4.600	5.360	−0.380	0.996	0.616	−1.376
IV	8.362	9.122	−0.380	2.914	2.534	−3.294
V	8.416	9.122	−0.353	2.923	2.570	−3.276
26"	8.362	9.158	−0.398	2.920	2.522	−3.318
VI	4.200	5.360	−0.580	0.956	0.376	−1.536
VII	4.536	5.360	−0.412	0.990	0.578	
VIII	4.536	5.404	−0.434	0.994	0.560	−1.428
IX	8.290	9.158	−0.434	2.908	2.474	−3.342

	GUESS									CHECK	
P_{III}	$\mathfrak{U}_{\text{III}} = \mathfrak{U}_{\text{IV}}$	$\mathfrak{A}_{\text{III}}$	$P_{\text{III}}/P_0 = P_{\text{IV}}/P_0$	$\mathfrak{A}_{\text{IV}}$	$P_{\text{IV}} = P_{26''}$	$\mathfrak{A}_{26''}$	$\mathfrak{U}_{26''}$	$Q_{26''}$	$\mathfrak{U}_{\text{IV}} - \mathfrak{A}_{\text{IV}}$	Q_{IV}	Q_{V}
4.600	−0.200	0.960	0.751	2.767	8.101	2.906	−0.617	9.335	−2.967	8.501	
	−0.400	1.000	1.000	2.930	8.390	2.920	−0.370	9.130	−3.330	9.190	
	−0.300	0.980	0.868	2.848	8.244	2.913	−0.495	9.234	−3.148	8.844	9.080
	−0.350	0.990	0.932	2.889	8.317	2.917	−0.434	9.185	−3.239	9.017	9.106
	−0.380	0.996	0.972	2.914	8.362	2.920	−0.398	9.158	−3.294	9.122	9.122

	GUESS				CHECK	
P_{VIII}	$\mathfrak{U}_{\text{VIII}} = \mathfrak{U}_{\text{IX}}$	$\mathfrak{A}_{\text{VIII}}$	Q_{IX}	$\mathfrak{A}_{\text{IX}}$	P_{VIII}/P_0	P_{IX}/P_0
4.536	−0.300	0.967	9.158	2.953	0.791	1.027
	−0.400	0.987		2.919	0.913	0.972
	−p.450	0.997		2.903	0.979	0.950
	−0.434	0.994		2.908	0.959	0.958

Fig. IX.b.7. Data sheet for Example IX.b.3 (abridged)

for $S_{\text{III}} = 0$ and the second equation is solved for $\mathfrak{A}_{\text{IV}}$ with $S_{\text{IV}} = S_E$. One obtains, thus, a value for $P_{\text{IV}} = P_{26''}$ and with the aid of Chart 1b also $\mathfrak{A}_{26''}$. A Q-wave plotted backward from IV to V with a slope corresponding to $\mathfrak{U}_{\text{IV}} - \mathfrak{A}_{\text{IV}}$ indicates the location of V, and the value of Q_{V} can then be found by interpolation between $Q_{18''}$ and $Q_{26''}$. The condition $Q_{\text{IV}} = Q_{\text{V}}$ constitutes the check that all variables are correct. Since Q_{IV} must lie between $Q_{18''}$ and $Q_{26''}$, it is obvious that the first two guesses (see table) must be wrong even without finding the resulting values of Q_V. Since the errors are in opposite directions, the correct value for $\mathcal{U}_{III}$ must lie somewhere between -0.200 and -0.400. In the example shown, three more steps led to the correct result.

The next points along the interface, VIII and IX, can be handled by means of the standard iteration procedure (Chapter VI.f). A sample of these calculations is listed in the third table of Fig. IX.b.7. From the estimated value of $\mathcal{U}_{VIII} = \mathcal{U}_{IX}$, one computes $\mathcal{A}_{VIII}$ and $\mathcal{A}_{IX}$, and the corresponding pressures should come out equal. In the illustrated example the correct answer was obtained in four attempts. The final guess was obtained by graphical interpolation which, probably, eliminated the need of more iterations.

IX.c. Expansion Wave Passing Through a Duct of Variable Cross Section

1. *Gradual Change of Area.* Two ducts of constant cross section are joined by a transition section of length L_0 and a cross-sectional area given by $A/A_0 = 1 + 0.7(x/L_0)^2$. A piston is located in the narrow duct at a distance of $0.2\,L_0$ from the junction with the cone. Both ducts extend to infinity. The gas is, initially, at rest and its speed of sound is a_0.

Let the piston be impulsively accelerated to a velocity $0.25\,a_0$ in the direction away from the conical section. How does the speed of sound vary with time at the junction of the cone with the larger duct? What is the final steady flow that would be established?

The obvious choice for the reference variables is a_0 and p_0 (the latter will not be required) and L_0.

If the simplified procedure for gradual area change is to be used (see Chapter V.b.1), the region of the transition must be approximated by a number of constant-area strips where the logarithmic area change between consecutive steps should be sufficiently small, perhaps of the order of 0.1 and, preferably,

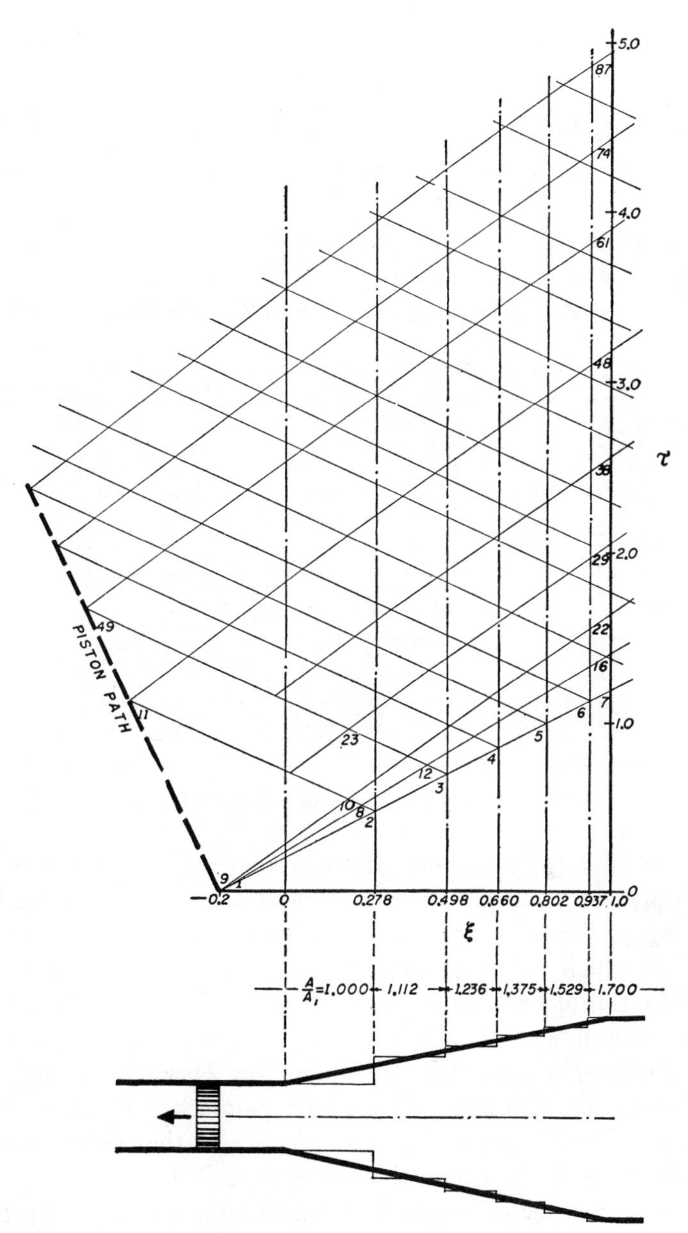

FIG. IX.c.1. Example IX.c.1. Expansion wave, created by a receding piston, passing through a duct of variable cross section

constant. Since ln 1.7 = 0.530, five steps should be adequate; the ratio of adjacent areas becomes then 1.112. The selected areas and locations of the steps are indicated below the wave diagram in Fig. IX.c.1.

The piston path can be immediately plotted through the point given by $\xi = -0.2$ and $\tau = 0$ with a slope corresponding to $\mathcal{V} = -0.250$. As a result of the sudden expansion that is

POINT	P	Q	$\mathcal{U}$	$\mathcal{A}$	$\mathcal{U} + \mathcal{A}$	$\mathcal{U} - \mathcal{A}$
1 - 7	5.000	5.000	0	1.000	1.000	-1.000
8	4.750	5.000	-0.125	0.975	0.850	-1.100
9, 10, 11	4.500	5.000	-0.250	0.950	0.700	-1.200
12	4.765	5.000	-0.117	0.976	0.859	-1.093
16	4.816	5.000	-0.092	0.982	0.890	-1.074
22	4.647	5.000	-0.176	0.965	0.789	-1.141
23	4.500	4.980	-0.240	0.948	0.708	-1.188
29	4.642	5.000	-0.179	0.964	0.785	-1.143
38	4.635	5.000	-0.182	0.964	0.782	-1.146
48	4.634	5.000	-0.183	0.963	0.780	-1.146
49	4.480	4.980	-0.250	0.946	0.696	-1.196
61	4.612	5.000	-0.194	0.961	0.767	-1.155
74	4.608	5.000	-0.196	0.961	0.765	-1.157
87	4.602	5.000	-0.199	0.960	0.761	-1.159

FIG. IX.c.2. Data sheet for Example IX.c.1 (abridged)

produced by the receding piston, a centered expansion wave originates at point 1. The head of this wave marks the beginning of the disturbance, and its intersections with the stations at which the area changes take place are denoted by 2 to 6. The tail of the expansion wave (point 9) is determined by $Q_9 = Q_1$ and $\mathcal{U}_9 = \mathcal{V}$. One additional wave of the expansion fan is plotted with a value of P halfway between P_1 and P_9. The data for a number of selected points are listed in Fig. IX.c.2. The computation for points in the variable-area region proceeds then with the aid of Eqs. (V.b.2).

The first reflected wave that reaches the piston at 11 terminates the region of steady flow adjacent to the latter. Its reflection P_{11} is found from the condition $\mathcal{U}_{11} = \mathcal{V}$. Since the region between the waves P_{10} and P_{11} is large, it is advisable to insert additional waves. For instance, the location of point 23 is arbitrarily selected on the Q-wave through 12. Its data are determined by $P_{23} = P_{10} = P_{11}$ (no interpolation is necessary in this case), and $Q_{23} = Q_{12} + \Delta_{-}Q$.

At all points of the wave diagram along the line $\xi = 1.0$, where the transition section joins the larger duct, the con-

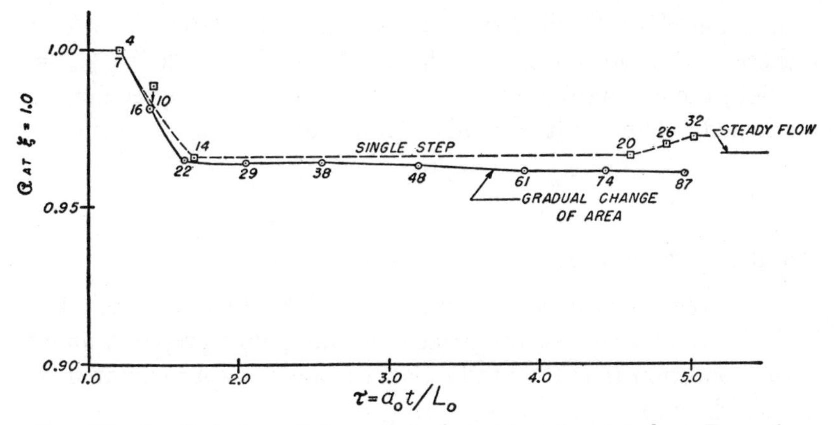

FIG. IX.c.3. Variation of the speed of sound at $\xi = 1.0$ from Examples IX.c.1 and 2

dition $Q = 5.000$ is always satisfied because no wave reflection occurs in the duct. The values of $\mathfrak{a}$ along this line, points 7, 16, 22, $\cdots$ are plotted in Fig. IX.c.3 as function of the non-dimensional time. It is seen that, after the initial expansion, the variations of $\mathfrak{a}$ due to wave reflections are rather small in this case.

If the wave diagram were sufficiently extended, one could study how the flow gradually approaches steady conditions as the waves become weaker and weaker. However, the final steady flow can be computed directly without preparing a wave diagram. Let subscripts n and w denote these conditions in the narrow and wide ducts, respectively. Since the flow is isentropic, there are four variables to be determined, namely, the speeds of sound $\mathfrak{a}_n$ and $\mathfrak{a}_w$, and the flow velocities $\mathfrak{u}_n$ and $\mathfrak{u}_w$ (or the Mach numbers M_n and M_w).

The boundary condition in the narrow duct determines immediately that $\mathfrak{u}_n = \mathfrak{v}$. The remaining unknowns are best found by means of a trial-and-error procedure. A guess for $\mathfrak{a}_n$ determines also $M_n = |\mathfrak{u}_n|/\mathfrak{a}_n$. With the aid of the continuity equation (VI.g.3), one obtains M_w; this and the energy equation (VI:g.1) yield $\mathfrak{a}_w$, so that $\mathfrak{u}_w$ is also determined.* If these values do not satisfy the condition $5\mathfrak{a}_w - \mathfrak{u}_w = Q_w = 5.000$, the estimate for $\mathfrak{a}_n$ must be modified until agreement is reached. The results for this example are:

$$\mathfrak{u}_n = -0.250, \quad \mathfrak{a}_n = 0.967$$
$$\mathfrak{u}_w = -0.144, \quad \mathfrak{a}_w = 0.971$$

and the value of $\mathfrak{a}_w$ is also shown in Fig. IX.c.3.

2. *Area Change Approximated by One Discontinuity.* Let the wave diagram for the preceding example be prepared under the condition that the gradual area change is approximated by a single step.

The wave diagram and corresponding data are shown in Figs. IX.c.4 and 5, respectively. The location of the discontinuous change of cross section is arbitrarily placed at $\xi = 0.659$ where $A/A_1 = A_2/A$.

The first step is to prepare the auxiliary chart for an area ratio of 1.7 as described in Chapter VI.g.1.2. This is reproduced as Chart 4a (see Chapter XI).

The initial, centered expansion wave is the same as in the preceding example. Its interaction with the area discontinuity will be described in detail for points 8 and 9. The iterations are here recorded in a separate table—the lower table in Fig. IX.c.5—where subscripts L and R are used to indicate whether a point is located on the left (narrow) or right side of the discontinuity. From the estimate for $\mathfrak{a}_9$ and the known value of

* A plot like Chart 4a for the correct area ratio speeds up these calculations, since M_w and $\mathfrak{a}_n/\mathfrak{a}_w$ can be read off directly for any value of M_n.

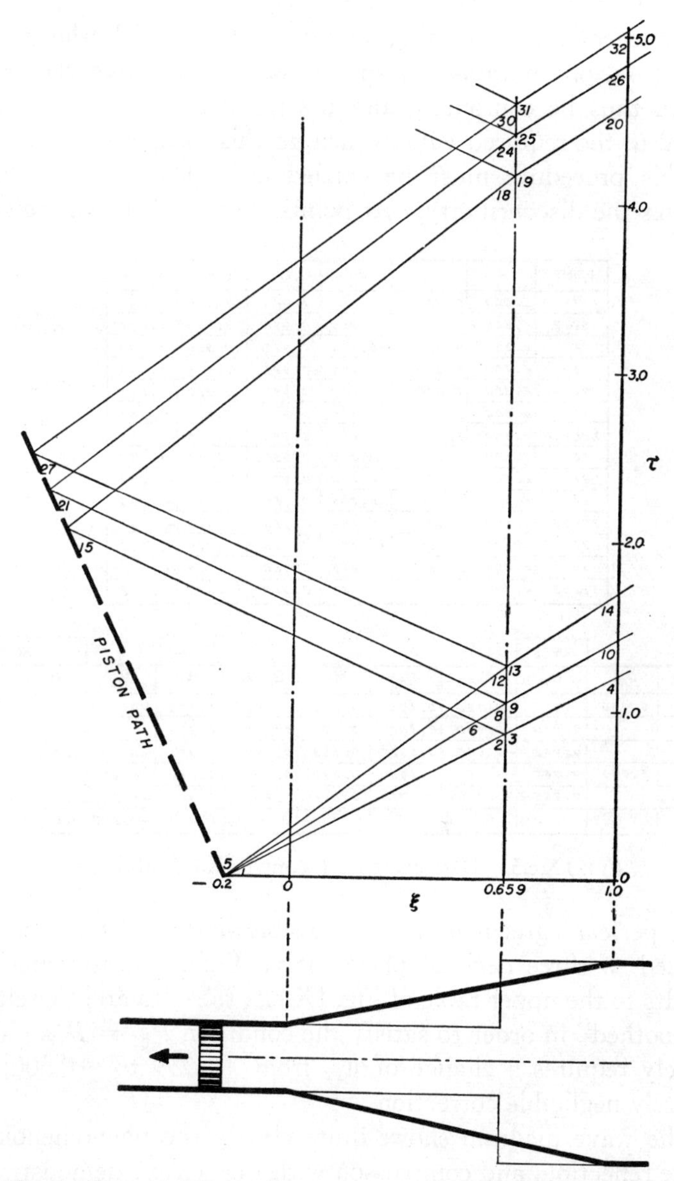

Fig. IX.c.4. Example IX.c.2. Expansion wave, created by a receding piston, passing through a duct of variable cross section (approximated by a single discontinuous area change)

Q_9, one computes $\mathcal{U}_9$ and, therefore, M_9. With the aid of Chart 4a, one obtains M_8 and $\mathcal{A}_8/\mathcal{A}_9$. The flow variables at 8 can thus be computed, and the resulting P_8 must become equal to the required value which, in this case, is P_6.

This procedure must be carried out whenever a P-wave crosses the discontinuity. At points 12 and 13, it will be seen

POINT	P	Q	$\mathcal{U}$	$\mathcal{A}$	$\mathcal{U}+\mathcal{A}$	$\mathcal{U}-\mathcal{A}$
1 to 4	5.000	5.000	0	1.000	1.000	−1.000
5, 15	4.500	5.000	−0.250	0.950	0.700	−1.200
6	4.750	5.000	−0.125	0.975	0.850	−1.100
8	4.750	5.060	−0.155	0.981	0.826	−1.136
9, 10	4.820	5.000	−0.090	0.982	0.892	−1.072
12	4.500	5.100	−0.300	0.960	0.660	−1.260
13, 14, 19, 20	4.660	5.000	−0.170	0.966	0.796	−1.136
18	4.500	5.100	−0.300	0.960	0.660	−1.260
21	4.560	5.060	−0.250	0.962	0.712	−1.212
24	4.560	5.090	−0.265	0.965	0.700	−1.230
25, 26	4.697	5.000	−0.152	0.970	0.818	−1.122
27	4.600	5.100	−0.250	0.970	0.720	−1.220
30	4.600	5.080	−0.240	0.968	0.728	−1.208
31, 32	4.720	5.000	−0.140	0.972	0.832	−1.112

		GUESS							CHECK	
POINTS	Q_R	$\mathcal{A}_R$	$\mathcal{U}_R$	M_R	M_L	$\mathcal{A}_L/\mathcal{A}_R$	$\mathcal{A}_L$	$\mathcal{U}_L$	P_L	$P_{requ.}$
8, 9	5.000	0.970	−0.150	0.155	0.270	0.995	0.965	−0.260	4.565	4.750
		0.980	−0.100	0.102	0.176	0.998	0.978	−0.172	4.718	
		0.982	−0.090	0.092	0.158	0.999	0.981	−0.155	4.750	
12, 13	5.000	0.968	−0.160	0.165	0.290	0.994	0.962	−0.279	4.531	4.500
		0.966	−0.170	0.176	0.311	0.993	0.960	−0.299	4.501	
							0.960	−0.300	4.500	

FIG. IX.c.5. Data sheet for Example IX.c.2 (abridged)

that perfect agreement is not possible if the calculations are carried to three decimal places only. Before transferring the results to the upper table of Fig. IX.c.5, the data are, therefore, "smoothed" in order to satisfy the condition $P_5 = P_{12}$. This merely requires a change of $\mathcal{U}_{12}$ from -0.299 to -0.300, an entirely negligible correction.

The wave diagram shows quite clearly the phenomenon of wave reflection, and comparison with Fig. IX.c.1 demonstrates the large saving that results from approximating a gradual area change by a discontinuous one. The variation of $\mathcal{A}$ with time at the junction of the conical section with the larger duct (points 4, 10, 14, $\cdots$) is plotted in Fig. IX.c.3 where the agreement between the two methods of wave diagram construction can be noted.

IX.d. Shock Tube

Consider a duct of constant cross section that is divided by a diaphragm into two sections, 1 and 2, in which different pressures are applied. If the diaphragm is suddenly broken, a system of pressure waves is established for which a wave diagram is to be drawn. It may be assumed that the diaphragm is instantaneously removed. There is ample experimental evidence (see the bibliography) that the flow disturbances that are caused by a properly chosen diaphragm are quite small.

Assume the following initial conditions in the shock tube: Both chambers are filled with air of equal temperature, and the pressure ratio is 20. Let the high-pressure chamber occupy one-quarter of the total length of the tube and the end of the low-pressure chamber be closed.

It is convenient to choose the initial state of the air in the low-pressure chamber as a reference state ($p_1 = p_0$ and $a_1 = a_0$). In view of this choice, one has then $\mathcal{A}_1 = \mathcal{A}_2 = 1.000$, $S_1 = 0$, and $S_2 = -\frac{1}{\gamma}\ln\frac{p_2}{p_1} = -2.140$. The length of either section of the shock tube or its entire length could serve equally well as reference length. Let L_0 be the total length of the shock tube. The wave diagram and the corresponding data are presented in Figs. IX.d.1 and 2, respectively.

It is necessary to establish the strength of the initial waves. To demonstrate a variation of the iteration procedure described in Chapter VII.g.3, let the guess be made for $\mathcal{A}_3$ (instead of $\mathcal{A}_4$). The progress of the calculations is shown in the second table of Fig. IX.d.2. The ratio $\mathcal{A}_3/\mathcal{A}_1$ determines the strength of the shock wave; $\mathfrak{U}_3$ and p_3/p_0 can then be obtained directly from Table 1a. From $P_4 = P_2$ and $\mathfrak{U}_4 = \mathfrak{U}_3$ one computes $\mathfrak{A}_4$ and then p_4/p_0. The estimate for $\mathfrak{A}_3$ must be

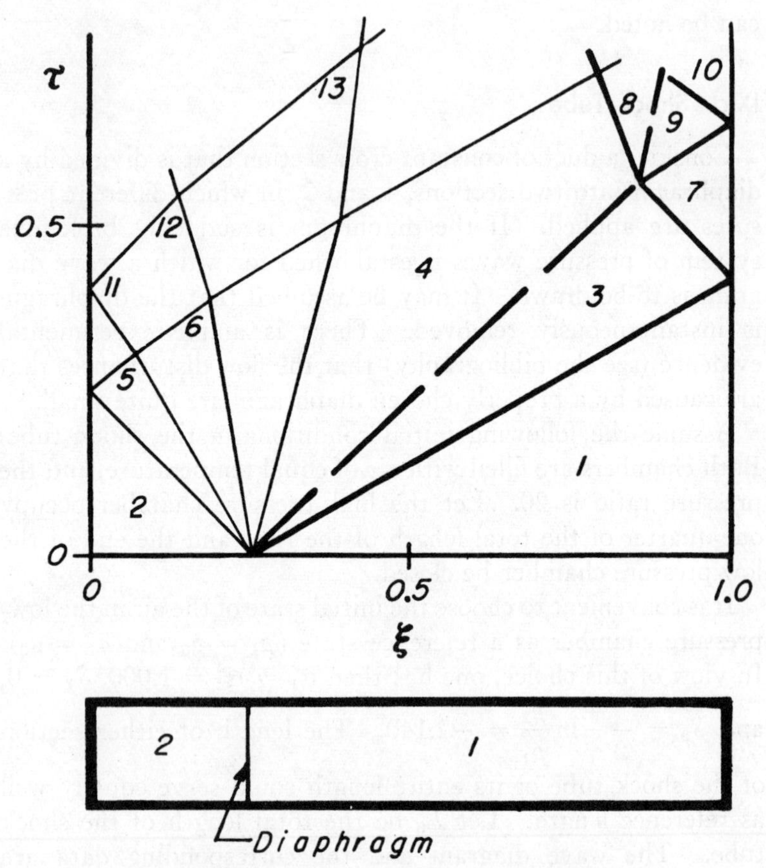

Fig. IX.d.1. Example IX.d. Shock tube

varied until p_3 and p_4 become equal. It is again necessary to smooth the best results that can be obtained within the accuracy of three decimal places, in order to achieve a perfect check.

It is seen that $\mathcal{U}_4$ turns out to be larger than $\mathcal{A}_4$—the flow in region 4 is supersonic and $\mathcal{U}_4 - \mathcal{A}_4$ becomes positive. This does not introduce any complications; one must only take care to

POINT	P	Q	$\mathcal{U}$	$\mathcal{A}$	$\mathcal{U}+\mathcal{A}$	$\mathcal{U}-\mathcal{A}$	$\mathcal{W}$	S	p/p_0	$\mathcal{A}'/\mathcal{A}$
1	5.000	5.000	0	1.000	1.000	−1.000		0	1.00	
2	5.000	5.000	0	1.000	1.000	−1.000		−2.140	20.00	
3	7.296	5.164	1.066	1.246	2.312	−0.180	1.825	0.158	3.73	1.246
4	5.000	2.868	1.066	0.787	1.853	+0.279		−2.140	3.73	
5	5.000	4.300	0.350	0.930	1.280	−0.580				
6	5.000	3.600	0.700	0.860	1.560	−0.160				
7	7.405	7.405	0	1.481	1.481	−1.481	−0.974	0.248	11.04	1.189
8	5.127	4.663	0.232	0.979	1.211	−0.747	−0.367	−1.906	13.80	1.244
9	7.872	7.408	0.232	1.528	1.760	−1.296	1.629	0.249	13.80	1.032
10	7.872	7.875	0	1.574	1.574	−1.574	−1.443	0.250	17.04	1.031
11	4.300	4.300	0	0.860	0.860	−0.860				
12	4.300	3.600	0.350	0.790	1.140	−0.440				
13	4.300	2.868	0.716	0.717	1.433	0				

GUESS					CHECK	
$\mathcal{A}_3/\mathcal{A}_1$	$\mathcal{U}_3=\mathcal{U}_4$	$P_4=P_2$	$\mathcal{A}_4$	$S_4=S_2$	p_4/p_0	p_3/p_0
1.260	1.118	5.000	0.776	−2.140	3.40	3.93
1.241	1.048		0.790		3.86	3.66
1.245	1.064		0.787		3.75	3.72
1.246	1.069		0.786		3.70	3.74
	1.066		0.787		3.73	3.73

GUESS						CHECK				
$\mathcal{A}_8/\mathcal{A}_4$	p_8/p_4	$-\frac{\mathcal{U}_8-\mathcal{U}_4}{\mathcal{A}_4}$	$\mathcal{U}_8=\mathcal{U}_9$	$\frac{\mathcal{U}_9-\mathcal{U}_7}{\mathcal{A}_7}$	p_9/p_7	p_9/p_0	p_8/p_0	$\mathcal{A}_9/\mathcal{A}_7$	S_9-S_7	S_8-S_4
1.230	3.51	1.010	0.271	0.183	1.29	14.24	13.09			
1.250	3.79	1.081	0.215	0.145	1.22	13.47	14.14			
1.243	3.69	1.056	0.235	0.159	1.25	13.80	13.76			
1.244	3.70	1.060	0.232	0.157	1.25	13.80	13.80	1.032	0.001	0.234

FIG. IX.d.2. Data sheet for Example IX.d

remember whether a wave with positive slope is a *P*- or a *Q*-wave (if desired, *P*- and *Q*-waves could be distinguished by different plotting).

The shock wave is reflected from the closed end of the duct, and the strength of the reflected shock is determined by $|\Delta\mathcal{U}|/\mathcal{A} = |\mathcal{U}_7 - \mathcal{U}_3|/\mathcal{A}_3$ with $\mathcal{U}_7 = 0$. For this value, Table 1a yields the speed of sound ratio across the shock

$\mathfrak{a}_7/\mathfrak{a}_3$ which is listed in the table at the top of Fig. IX.d.2, and the shock Mach number. From the latter and the data for region 3, one can compute the propagation velocity of the reflected shock wave.

The iteration procedure for the interaction of the reflected shock wave with the contact surface is presented in the third table of Fig. IX.d.2. A guess is made for the strength of the transmitted shock wave, $\mathfrak{a}_8/\mathfrak{a}_4$. Table 1a yields then both p_8/p_4 and $|\mathfrak{u}_8 - \mathfrak{u}_4|/\mathfrak{a}_4$ from which $\mathfrak{u}_8$ and p_8/p_0 can be determined. Since $\mathfrak{u}_9(=\mathfrak{u}_8)$ turns out to be positive and $\mathfrak{u}_7 = 0$, it follows that the reflected wave is also a shock wave of a strength determined by $|\mathfrak{u}_9 - \mathfrak{u}_7|/\mathfrak{a}_7$. This yields p_9/p_7 and, as a check of the calculation, the resulting p_9/p_0 must be equal to p_8/p_0. For the final values only, one obtains, again from Table 1a, $\mathfrak{a}_9/\mathfrak{a}_7$, $S_9 - S_7$, and $S_8 - S_4$, so that all variables for points 8 and 9 have now been determined.

Construction of the wave diagram could be extended as long as desired, but no additional types of procedures would be required. From diagrams like this, all information may be obtained that is required for the design of a shock tube. For instance, if the shock tube is to be employed as a short-duration wind tunnel using the flow in region 3, the best location of the test section would be at the point where the first reflected shock wave meets the contact surface since maximum duration of the flow is available there.

IX.e. Interruption of a Supersonic Flow

Let a steady supersonic air stream of $M = 2.0$ in a duct of constant cross section be discharged into the atmosphere at atmospheric pressure and temperature. Assume that the exit of the duct is instantaneously closed and, some time later, opened again. How far upstream can this disturbance be observed and after what time is the initial flow reestablished?

It is again convenient to let the atmospheric conditions serve as a reference state so that the flow variables of the

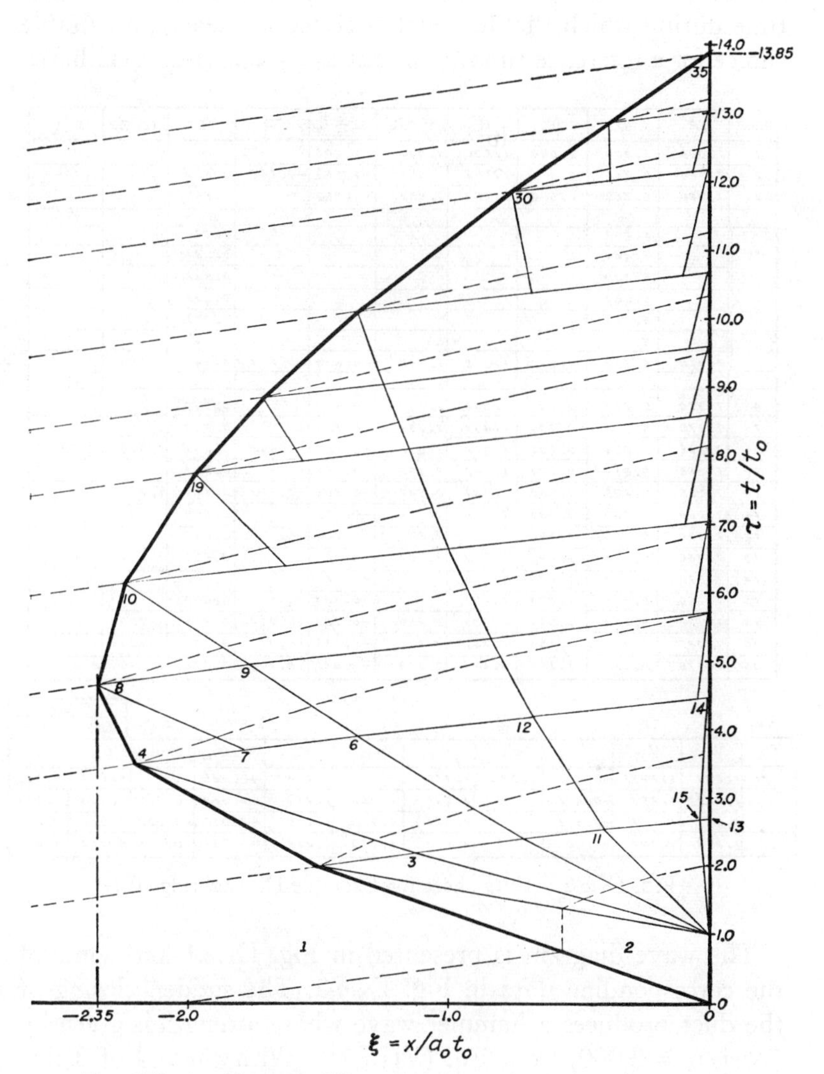

FIG. IX.e.1. Example IX.e. Interruption of a supersonic flow

initial flow are $\mathcal{A}_1 = 1.000$, $\mathcal{U}_1 = 2.000$, and $S_1 = 0$. The time during which the duct exit is closed is, clearly, a suitable choice for a reference time t_0, so that $L_0 = a_0 t_0$ (Eqs. (III.b.3)).

POINT	P	Q	$\mathcal{U}$	$\mathcal{A}$	$\mathcal{U}+\mathcal{A}$	$\mathcal{U}-\mathcal{A}$	$\mathcal{W}$	S	$\mathcal{A}'/\mathcal{A}$	p/p_0
1	7.000	3.000	2.000	1.000	3.000	1.000		0		1.00
2	7.760	7.760	0	1.552	1.552	−1.552	−0.762	0.650	1.552	8.73
3	7.760	6.760	0.500	1.452	1.952	−0.952		0.650		
4		6.760						0.432		
4		6.444						0.367		
4		6.349						0.348		
4		6.322						0.342		
4		6.313						0.341		
4		6.311						0.340		
4	7.370	6.310	0.530	1.368	1.289	−0.838	−0.216	0.340	1.368	
6	7.370	5.632	0.869	1.300	2.169	−0.431		0.340		
7	7.370	5.970	0.700	1.334	2.034	−0.634		0.340		
8	7.266	5.854	0.706	1.312	2.018	−0.606	−0.042	0.253	1.312	
9	7.266	5.519	0.874	1.278	2.152	−0.404		0.253		
10	7.185	5.425	0.880	1.261	2.142	−0.381	+0.123	0.179	1.261	
11	7.760	5.600	1.080	1.336	2.416	−0.256		0.650		
12	7.370	5.186	1.092	1.256	2.348	−0.164		0.340		
13	7.760	5.173	1.293	1.293	2.587	0		0.650		2.44
14	7.370	4.813	1.279	1.218		+0.061		0.340		2.47
15	7.760	5.214	1.273	1.297		−0.024		0.650		
19	7.136	5.154	0.991	1.229	2.220	−0.238	+0.227	0.138	1.229	
30	7.089	4.781	1.154	1.187	2.341	−0.033	+0.371	0.088	1.187	
35	7.078	4.688	1.198	1.176	2.374	+0.022	+0.406	0.078	1.176	2.81

										CHECK	
										GUESS	INTERPOL
P_{15}	$\mathcal{U}_{15}$	$\mathcal{A}_{15}$	$\Delta_- Q$	P_{14}	Q_{14}	$\mathcal{U}_{14}$	$\mathcal{A}_{14}$	$(\mathcal{U}-\mathcal{A})_{14}$	$(\mathcal{U}-\mathcal{A})_{15}$	Q_{15}	Q_{15}
7.760	1.293	1.293	−0.401	7.370	4.772	1.299	1.214	+0.085	0	5.173	5.331
	1.280	1.296	−0.401		4.799	1.286	1.217	+0.069	−0.016	5.200	5.224
	1.274	1.297	−0.401		4.811	1.280	1.218	+0.062	−0.023	5.212	5.216
	1.273	1.297	−0.401		4.813	1.279	1.218	+0.061	−0.024	5.214	5.214

FIG. IX.e.2. Data sheet for Example IX.e (abridged)

The wave diagram is presented in Fig. IX.e.1 and some of the corresponding data in Fig. IX.e.2. The sudden closing of the duct produces a hammer wave whose strength is given by $|\Delta\mathcal{U}|/\mathcal{A} = 2.000$, from Eq. (VII.f.2). With the aid of Table 1a, one obtains then all flow variables in region 2 behind the hammer wave. At $\tau = 1.0$, the exit is opened again and a centered expansion wave is created that follows the hammer wave into the duct.

In region 2, the pressure is 8.73 times atmospheric and, if the expansion wave were to reduce this to atmospheric, the speed of sound would have to drop from $\mathfrak{a}_2 = 1.552$ to $1.552 \times 8.73^{1/7} = 1.139$. However, from this and $P_2 = 7.760$, one would obtain a supersonic exit velocity which could not be produced by an expansion wave (see Chapter VI.d.2). The flow accelerates, therefore, only until it becomes sonic at the exit, point 13, where the flow variables are determined by $P_{13} = P_2$, $\mathfrak{u}_{13} = \mathfrak{a}_{13}$, and $S_{13} = S_2$.

The waves of the expansion fan catch up with the hammer wave, and the latter becomes weaker as a result of the interaction. The flow field behind the hammer wave is, therefore, multi-isentropic, and the appropriate changes of the Riemann variables must be taken into account. It is most convenient to select the particle paths through those shock points at which the interaction with the expansion wave is computed as boundaries of strips within which the entropy level is assumed to be constant, as discussed in Chapter V.c.1. According to the procedures described there, one has, for instance, $Q_4 = Q_3 + \mathfrak{a}_3(S_4 - S_3)$. Since S_4 depends on the strength of the shock at 4, one may assume at first $S_4 = S_3$, so that $Q_4 = Q_3$. The resulting shock strength is given by $(Q_4 - Q_1)/\mathfrak{a}_1$ and Table 1a yields then a value for S_4. With this, one computes an improved value for Q_4; iteration must be continued until S_4 remains unchanged. From Table 1a, one obtains then also the remaining variables at 4. The iteration steps for point 4 are listed in Fig. IX.e.2. It is interesting to note that if one had used the mean value $\mathfrak{a}_{3,4}$ instead of $\mathfrak{a}_3$ for the calculation of Q_4, the final result would have been $\mathfrak{a}_4 = 1.369$, compared to 1.368 for the simpler procedure. Therefore, the extra work of using mean values of $\mathfrak{a}$ does not seem warranted in this case. For the remaining points that are listed in Fig. IX.e.2, only the final results of the iterations are indicated. It is seen in Fig. IX.e.1 that, several times, Q-waves have been entered by interpolation to prevent the steps between points from becoming too large.

The behavior of the flow at the exit requires some further discussion. Immediately after opening, the flow is steady and sonic up to point 13, where the reflections of the expansion wave from the shock wave begin to arrive. From then on, the flow cannot remain sonic because a pressure wave does not produce equal changes of $\mathcal{U}$ and $\mathcal{A}$. It was pointed out in Chapter VII.c.2 that, in the case of one shock wave overtaking another, the reflected wave is a weak expansion wave. One should, therefore, expect a weak compression wave to be reflected when an expansion wave catches up with a shock. Such a wave would cause the sonic flow at the exit to become supersonic. The calculations are illustrated for point 14 (see the second table in Fig. IX.e.2). The location of this point is found by plotting the P-wave through 12 with a slope corresponding to $(\mathcal{U} + \mathcal{A})_{12}$. The wave Q_{14} must intersect the P-wave through 11 and 13 at some point 15 whose location is not known. From an estimated value of Q_{15}, one obtains $Q_{14} = Q_{15} + \Delta_{-}Q$, where $\Delta_{-}Q = \mathcal{A}_{15}(S_{14} - S_{15})$. Since both 14 and 15 are, clearly, located within strips of entropy levels that are already known ($S_{14} = S_4$ and $S_{15} = S_2$), the increment $\Delta_{-}Q$ is given by $-0.310\mathcal{A}_{15}$.

As a first guess, let 15 coincide with 13, which determines an estimate for the flow variables at 14 and 15. If one plots a Q-wave through 14 with the slope $(\mathcal{U} - \mathcal{A})_{14,15}$, one obtains an approximate location of 15. The use of mean slopes is important here, since $(\mathcal{U} - \mathcal{A})_{15}$ is always negative and becomes, at most, zero when 15 coincides with 13. For the position of 15 thus obtained, one can determine Q_{15} by interpolation between Q_{11} and Q_{13}. If this interpolated value which is entered in the last column of the second table of Fig. IX.e.2 does not agree with the guess for Q_{15}, the procedure must be repeated with an improved guess. The corresponding iteration steps are listed in the table. It is seen that, as anticipated, the pressure at 14 is higher than at 13 and the flow is slightly supersonic. This procedure must be repeated every time a *P*-wave reaches the open end.

The Mach number of the hammer wave is 2.762. This wave is weakened by its interaction with the expansion wave so that its propagation velocity decreases. The latter becomes zero when the shock Mach number equals that of the initial supersonic flow ($M_1 = 2.000$), near point 8; upon further weakening, the shock is no longer capable of advancing upstream and is swept downstream. Finally, when it reaches the exit of the duct at point 35, the initial flow is reestablished. It is seen from the wave diagram that the greatest distance from the exit at which the temporary closing of the latter can be felt is $2.35a_0t_0$, and the initial flow is reestablished after a time $13.85t_0$ has elapsed since the closing of the duct.

IX.f. Heat Addition

Consider a duct of constant cross section that is closed at one end and extends to infinity at the other. A region adjacent to the closed end contains a combustible air/fuel mixture, while the rest of the duct is filled with air. Initially, the gas is at rest, and pressure and temperature are constant throughout the duct.

Construct the wave diagram under the assumption that the rate of heat release is constant everywhere in the combustion region and for the following conditions:

Heat content of the fuel.........	$H = 19{,}000$ Btu/lb
Air/fuel ratio..................	$\alpha = 30$
Combustion efficiency..........	$\eta_c = 90\%$
Length of combustion region....	$L = 1$ ft
Combustion time..............	$t_c = 0.5$ ms
Initial speed of sound in duct...	$a_1 = 1140$ ft/sec
Ratio of specific heats..........	$\gamma = 1.4 =$ const.

The initial state of the gas is the natural choice for the reference conditions ($a_0 = a_1$). In addition, either L or t_c may be selected to make the variables nondimensional. Let $L_0 = L$, so that $t_0 = L_0/a_0$ and $\tau_c = t_c/t_0 = 0.57$.

Before one can start to plot the wave diagram, one must determine the rate of entropy rise in the combustion region. Because of the assumption of constant rate of heat release, one has $dq/dt = q/t_c$, where q is computed from Eq. (V.c.4). From Eq. (V.c.3), one obtains then

$$\frac{DS}{D\tau} = \frac{gJL_0q}{a_0{}^3t_c} \cdot \frac{1}{\mathcal{A}^2} = \frac{K}{\mathcal{A}^2} = \frac{18.65}{\mathcal{A}^2}$$

for the given numerical values. Note that, in this example, contrary to the previous ones, the numerical values of the reference variables must be prescribed in order to allow the computation of the constant K. The wave diagram procedures are then based on the following relations:

$$\Delta S = \frac{K}{\mathcal{A}^2}\Delta\tau \quad \text{from Eq. IV.a.1}$$

$$\Delta_+ P = (\gamma - 1)K\frac{\Delta_+\tau}{\mathcal{A}} + \mathcal{A}\Delta_+ S \quad \text{from Eqs. (IV.a.2 \& III.d.12)}$$

$$\Delta_- Q = (\gamma - 1)K\frac{\Delta_-\tau}{\mathcal{A}} + \mathcal{A}\Delta_- S \quad \text{from Eqs. (IV.a.3 \& III.d.13)}$$

The wave diagram and table of the corresponding data for a few selected points are presented in Figs. IX.f.1 and 2. Before the arrival of the first wave Q_1, the gas in the combustion region remains at rest. Although it would be possible to calculate the changes of state in this region of constant-volume combustion by means of the foregoing wave diagram relations, this is unnecessarily complicated, since, at any point within this region, the temperature is given as a function of time by

$$T = T_0 + \frac{qt}{t_c c_v} = T_0\left(1 + \frac{qL_0\gamma R}{t_c c_v a_0{}^3}\tau\right)$$

and the speed of sound is, therefore, obtained as

$$\alpha^2 = \frac{T}{T_0} = 1 + \gamma(\gamma - 1)K\tau$$

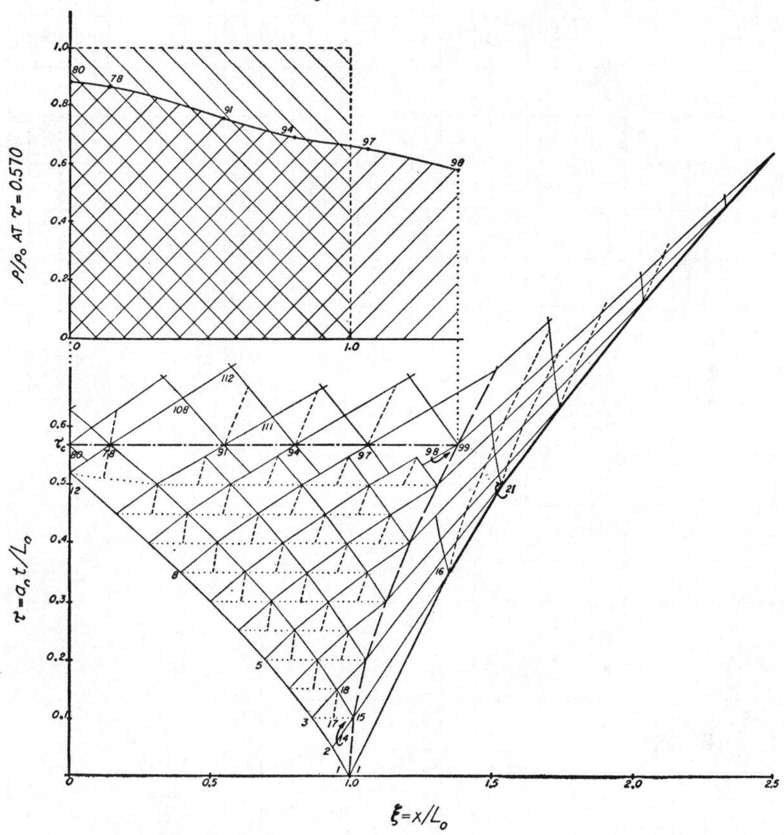

FIG. IX.f.1. Example IX.f. Combustion in a region adjacent to a closed end of a duct

where K is the same constant as in the preceding wave diagram relations. Since the density is constant in the region of constant volume combustion, the entropy level is determined by

$$S = \frac{2}{\gamma(\gamma - 1)} \ln \alpha$$

POINT	P	Q	u	a	$u+a$	$u-a$	w	S	τ	$\Delta_+ P$: a_{MEAN}	$\Delta_+ P$: $a_{MEAN}\Delta_+ S$	$\Delta_+ P$: $\frac{(\gamma-1)K\Delta_+\tau}{a_{MEAN}}$	$\Delta_+ P$: $\Delta_+ P$	$\Delta_- Q$: a_{MEAN}	$\Delta_- Q$: $a_{MEAN}\Delta_- S$	$\Delta_- Q$: $\frac{(\gamma-1)K\Delta_-\tau}{a_{MEAN}}$	$\Delta_- Q$: $\Delta_- Q$	ΔS: a^2_{MEAN}	ΔS: ΔS	INTERFACE: a_R/a_L	INTERFACE: $\frac{\gamma-1}{2}\left(1+\frac{a_R}{a_L}\right)$
1	5.000	5.000	0	1.000	1.000	−1.000		0	0												
2	6.170	6.170	0	1.234	1.234	−1.234		0.750	0.050												
3	7.150	7.150	0	1.430	1.430	−1.430		1.277	0.100												
5	8.790	8.790	0	1.578	1.578	−1.578		2.014	0.200												
8	10.790	10.790	0	2.158	2.747	−2.747		2.747	0.350												
12	12.680	12.680	0	2.536	2.536	−2.536		3.324	0.520												
14	7.848			1.522				1.865	0.100	1.234	1.376	0.302	1.678					1.000	1.865	1.452	12.260
15		5.000		1.048				0	0.100												
14	7.024			1.343				1.173	0.100	1.378	0.583	0.271	0.854					1.590	1.173	1.264	11.322
15		5.000		1.062				0	0.100												
14	7.244			1.390				1.359	0.100	1.288	0.784	0.290	1.074					1.372	1.359	1.312	11.562
15		5.000		1.059				0	0.100												
14	7.183			1.376				1.306	0.100	1.312	0.729	0.284	1.013					1.428	1.306	1.298	11.492
15		5.000		1.060				0	0.100												
14	7.201			1.380				1.321	0.100	1.305	0.745	0.286	1.031					1.411	1.321	1.303	11.512
15		5.000		1.060				0	0.100												
14	7.196			1.379				1.317	0.100	1.307	0.741	0.285	1.026					1.416	1.317	1.301	11.506
15		5.000		1.060				0	0.100												
14	7.198	6.600	0.300	1.380	1.680	−1.080		1.318	0.100	1.306	0.742	0.286	1.028					1.415	1.318	1.302	11.508
15	5.600	5.000	0.300	1.060	1.360	−0.760		0	0.100												
16	5.605	5.000	0.305	1.060	1.365	−0.755	1.195	0.005													
17				1.403				1.292	0.100												
18	8.110	7.488	0.311	1.560				1.766	0.150	1.430	0.699	0.261	0.960	1.380	0.618	0.270	0.888	1.968	0.474		
18	8.058	7.440	0.309	1.550				1.717	0.150	1.495	0.658	0.250	0.908	1.470	0.586	0.254	0.840	2.195	0.425		
18	8.060	7.444	0.308	1.550	1.858	−1.242		1.720	0.150	1.490	0.660	0.250	0.910	1.465	0.589	0.255	0.844	2.180	0.428		
21	6.056	5.020	0.518	1.108	1.626	−0.590	1.357	0.023													
78	13.206	12.504	0.351	2.571	2.922	−2.220		3.473													
80	12.877	12.877	0	2.575	2.575	−2.575		3.467													
91	13.307	11.991	0.658	2.530	3.188	−1.872		3.517													
94	13.400	11.781	0.810	2.518	3.328	−1.708		3.560													
97	13.506	11.757	0.874	2.526	3.400	−1.652		3.614	} STEPS OF ITERATION NOT LISTED												
98	13.659	11.595	1.035	2.526	3.561	−1.491		3.693													
99	7.070	5.000	1.035	1.207	2.242	−0.172		0													
108	13.319	11.991	0.684	2.531	3.195	−1.867		3.517													
111	13.416	11.781	0.818	2.520	3.338	−1.702		3.560													
112	13.319	11.673	0.823	2.499	3.322	−1.676		3.517							2.520	−0.108		−0.108				

FIG. IX.f.2. Data sheet for Example IX.f. (abridged)

The data for the points along the first Q-wave can, therefore, be directly computed since $\mathcal{U} = 0$ along this wave.

After plotting the first Q-wave, one must evaluate a point along the interface between the burning gas and air, point 14. In view of the rapid changes of the flow variables, one should keep the steps between points rather small. From an estimate of τ_{14}, one computes $S_{14} = S_1 + \Delta S$ and then $P_{14} = P_2 + \Delta_+ P$ using for $\mathcal{A}$ the values of $\mathcal{A}_1$ and $\mathcal{A}_2$, respectively. With some experience, one may speed up the calculations somewhat by using better estimates for $\mathcal{A}$ in the evaluation of ΔS and $\Delta_+ P$. On the other side of the interface, one has $Q_{15} = Q_1$. Eqs. (VI.f.5) can then be solved for $\mathcal{A}_{14}$ and $\mathcal{A}_{15}$. P_{14} and $\mathcal{A}_{14}$ determine $\mathcal{U}_{14}$, so that one can plot the segments 1–14 and 2–14 with the appropriate mean slopes and establish a better estimate of τ_{14}, if necessary. The increments ΔS and $\Delta_+ P$ can then be recomputed with the improved mean values of $\mathcal{A}$, and this procedure must be continued until no further changes of the flow variables occur.

Having obtained a point on the interface, one may proceed into the interior of the combustion region, point 18. Again, one must first guess the location of this point. The estimated particle path through 18 intersects the line that connects points 3 and 14 at 17. S_{17} is found by interpolation and, from this and ΔS, one obtains the first estimate for S_{18}. This allows determination of $\Delta_+ P$ and $\Delta_- Q$, using again reasonable estimates for $\mathcal{A}$. From P_{18} and Q_{18}, all variables at 18 can be found. This enables one to compute improved values of the increments ΔS, $\Delta_+ P$, and $\Delta_- Q$ and, if necessary, also of τ_{18}, and so forth.

It is seen that the procedures are rather lengthy. They must be repeated for every point to the left of the interface until the line $\tau = \tau_c$ is reached when combustion is completed. After that, the particle paths through the known points along $\tau = \tau_c$ may be used to divide the flow field into strips of constant entropy as described in Chapter V.c.1. For instance,

$P_{112} = P_{108}$ and $Q_{112} = Q_{111} + \mathfrak{a}_{111,\,112}\,\Delta_{-}S$ where $\Delta_{-}S$ is the entropy change between the two strips.

On the right side of the interface, the expansion of the burning gas produces a compression wave which steepens to a shock at 16. In the undisturbed air to the right of the shock, one has always $P = P_1$ and the P-wave that catches up with the shock determines $\Delta P/\mathfrak{a}$, which describes the strength of the shock (Table 1a). As the shock grows stronger, the entropy behind it varies, and one may choose again the paths of the particles through the known points along the shock to divide the region behind the latter into strips of constant entropy. As the P- and Q-waves cross these strips, their Riemann variables must be modified in the appropriate manner.

In the case of a wave diagram as complicated as the one discussed here, it is advisable to check the accuracy of the calculations. For instance, the mass between the closed end of the duct and the interface must remain constant. The values of ρ/ρ_0 at the beginning and end of combustion as function of ξ are plotted above the wave diagram in Fig. IX.f.1, and the two shaded areas should be equal. In this example, an agreement within one percent was obtained.

X

BIBLIOGRAPHY

Publications of Historical Interest (See also references 67, 81, 82, and 96)

1. S. D. Poisson, Mémoire sur la théorie du son. *Journ. de l'École Polytechn.* **7,** 319–392, 1808 (14e cahier).

2. Sir G. G. Stokes, On a difficulty in the theory of sound. *Phil. Mag.* **23,** 349–356, 1848.

3. S. Earnshaw, On the mathematical theory of sound. *Phil. Trans. Roy. Soc.* (London) **150,** 133–148, 1860.

4. B. Riemann, Über die Fortpflanzung ebener Luftwellen von endlicher Schwingungsweite. *Gött. Abh.* **8** (Math.), 43–65, 1858/59; see also *Gesammelte Mathematische Werke*, Teubner, Leipzig, 1892, 2nd Ed., pp. 156–175.

5. Lord Rayleigh, Aerial plane waves of finite amplitude. *Proc. Roy. Soc.* **A84,** 247–284, 1910. See also *The Theory of Sound*, Dover, New York, 1945, 2nd Ed., Vol. II, Chapter XI.

6. P. Vieille, Sur les discontinuités produites par la détente brusque de gas comprimés. *Comptes Rendus* **129,** 1228–1230, 1899.

General Descriptions of Various Forms of Wave Diagram Procedures (See also references 44-47, 55, 61, 63, 64, 99, and 130)

7. J. Aschenbrenner, Nichtstationäre Gasströmungen in Leitungen mit veränderlichem Querschnitt. *Forsch. Geb. Ing. Wes.* **8,** 118–130, 1937.

8. F. Schultz-Grunow, Nichtstationäre eindimensionale Gasbewegung. *Forsch. Geb. Ing. Wes.* **13,** 125–134, 1942.

9. R. Sauer, Charakteristikenverfahren für die eindimensionale instationäre Gasströmung. *Ing. Arch.* **13,** 79–89, 1942.

10. R. Sauer, Theory of nonstationary gas flow. A.F. Transl. from *Deutsche Luftfahrtforschung FB1675/1–4*, 1942/3.

I. Laminar steady gas waves. *F-TS-763-RE*, August 1946.

II. Plane gas waves with compression shocks. *F-TS-758-RE*, September 1946.

III. Laminar flow in tubes of variable cross section in particular spherical and cylindrical waves. *F-TS-770-RE*, February 1947.

IV. Plane gas flow with friction, heat transfer, and temperature differences. *F-TS-949-RE*, March 1947.

11. F. Schultz-Grunow, Nichtstationäre kugelsymmetrische Gasbewegung und nichtstationäre Gasströmung in Düsen und Diffusoren. *Ing. Arch.* **14**, 21–29, 1943.

12. P. de Haller, On a graphical method of gas dynamics. *R.T.P. Translation No. 2555 from Technische Rundschau Sulzer No. 1*, 6–24, 1945.

13. R. Courant and K. O. Friedrichs, *Supersonic Flow and Shock Waves*. Interscience Publishers, Inc., New York, 1948.

14. A. Kahane and L. Lees, Unsteady one-dimensional flows with heat addition or entropy gradients. *J. Aeron. Sci.* **15**, 665–670, 1948.

15. G. Rudinger and L. D. Rinaldi, The construction of wave diagrams for the study of one-dimensional, nonsteady gas flow. *Project SQUID Technical Memorandum No. CAL-14* (Cornell Aeronautical Laboratory, Inc.), March 10, 1948.

16. W. Weibull, Waves in compressible media. I. Basic equations. II. Plane continuous waves. *Acta Polytechnica*, Physics and Applied Mathematics Series, Stockholm, Sweden, 1, No. 5, 1–37, 1948.

17. G. Guderley, Nonsteady gas flows in thin tubes of variable cross-section. *ZWB FB 17444*, October 22, 1942. Translation T.T. 82 of Nat. Res. Coun. Can., Div. of Mech. Engng. Ottawa, October 15, 1948.

18. F. Cap, Über zwei Verfahren zur Lösung eindimensionaler instationärer gasdynamischer Probleme. *Acta Phys. Austr.* **2**, 224–238, 1949.

19. R. V. Hess, Study of unsteady flow disturbances of large and small amplitudes moving through supersonic or subsonic steady flows. *NACA TN 1878*, May 1949.

20. J. Kestin and J. S. Glass, Application of the method of characteristics to the transient flow of gases. *Proc. Inst. Mech. Eng.* **161**, 250–257, 1949.

21. N. A. Hall, An introduction to the analysis of one-dimensional nonsteady flow. *United Aircraft Corp. Meteor Rept. UAC-44*, January 1950.

22. H. H. Korst, Analysis of some thermodynamic processes by the method of characteristics for nonsteady, one-dimensional flow.

Proc. First Midwest Conf. Fluid Dyn. Univ. of Ill., May 1950; J. W. Edwards, Ann Arbor 1951, pp. 320–339.

23. C. C. Lin, Note on the characteristics in unsteady one-dimensional flows with heat addition. *Quart. Appl. Math.* **7**, 443–445, 1950.

24. W. H. Heybey, A general solution for one-dimensional non-steady flow of a perfect gas. *Nav. Ord. Lab. NAVORD Rept. No. 2210* (*Aeroball. Res. Rept. No. 52*), September 27, 1951.

25. R. Sauer, *Écoulements des fluides compressibles.* Béranger, Paris 1951.

26. S. N. B. Murthy, On the characteristic method for unsteady compressible flow. *Proc. Indian Acad. Sci.* **A37**, 664–680, 1953.

27. A. Kantrowitz, *Nonsteady One-dimensional Gas Dynamics.* High Speed Aerodynamics and Jet Propulsion Series, Vol. 3, Section C, Princeton University Press, 1958 (H. W. Emmons, Editor).

Procedures for Special Cases
(See also references 49, 51, 56, 61, and 95)

28. H. Pfriem, Die ebene ungedämpfte Druckwelle grosser Schwingungsweite. *Forsch. Geb. Ing. Wes.* **12**, 51–64, 1941.

29. J. von Neumann, Progress report on the theory of shock waves. National Defense Research Committee, *OSRD Report No. 1140*, 1943.

30. S. Paterson, The reflection of a plane shock wave at a gaseous interface. *Proc. Roy. Soc.* (*London*) **61**, 119–121, 1948.

31. G. Guderley and K. Schlagentweite, Nonstationary one-dimensional flow with combustion. *A.F. Transl. F-TS-3423-RE, Reel C724, ATI 19199.*

32. F. Cap, Zum Problem der instationären Stosspolaren. *Helv. Phys. Acta* **21**, 505–512, 1948.

33. G. Rudinger and L. D. Rinaldi, The construction of wave diagrams to study the propagation of flame fronts in ducts. *Project SQUID Technical Memorandum No. CAL-23* (Cornell Aeronautical Laboratory, Inc.), November 1948.

34. F. Schultz-Grunow, Der Carnotsche Stossverlust in nichtstationärer Gasströmung. *Z. angew. Math. Phys.* **29**, 257–267, 1949.

35. R. Sauer, Shock waves in one-dimensional nonsteady gas flow. *Helv. Phys. Acta* **22**, 467–472, 1949.

36. D. Bitondo, I. I. Glass, and G. N. Patterson, One-dimensional theory of absorption and amplification of a plane shock wave by a gaseous layer. *Univ. of Toronto, Inst. of Aerophysics Rept. No. 5*, June 1950.

37. E. Mayenfisch, Plötzliches Vordringen einer bewegten Gassäule gegen eine ruhende. *Schweiz. Arch.* **17,** 119–126, 1951.

38. N. A. Hall, The action of friction in nonsteady flow of fluids. *Proc. First Midwest. Conf. Fluid Dyn. Univ. of Ill.,* May 1950; J. W. Edwards, Ann Arbor 1951, pp. 340–353.

39. C. I. H. Nicholl, The head-on collision of shock and rarefaction waves. *Univ. of Toronto, Inst. of Aerophysics Rept. No. 10,* October 1951.

40. D. G. Gould, The head-on collision of two shock waves and a shock and rarefaction wave in one-dimensional flow. *Univ. of Toronto, Inst. of Aerophysics Rept. No. 17,* May 1952.

41. W. R. Warren, Jr., Interaction of plane waves of finite amplitude with channels of varying cross section. *Princeton Univ., Aero. Eng. Dept. Rept. No. 206,* June 1952. An extension of this report by A. Kahane, W. R. Warren, W. C. Griffith, and A. A. Marino, entitled "A theoretical and experimental study of finite amplitude wave interactions with channels of varying area," was published in *J. Aeron. Sci.* **21,** 505–524 and 565, 1954.

42. P. W. H. Howe, Reflection of a small pressure pulse by distributed friction in one-dimensional gas flow. *National Gas Turbine Establishment (Gt. Britain) Memorandum No. M. 168,* July 1953.

43. G. V. Bull, L. R. Fowell, and D. H. Henshaw, The interaction of two similarly facing shock waves. *Univ. of Toronto, Inst. of Aerophysics Rept. No. 25,* January 1953. An abridged version of this report is published in *J. Aeron. Sci.* **21,** 210–212, 1954.

Nonsteady-Flow Engines

(See also references 22, 84, 87, 96, 107, 108, 110, 112, 113, 122, 123, 129, 131, and 134 regarding wave engines, and references 22, 86, 87, and 96 regarding the pulsejet)

44. A. Capetti, Contributo allo studio del flusso nei cilindri dei motori veloci. *Ingegneria* **2,** 206–210, 1923.

45. F. Schultz-Grunow, Gasdynamic investigation of the pulsejet tube. *NACA TM 1131,* Parts I and II, February 1947.

46. A. Kantrowitz, Heat engines based on wave processes. Paper presented at the annual meeting of the ASME, November 1948.

47. E. Jenny, Berechnungen und Modellversuche über Druckwellen grosser Amplituden in Auspuff-Leitungen, Thesis, Eidgen. Techn. Hochsch. Zürich, Ameba-Druck, Basel, 1949. A review of this is presented in "Unidimensional transient flow with consideration of friction, heat transfer and change of section." *Brown Boveri Rev.* **37,** 447–461, 1950.

48. D. G. Stewart, A gasdynamic method for the solution of the pulsejet problem. *Austral. Dept. of Supply and Development. Div. of Aeronautics Rept. E64* (*ATI 74306*), July 1949.

49. S. N. Yen, H. H. Korst, and R. W. McCloy, Gasdynamic investigation of a valveless pulsejet tube. *Proc. Sec. Midwest. Conf. Fluid Mech.*, Ohio State Univ. Press, Columbus, 1952, pp. 507–520.

50. G. Rudinger, A problem arising in the performance analysis of the ducted pulsejet. *Proc. First Nat. Congr. Appl. Mech., Chicago, June 1950*, ASME, New York 1952, pp. 935–939.

51. A. Kahane, A. A. Marino, C. W. Messinger, Jr., and H. J. Shafer, A theoretical and experimental investigation of the feasibility of the intermittent ramjet engine. *Project SQUID Technical Report No. 35*, Princeton Univ., August 1, 1951.

Shock Tube

(See also references 6, 36, 39, 40, 72-74, 88, 89, 91-93, 95, 102, 104, 111, 120, 127, and 131)

52. G. N. Patterson, Theory of the shock tube. *Naval Ordnance Lab. Memo 9903*, September 21, 1948.

53. F. W. Geiger, C. W. Mautz, and R. N. Hollyer, The shock tube as an instrument for the investigation of transonic and supersonic flow patterns. *Univ. of Michigan, Eng. Res. Inst. Project M720–4*, June 1949.

54. P. W. Huber, C. E. Fitton, Jr., and F. Delpino, Experimental investigation of moving pressure disturbances and shock waves and correlation with one-dimensional unsteady-flow theory. *NACA TN 1903*, July 1949.

55. J. Lukasiewicz, Shock tube theory and applications. *Nat. Res. Coun. Can. Rept. No. MT-10*, January 18, 1950.

56. J. Lukasiewicz, Flow in a shock tube of nonuniform cross-section. *Nat. Res. Coun. Can. Rept. No. MT-11*, January 18, 1950.

57. R. K. Lobb, A study of supersonic flows in a shock tube. *Univ. of Toronto, Inst. of Aerophysics Rept. No. 8*, May 1950.

58. R. K. Lobb, On the length of a shock tube. *Univ. of Toronto, Inst. of Aerophysics Rept. No. 4*, July 1950.

59. A. Hertzberg and A. Kantrowitz, Studies with an aerodynamically instrumented shock tube. *J. Appl. Phys.* **21**, 874–878, 1950.

60. A. Hertzberg, A shock tube method of generating hypersonic flows. *J. Aeron. Sci.* **18**, 803–805, 1951.

61. J. A. Steketee, On the interaction of rarefaction waves in a shock tube. *Univ. of Toronto, Inst. of Aerophysics Review No. 4*, March 1952.

62. I. I. Glass, W. Martin, and G. N. Patterson, A theoretical and experimental study of the shock tube. *Univ. of Toronto, Inst. of Aerophysics Rept. No. 2*, November 1953.

Pulse Starting of Supersonic Wind Tunnels

63. A. Kantrowitz, R. Perry, and E. E. McDonald, Final report to Sverdrup & Parcel of theoretical study of pulse starting mechanism for supersonic wind tunnel. *Cornell Univ., Grad. School of Aero. Eng.* (No number) (*ATI 93908*), January 22, 1948.

64. G. V. Bull, Starting processes in an intermittent supersonic wind tunnel. *Univ. of Toronto, Inst. of Aerophysics Rept. No. 12*, February 1951.

Internal Ballistics
(See also reference 18)

65. P. Carrière, The method of characteristics applied to problems of internal ballistics. *Proc. 7th Int. Congr. Appl. Mech. London, 1948*, Vol. 3, 139–153.

66. J. Corner, *Theory of the interior ballistics of guns.* John Wiley & Sons, Inc., New York, 1950, Chapter 9.

Miscellaneous Applications
(See also references 91, 103, 106, 109, 115-117, and 125)

67. K. Kobes, Die Durchschlaggeschwindigkeit bei den Luftsauge- und Druckluftbremsen; Studien über unstetige Gasbewegungen. *Z. Öst. Ing. Arch. Ver.* **62,** 553–579, 1910.

68. C. De Prima and A. Leopold, On the effectiveness of various modes of detonation or combustion. *Appl. Math. Group—New York Univ. Rept. No. 145*, February 1946.

69. J. V. Foa and G. Rudinger, On the addition of heat to a gas flowing in a pipe at subsonic speed. *J. Aeron. Sci.* **16,** 84–95, 1949.

70. S. F. Neice, A method for stabilizing shock waves in channel flow by means of a surge chamber. *NACA TN 2694*, July 1953.

71. J. C. Freeman, The solution of nonlinear meterological problems by the method of characteristics. In: *Compendium of Meteorology*, Am. Meteorol. Soc., Boston, 1951, pp. 421–433.

72. E. L. Resler, S. C. Lin, and A. Kantrowitz, The production of high temperature gases in shock tubes. *J. Appl. Phys.* **23,** 1390–1399, 1952.

73. A. Hertzberg and W. Squire, The use of the shock tube for studying chemical kinetics. Paper presented at the Washington Meeting of the American Physical Society, April 1953.

74. H. S. Glick, W. Squire, and A. Hertzberg, A new shock tube technique for the study of high temperature gas phase reactions. Fifth Symposium (International) on Combustion, Reinhold Publishing Corp., New York, 1955, pp. 393–402.

Useful Tables and Charts
(See also references 17, 19, 56, 62, 104, and 131)

75. F. O. Ellenwood, N. Kulik, and N. R. Gay, The specific heats of certain gases over wide ranges of pressures and temperatures. *Cornell Univ. Engng. Exp. Sta., Bull. No. 30*, October 1942.

76. H. W. Emmons, *Gas Dynamic Tables for Air.* Dover, New York, 1947.

77. M. A. Burcher, Compressible flow tables for air. *NACA TN 1592*, August 1948.

78. J. V. Foa, *Mach number functions for ideal diatomic gases.* Cornell Aeronautical Laboratory, Inc., October 1949.

79. Handbook of supersonic aerodynamics. *NAVORD Rept. 1488*, Vol. 1, April 1, 1950, and Vol. 2, October 1, 1950. Superintendent of Documents, U. S. Government Printing Office, Washington 25, D. C.

Other References

80. F. Schultz-Grunow, Pulsierender Durchfluss durch Rohre. *Forsch. Geb. Ing. Wes.* **11,** 170–187, 1940.

81. Sir H. Lamb, *Hydrodynamics.* Dover, New York, 1945, 6th Ed., p. 481.

82. A. Stodola, *Steam and Gas Turbines.* Vol. II, Peter Smith, New York, 1945.

83. W. Jost, *Explosion and Combustion Processes in Gases.* (Translation) McGraw-Hill Book Co., Inc., New York, 1946.

84. C. Seippel, Pressure Exchanger. U. S. Patent 2,399,394, April 30, 1946.

85. H. W. Liepmann and A. E. Puckett, *Introduction to Aerodynamics of a Compressible Fluid.* John Wiley & Sons, Inc., New York, 1947.

86. L. B. Edelman, The pulsating jet engine—its evolution and future prospects. *SAE Quart. Trans.* **1,** 204–216, 1947.

87. F. W. Godsey, Jr., and L. A. Young, *Gas Turbines for Aircraft.* McGraw-Hill Book Co., Inc., New York, 1949.

88. C. du P. Donaldson and R. D. Sullivan, The effect of wall friction on the strength of shock waves in tubes and hydraulic jumps in channels. *NACA TN 1942*, 1949.

89. I. I. Glass, The design of a wave interaction tube. *Univ. of Toronto, Inst. of Aerophysics Rept. No. 6*, May 1950.

90. Sir G. I. Taylor, The instability of liquid surfaces when accelerated in a direction perpendicular to their planes. Part 1. *Proc. Roy. Soc. (London)* **A201,** 192, 1950; and D. J. Lewis, same title, Part II, *ibid.* **A202,** 81, 1950.

91. T. Carrington and N. Davidson, Photoelectric observation of the rate of dissociation of N_2O_4 by a shock wave. *J. Chem. Phys.* **19,** 1313, 1951.

92. J. D. Calhoun and A. Kogan, A preliminary investigation of the flow and reflection conditions associated with a uniform normal shock wave emerging from a two-dimensional sharp-edged channel. *Princeton Univ. Aeron. Engng. Lab. Rept. No. 209*, June 1, 1952.

93. F. K. Elder, Jr., and N. de Haas, Experimental study of the formation of a vortex ring at the open end of a cylindrical shock tube. *J. Appl. Phys.* **23,** 1065–1069, 1952.

94. L. Prandtl, *Essentials of Fluid Dynamics.* Hafner, New York, 1952.

95. G. Rudinger, On the reflection of shock waves from an open end of a duct. *J. Appl. Phys.* **26,** 981–993, 1955.

96. J. V. Foa, *Intermittent Jets.* High Speed Aerodynamics and Jet Propulsion Series, Vol. 12, Section F, Princeton University Press, 1959 (O. E. Lancaster, Editor).

97. A Kantrowitz, The formation and stability of normal shock waves in channel flow. *NACA TN 1225*, 1947.

98. Th. v. Kármán and M. A. Biot, *Mathematical Methods in Engineering.* McGraw-Hill Co., New York, 1940, pp. 403–405.

99. R. C. Roberts, The method of characteristics in compressible flow. Part II (Unsteady flow). *Wright Field Air Materiel Command, Dayton, Ohio, Technical Report No. F-TR-1173 D-ND*, December 1947.

100. J. Altenhoff, A test of the uniqueness of solutions for problems of nonsteady flow by the method of characteristics. *Industrial Mathematics* **8,** 83–92, 1957.

101. G. H. Markstein, Flow disturbances induced near a slightly wavy contact surface or flame front, traversed by a shock wave. *J. Aeron. Sci.* **24,** 238, 1957.

102. H. Mirels and W. H. Braun, Nonuniformities in shock-tube flow due to unsteady-boundary-layer action. *NACA TN 4021*, May, 1957.

103. G. Rudinger, The reflection of pressure waves of finite amplitude from an open end of a duct. *J. Fluid Mech.* **3**, 48–66, 1957.

104. I. J. Glass, Shock tubes. Part I. Theory and performance of simple shock tubes. *Institute of Aerophysics, University of Toronto, UTIA Review No.* **12,** 1958.

105. G. Rudinger, Shock wave and flame interactions. *Combustion and Propulsion*, Third AGARD Colloquium, Pergamon Press, London, 1958, pp. 153–182.

106. G. Rudinger, The reflection of shock waves from an orifice at the end of a duct. *J. Appl. Math. Phys.* (*ZAMP*), **9b,** 570–585, 1958 (Festschrift Jakob Ackeret).

107. R. S. Benson, Experiments on two-stroke engine exhaust ports under steady and unsteady flow conditions. *Proc. Inst. Mech. Engrs.* **173,** 511–546, 1959.

108. M. Berchtold, The Comprex Diesel supercharger. *SAE Trans.* **67,** 5–14, 1959.

109. A. K. Oppenheim, P. A. Urtiew, and R. A. Stern, Peculiarity of shock impingement on area convergence. *Phys. Fluids* **2,** 427–431, 1959.

110. R. C. Weatherston, W. E. Smith, A. L. Russo, and P. V. Marrone, Gasdynamics of a wave superheater facility for hypersonic research and development. *Cornell Aeronautical Laboratory Report No. AD-1118-A-1*, 1959, (DDC 210223, PB-142603).

111. C. E. Wittliff, M. R. Wilson, and A. Hertzberg, The taylored-interface hypersonic shock tunnel. *J. Aero/Space Sci.* **26,** 219–228, 1959.

112. R. S. Benson and W. A. Woods, Wave action in the exhaust system of a supercharged two-stroke engine model. *Int. J. Mech. Sci.* **1,** 253–281, 1960.

113. J. V. Foa, *Elements of Flight Propulsion.* John Wiley and Sons, Inc., New York, 1960.

114. G. E. Forsythe and W. R. Wasow, *Finite-Difference Methods for Partial Differential Equations.* John Wiley and Sons, Inc., New York, 1960.

115. G. Rudinger, Passage of shock waves through ducts of variable cross section. *Phys. Fluids* **3,** 449–455, 1960.

116. W. B. Brower, Jr., An investigation of the flow due to shock impingement on a constriction. *Rensselaer Polytechnic Institute, Report TR AE 6107,* 1961.

117. G. Rudinger, Nonsteady discharge of subcritical flow. *J. Basic Engng.* (*Trans. ASME*) **83,** Ser. D, 341–348, 1961.

118. J. E. Dove and H. Gg. Wagner, A photographic investigation of the mechanism of spinning detonation. *Eighth Symposium* (*International*) *on Combustion,* Williams & Wilkins Co., Baltimore, 1962, pp. 589–600.

119. L. Fox. *Numerical Solution of Ordinary and Partial Differential Equations.* Addison-Wesley Publishing Co., Inc., Reading, Mass., 1962.

120. A. Hertzberg, C. E. Wittliff, and J. G. Hall, Development of the shock tunnel and its application to hypersonic flight. *Hypersonic Flow Research. Progress in Astronautics and Rocketry,* Vol. 7, (F. R. Riddell, Editor), Academic Press, New York, 1962, pp. 701–758.

121. A. K. Oppenheim, N. Manson, and H. Gg. Wagner, Recent progress in detonation research. *AIAA J.* **1,** 2243–2252, 1963.

122. D. B. Spalding, Pressure exchangers. *U.S. Patent* 3,082,934, March 26, 1963.

123. R. S. Benson, R. D. Garg, and D. Woollatt, A numerical solution of unsteady flow problems. *Int. J. Mech. Sci.* **6,** 117–144, 1964.

124. G. H. Markstein, (Editor), *Nonsteady Flame Propagation.* AGARDograph 75, Macmillan Co., New York, 1964.

125. H. Rosenberg, Instationäre Strömungsvorgänge in Leitungssystemen mit flexibel-elastischen Rohrwänden. *Forschungsbericht des Landes Nordrhein-Westfalen, Nr. 1450,* Westdeutscher Verlag, Köln und Opladen, 1964.

126. V. L. Streeter, W. F. Keitzer, and D. F. Bohr, Energy dissipation in pulsatile flow through distensible tapered vessels. *Pulsatile Blood Flow* (E. O. Attinger, Editor), Blakiston Division, McGraw-Hill Book Co., New York, 1964, pp. 149–177.

127. F. J. Stoddard, A. Hertzberg, and J. G. Hall, The isentropic compression tube: a new approach to generating hypervelocity flows with low dissociation. *Proc. 4th Hypervelocity Impact Symposium, Arnold Engineering Development Center*, Tullahoma, Tenn., November 1965.

128. W. G. Vincenti and Ch. H. Kruger, Jr., *Introduction to Physical Gas Dynamics*. John Wiley and Sons, Inc., New York, 1965, pp. 300–305.

129. P. H. Azoury, An introduction to the dynamic pressure exchanger. *Proc. Inst. Mech. Engrs.* **180,** 451–480, 1965–66.

130. M. B. Abbott, *An Introduction to the Method of Characteristics*. American Elsevier Publishing Co., Inc., New York, 1966.

131. H. Oertel, *Stossrohre*. Springer Verlag, Wien-New York, 1966.

132. G. Rudinger, Review of current mathematical methods for the analysis of blood flow. *Biomedical Fluid Mechanics Symposium, ASME*, New York, 1966, pp. 1–33.

133. R. Skalak, Wave propagation in blood flow. *Biomechanics* (Y. C. Fung, Editor), ASME, New York, 1966, pp. 20–46.

134. R. C. Weatherston and A. Hertzberg, The energy exchanger, a new concept for high-efficiency gas turbine cycles. *J. Engng. for Power* (*Trans. ASME*) **89,** Ser. A, 217–228, 1967.

XI

CHARTS AND TABLES

Certain computations can be speeded up considerably if suitable charts or tables are available. A number of these are collected here for different values of γ which are indicated by the letter following the identification number. Thus,

a corresponds to $\gamma = 1.4$
b corresponds to $\gamma = \frac{5}{3}$
c corresponds to $\gamma = \frac{4}{3}$

The presented data are based on relations derived in:

Chapter VI.d.3 for Chart 1;
Chapter VI.d.5.2 for Chart 2;
Chapter VI.e.3 for Chart 3;
Chapter VI.g.2 for Chart 4;
Chapter VI.e.3 for Table 1.

Table 1 contains the same information as Chart 3 but in a form that is more accurate and convenient for actual use. Table entries are arranged in equal increments of 0.01 for $\Delta P/\mathfrak{a}$ or $\Delta Q/\mathfrak{a}$. Linear interpolation is permissible, and the increments are so small that this can be done mentally without difficulty. The obtainable accuracy is entirely adequate for all wave diagram procedures.

Although values for both ΔP and ΔQ exist for any shock wave, it is important to remember that the first column of the table refers to $\Delta P/\mathfrak{a}$ of a P-shock or $\Delta Q/\mathfrak{a}$ of a Q-shock.

The setting up and evaluation of wave diagrams require the knowledge of certain properties of the gases involved. A number of such quantities for several important gases are listed in the following table for convenient reference.

PROPERTIES OF SOME IMPORTANT GASES *

Gas	γ	$\frac{2}{\gamma - 1}$	$\frac{a}{a_{air}}$	R ft lb/slug, °R	c_p Btu/lb °R	c_v Btu/lb °R
Air	$\frac{7}{5}$	5	1.00	1715	0.24_0	0.17_1
Hydrogen	$\frac{7}{5}$	5	3.79	24630	3.44	2.46
Helium	$\frac{5}{3}$	3	2.93	12410	1.24	0.74_5
Neon	$\frac{5}{3}$	3	1.31	2461	0.24_6	0.14_7
Oxygen	$\frac{7}{5}$	5	0.95	1552	0.21_7	0.15_5
Nitrogen	$\frac{7}{5}$	5	1.02	1772	0.24_8	0.17_7
Argon	$\frac{5}{3}$	3	0.93	1244	0.12_4	0.74_8
Hot combustion gases †	$\frac{4}{3}$	6	‡	1715	0.27_4	0.20_6

* The values of γ listed in this table are given to the nearest simple fraction. The effects of small deviations of the actual values from those listed are, in general, unimportant compared to the errors due to other simplifying assumptions that must be made. On the other hand, a substantial time saving results if a simple number for $2/(\gamma - 1)$ can be used in a wave diagram.

† It was pointed out in Chapter VI.h.2.1 that, in the case of burning of the usually employed air/fuel mixtures, the change of R need not be considered. If one wishes to make allowance for the changes of the specific heats and γ, some reasonable values must be selected. The choice $\gamma = \frac{4}{3}$ has the advantage that most of the algebraic expressions in gas-dynamics relations that involve γ then assume simple numerical values. This choice is, therefore, convenient not only for use in wave diagrams but also in general analytical studies (e.g.96). Once R and γ are chosen, the values for c_p and c_v are, of course, also determined.

‡ Depending on temperature.

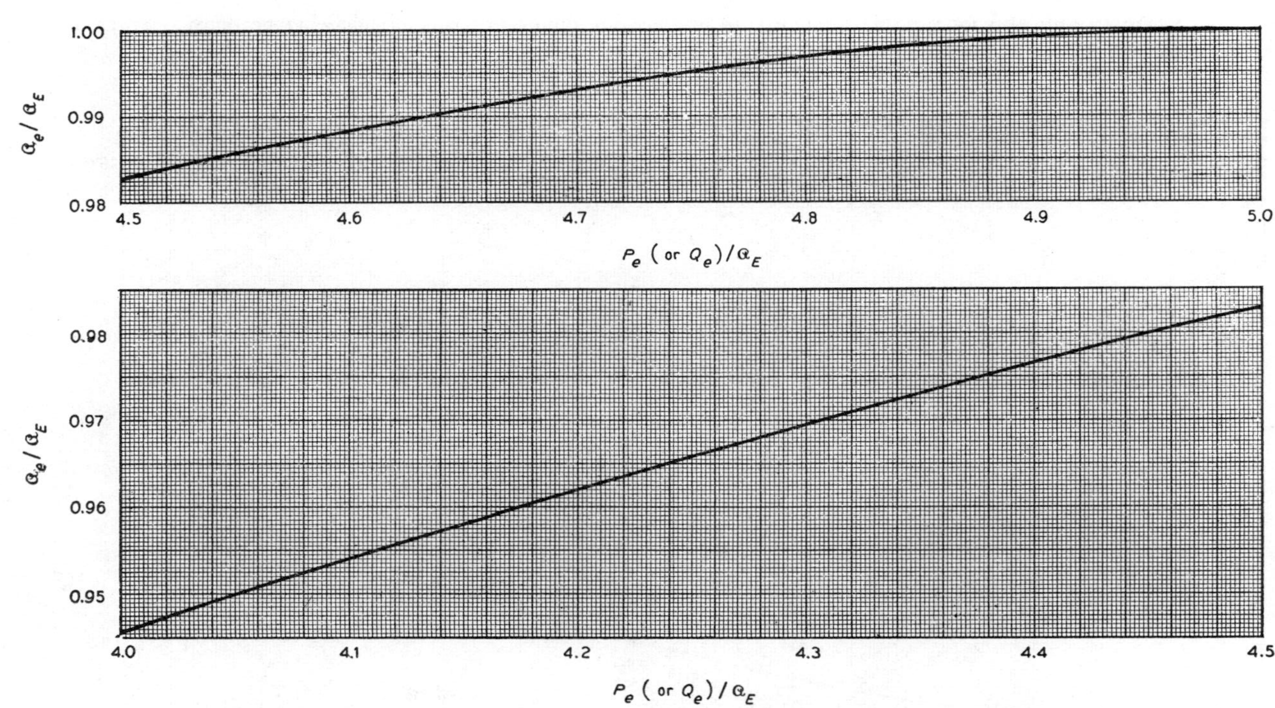

CHART 1a. Relation between the speed of sound and the known Riemann variable for flow into a duct; $\gamma = 1.4$

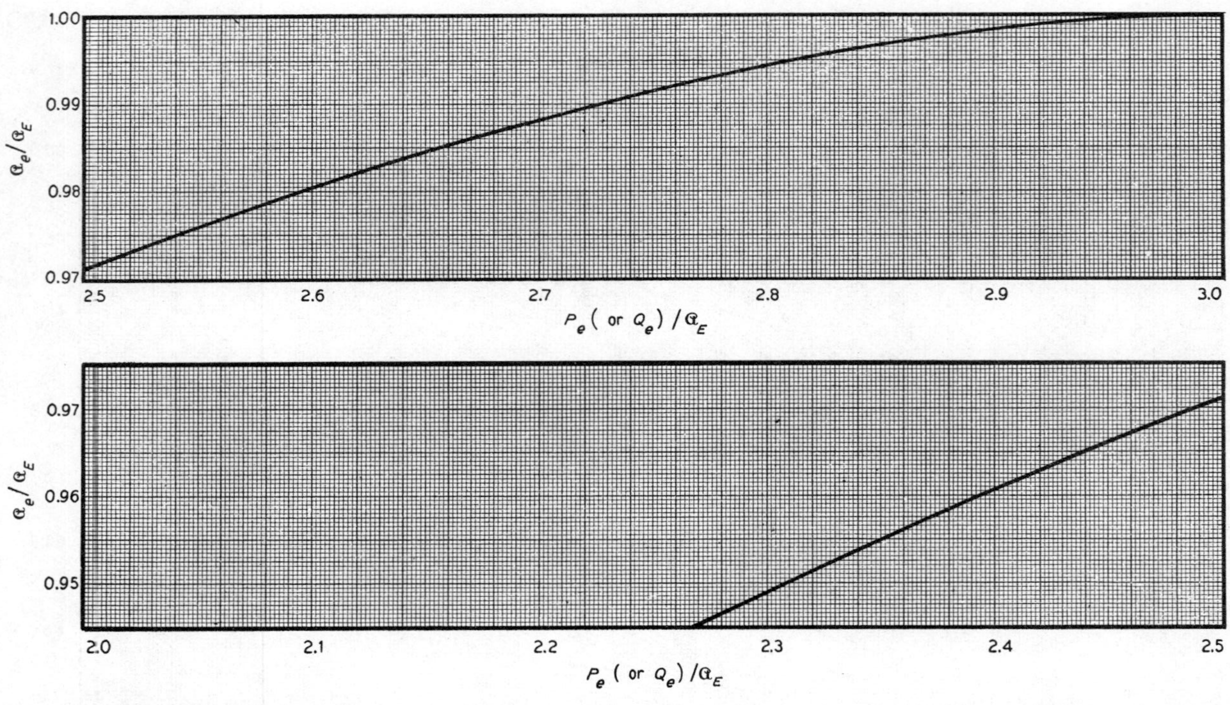

CHART 1b. Relation between the speed of sound and the known Riemann variable for flow into a duct; $\gamma = 5/3$

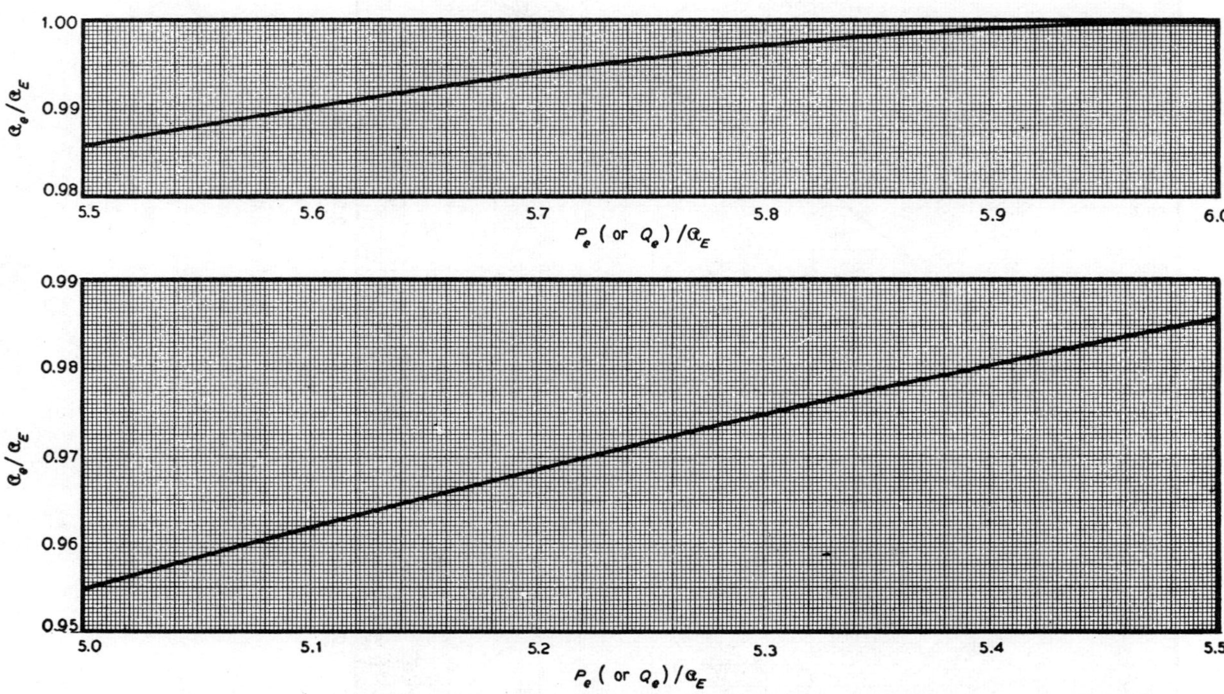

CHART 1c. Relation between the speed of sound and the known Riemann variable for flow into a duct; $\gamma = 4/3$

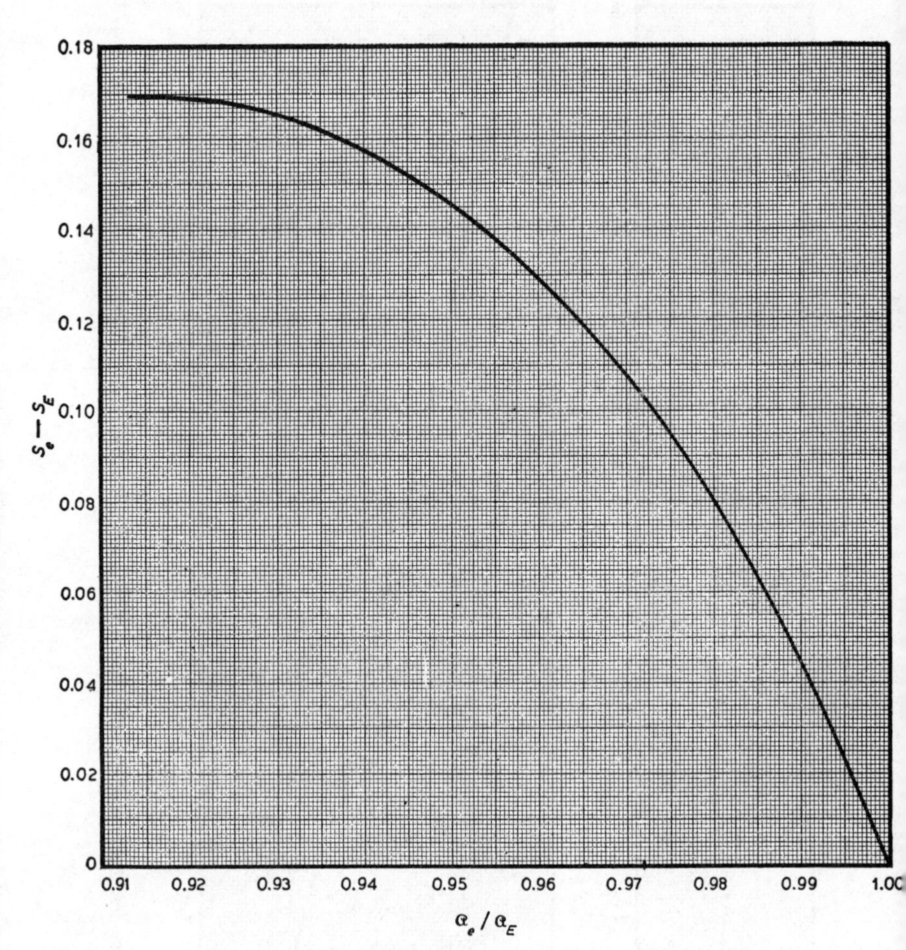

CHART 2a. Entropy rise in a Borda nozzle; $\gamma = 1.4$

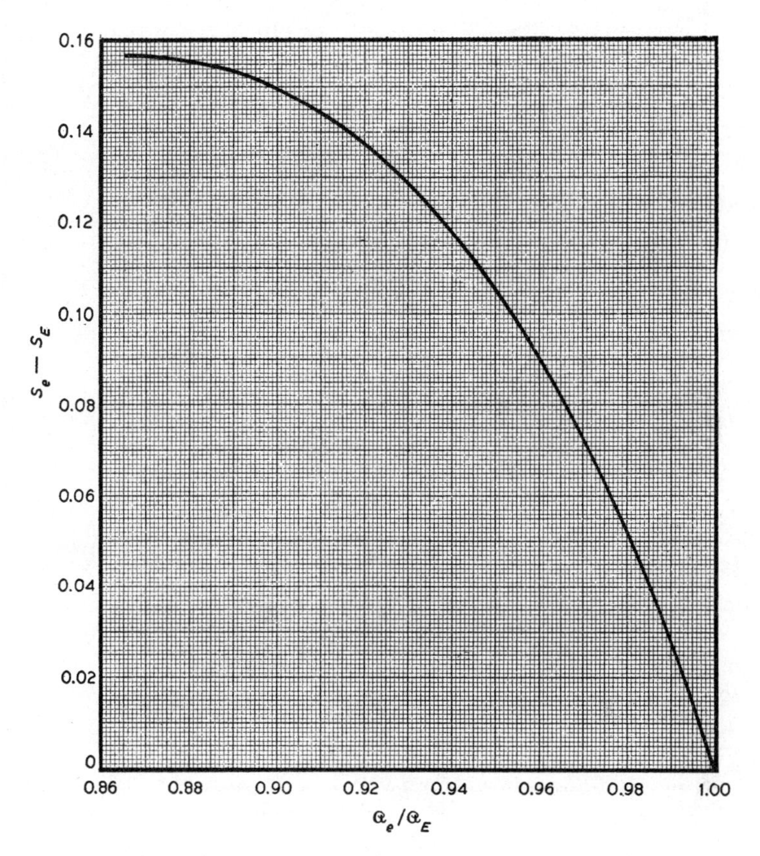

CHART 2b. Entropy rise in a Borda nozzle; $\gamma = 5/3$

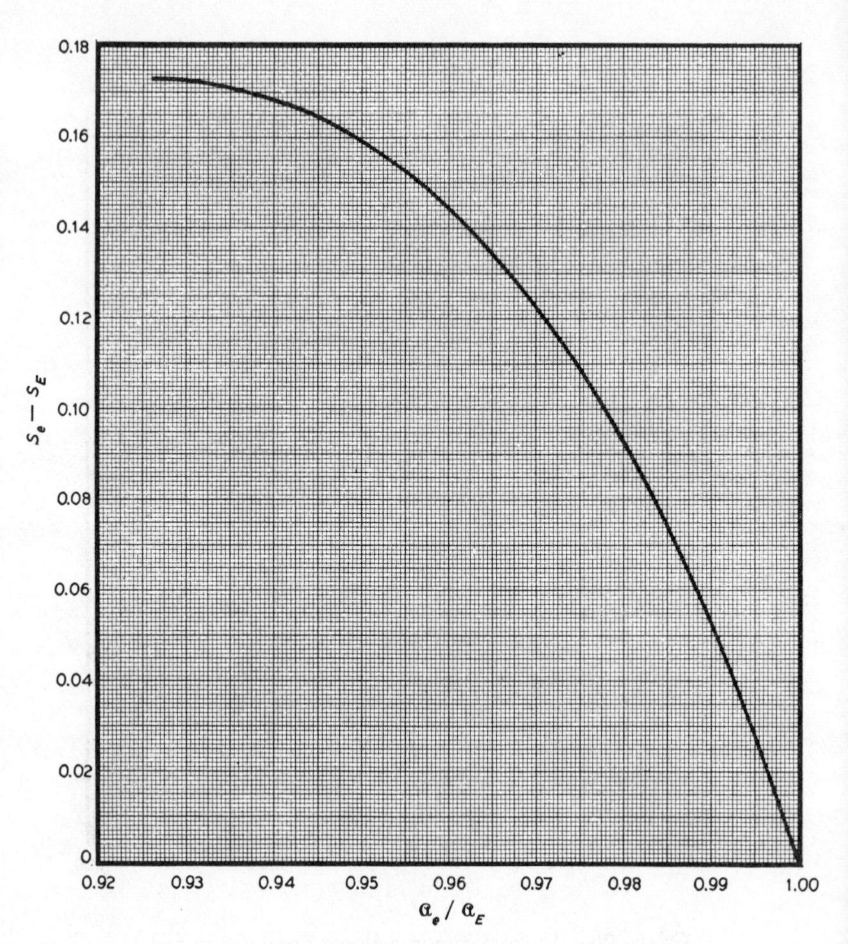

CHART 2c. Entropy rise in a Borda nozzle; $\gamma = 4/3$

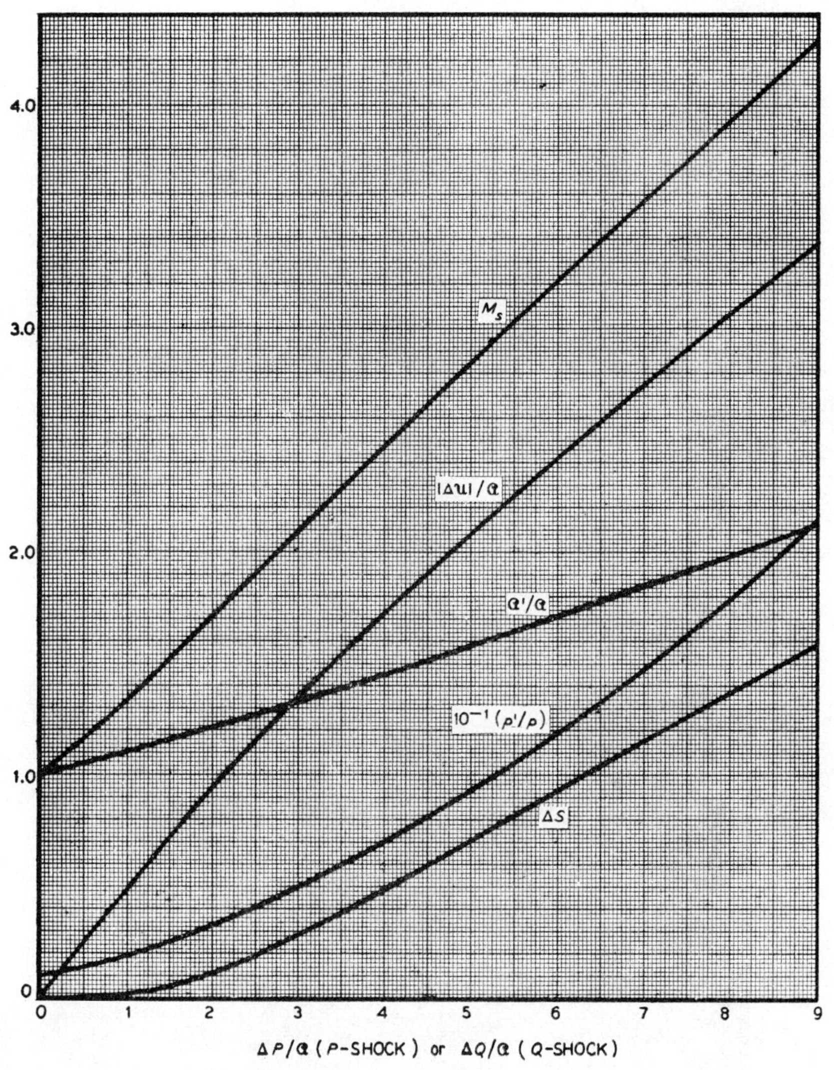

CHART 3a. Shock wave relations; $\gamma = 1.4$

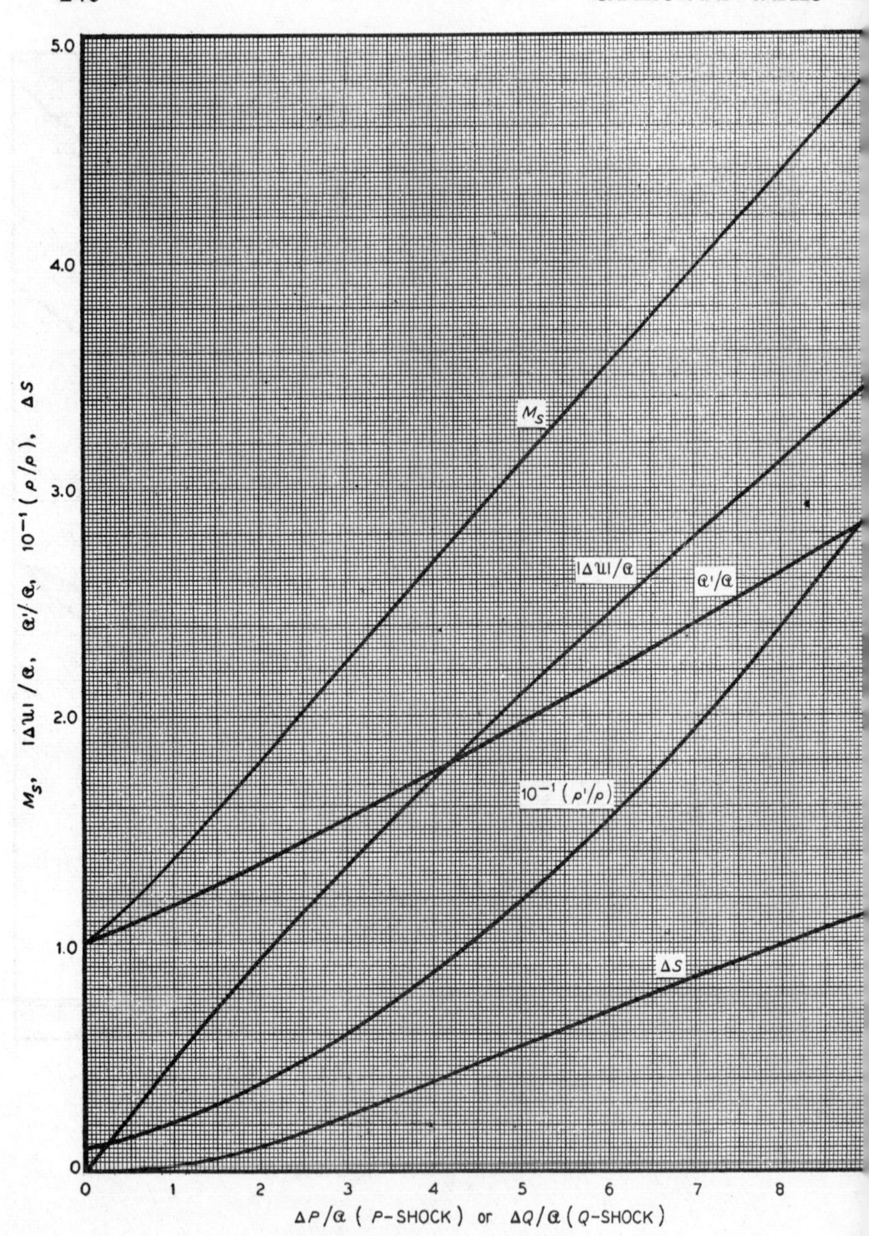

CHART 3b. Shock wave relations; $\gamma = 5/3$

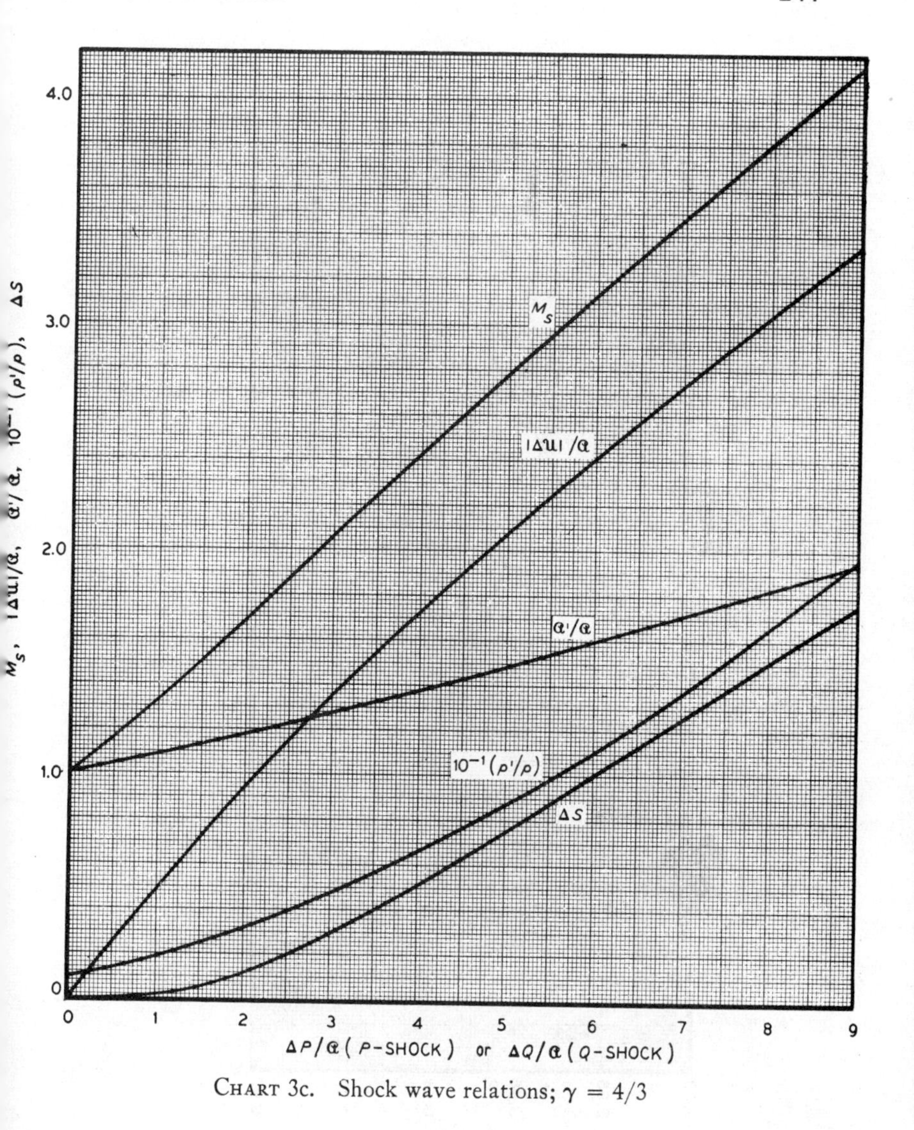

CHART 3c. Shock wave relations; $\gamma = 4/3$

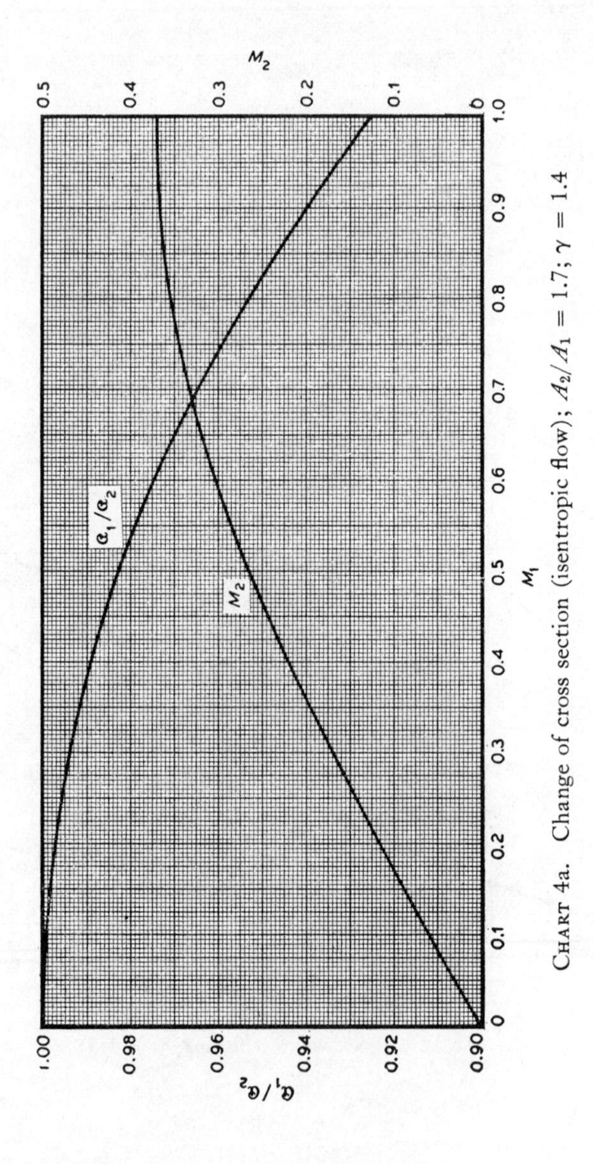

CHART 4a. Change of cross section (isentropic flow); $A_2/A_1 = 1.7$; $\gamma = 1.4$

TABLE 1a. SHOCK WAVE RELATIONS FOR $\gamma = 1.4$

$\frac{\Delta P}{\mathfrak{a}}, \frac{\Delta Q}{\mathfrak{a}}$	M_s	$\frac{\lvert\Delta \mathfrak{U}\rvert}{\mathfrak{a}}$	$\frac{\mathfrak{a}'}{\mathfrak{a}}$	$\frac{p'}{p}$	ΔS	$\frac{\Delta P}{\mathfrak{a}}, \frac{\Delta Q}{\mathfrak{a}}$	M_s	$\frac{\lvert\Delta \mathfrak{U}\rvert}{\mathfrak{a}}$	$\frac{\mathfrak{a}'}{\mathfrak{a}}$	$\frac{p'}{p}$	ΔS
.00	1.000	.000	1.000	1.00	.000	.50	1.160	.249	1.050	1.40	.003
.01	1.003	.005	1.001	1.01	.000	.51	1.164	.254	1.051	1.41	.003
.02	1.006	.010	1.002	1.01	.000	.52	1.167	.259	1.052	1.42	.003
.03	1.009	.015	1.003	1.02	.000	.53	1.170	.264	1.053	1.43	.004
.04	1.012	.020	1.004	1.03	.000	.54	1.174	.269	1.054	1.44	.004
.05	1.015	.025	1.005	1.04	.000	.55	1.177	.273	1.055	1.45	.004
.06	1.018	.030	1.006	1.04	.000	.56	1.181	.278	1.056	1.46	.004
.07	1.021	.035	1.007	1.05	.000	.57	1.185	.283	1.057	1.47	.004
.08	1.024	.040	1.008	1.06	.000	.58	1.188	.288	1.058	1.48	.005
.09	1.027	.045	1.009	1.06	.000	.59	1.191	.293	1.059	1.49	.005
.10	1.031	.050	1.010	1.07	.000	.60	1.195	.298	1.060	1.50	.005
.11	1.034	.055	1.011	1.08	.000	.61	1.198	.303	1.061	1.51	.005
.12	1.037	.060	1.012	1.09	.000	.62	1.201	.308	1.062	1.52	.005
.13	1.040	.065	1.013	1.09	.000	.63	1.205	.313	1.064	1.53	.006
.14	1.043	.070	1.014	1.10	.000	.64	1.209	.318	1.065	1.54	.006
.15	1.046	.075	1.015	1.11	.000	.65	1.212	.322	1.066	1.55	.006
.16	1.049	.080	1.016	1.12	.000	.66	1.215	.327	1.067	1.56	.006
.17	1.052	.085	1.017	1.13	.000	.67	1.219	.332	1.068	1.57	.007
.18	1.055	.090	1.018	1.13	.000	.68	1.222	.337	1.069	1.58	.007
.19	1.058	.095	1.019	1.14	.000	.69	1.226	.342	1.070	1.59	.007
.20	1.061	.100	1.020	1.15	.000	.70	1.230	.347	1.071	1.60	.008
.21	1.065	.105	1.021	1.16	.000	.71	1.233	.352	1.072	1.61	.008
.22	1.068	.110	1.022	1.16	.000	.72	1.237	.357	1.073	1.62	.008
.23	1.071	.115	1.023	1.17	.000	.73	1.240	.361	1.074	1.63	.008
.24	1.074	.120	1.024	1.18	.000	.74	1.244	.366	1.075	1.64	.009
.25	1.077	.125	1.025	1.19	.000	.75	1.247	.371	1.076	1.65	.009
.26	1.081	.130	1.026	1.20	.001	.76	1.250	.375	1.077	1.66	.009
.27	1.084	.135	1.027	1.20	.001	.77	1.254	.380	1.078	1.67	.010
.28	1.088	.140	1.028	1.21	.001	.78	1.257	.385	1.079	1.68	.010
.29	1.091	.145	1.029	1.22	.001	.79	1.261	.390	1.080	1.69	.010
.30	1.094	.150	1.030	1.23	.001	.80	1.265	.395	1.081	1.70	.011
.31	1.097	.154	1.031	1.24	.001	.81	1.268	.400	1.082	1.71	.011
.32	1.100	.159	1.032	1.25	.001	.82	1.272	.405	1.083	1.72	.011
.33	1.103	.164	1.033	1.25	.001	.83	1.275	.410	1.084	1.73	.012
.34	1.107	.169	1.034	1.26	.001	.84	1.279	.414	1.085	1.74	.012
.35	1.110	.174	1.035	1.27	.001	.85	1.283	.419	1.086	1.75	.012
.36	1.114	.179	1.036	1.28	.001	.86	1.287	.424	1.087	1.76	.013
.37	1.117	.184	1.037	1.29	.001	.87	1.290	.429	1.088	1.78	.013
.38	1.120	.189	1.038	1.30	.001	.88	1.293	.434	1.089	1.79	.014
.39	1.123	.194	1.039	1.31	.002	.89	1.297	.439	1.090	1.80	.014
.40	1.126	.199	1.040	1.31	.002	.90	1.301	.443	1.091	1.81	.015
.41	1.130	.204	1.041	1.32	.002	.91	1.304	.448	1.093	1.82	.015
.42	1.133	.209	1.042	1.33	.002	.92	1.308	.453	1.094	1.83	.016
.43	1.137	.214	1.043	1.34	.002	.93	1.312	.458	1.095	1.84	.016
.44	1.140	.219	1.044	1.35	.002	.94	1.315	.462	1.096	1.85	.017
.45	1.144	.224	1.045	1.36	.002	.95	1.318	.467	1.097	1.86	.017
.46	1.147	.229	1.046	1.37	.002	.96	1.322	.472	1.098	1.87	.018
.47	1.150	.234	1.047	1.38	.003	.97	1.326	.476	1.099	1.88	.018
.48	1.153	.239	1.048	1.39	.003	.98	1.329	.481	1.100	1.90	.019
.49	1.157	.244	1.049	1.40	.003	.99	1.333	.486	1.101	1.91	.020
.50	1.160	.249	1.050	1.40	.003	1.00	1.337	.491	1.102	1.92	.020

TABLE 1a (*Continued*). $\gamma = 1.4$

$\frac{\Delta P}{\mathfrak{a}}, \frac{\Delta Q}{\mathfrak{a}}$	M_s	$\frac{\lvert\Delta \mathfrak{u}\rvert}{\mathfrak{a}}$	$\frac{\mathfrak{a}'}{\mathfrak{a}}$	$\frac{p'}{p}$	ΔS
1.00	1.337	.491	1.102	1.92	.020
1.01	1.341	.495	1.103	1.93	.021
1.02	1.344	.500	1.104	1.94	.021
1.03	1.348	.505	1.105	1.95	.022
1.04	1.351	.509	1.106	1.96	.022
1.05	1.355	.514	1.107	1.98	.023
1.06	1.359	.519	1.108	1.99	.023
1.07	1.362	.523	1.109	2.00	.024
1.08	1.366	.528	1.110	2.01	.025
1.09	1.370	.533	1.112	2.02	.026
1.10	1.373	.538	1.113	2.03	.026
1.11	1.377	.542	1.114	2.05	.027
1.12	1.381	.547	1.115	2.06	.027
1.13	1.384	.552	1.116	2.07	.028
1.14	1.388	.556	1.117	2.08	.028
1.15	1.392	.561	1.118	2.09	.029
1.16	1.395	.566	1.119	2.11	.030
1.17	1.399	.570	1.120	2.12	.031
1.18	1.403	.575	1.121	2.13	.031
1.19	1.407	.580	1.122	2.14	.032
1.20	1.411	.585	1.123	2.15	.032
1.21	1.414	.589	1.124	2.17	.033
1.22	1.418	.594	1.125	2.18	.034
1.23	1.422	.599	1.126	2.19	.035
1.24	1.425	.603	1.127	2.20	.035
1.25	1.429	.608	1.129	2.22	.036
1.26	1.433	.612	1.130	2.23	.037
1.27	1.436	.617	1.131	2.24	.037
1.28	1.440	.621	1.132	2.25	.038
1.29	1.444	.626	1.133	2.27	.039
1.30	1.448	.631	1.134	2.28	.040
1.31	1.451	.635	1.135	2.29	.041
1.32	1.455	.640	1.136	2.30	.042
1.33	1.459	.644	1.137	2.32	.042
1.34	1.463	.649	1.138	2.33	.043
1.35	1.467	.654	1.139	2.34	.044
1.36	1.470	.658	1.140	2.36	.045
1.37	1.474	.663	1.142	2.37	.046
1.38	1.478	.667	1.143	2.38	.047
1.39	1.481	.672	1.144	2.39	.047
1.40	1.485	.676	1.145	2.41	.048
1.41	1.489	.681	1.146	2.42	.049
1.42	1.492	.685	1.147	2.43	.050
1.43	1.496	.690	1.148	2.45	.051
1.44	1.500	.694	1.149	2.46	.052
1.45	1.504	.699	1.150	2.47	.053
1.46	1.507	.703	1.151	2.48	.054
1.47	1.511	.708	1.152	2.50	.055
1.48	1.515	.713	1.154	2.51	.056
1.49	1.519	.717	1.155	2.53	.057
1.50	1.523	.722	1.156	2.54	.057

$\frac{\Delta P}{\mathfrak{a}}, \frac{\Delta Q}{\mathfrak{a}}$	M_s	$\frac{\lvert\Delta \mathfrak{u}\rvert}{\mathfrak{a}}$	$\frac{\mathfrak{a}'}{\mathfrak{a}}$	$\frac{p'}{p}$	ΔS
1.50	1.523	.722	1.156	2.54	.057
1.51	1.527	.726	1.157	2.55	.058
1.52	1.530	.731	1.158	2.57	.059
1.53	1.534	.735	1.159	2.58	.060
1.54	1.538	.740	1.160	2.59	.061
1.55	1.542	.744	1.161	2.61	.062
1.56	1.546	.749	1.162	2.62	.063
1.57	1.549	.753	1.163	2.63	.064
1.58	1.553	.758	1.164	2.65	.065
1.59	1.557	.762	1.166	2.66	.066
1.60	1.561	.767	1.167	2.68	.067
1.61	1.565	.771	1.168	2.69	.069
1.62	1.568	.776	1.169	2.70	.070
1.63	1.572	.780	1.170	2.72	.071
1.64	1.576	.784	1.171	2.73	.072
1.65	1.580	.789	1.172	2.75	.073
1.66	1.584	.794	1.173	2.76	.074
1.67	1.587	.798	1.174	2.77	.075
1.68	1.591	.802	1.176	2.79	.077
1.69	1.595	.807	1.177	2.80	.078
1.70	1.599	.811	1.178	2.82	.079
1.71	1.603	.816	1.179	2.83	.080
1.72	1.607	.820	1.180	2.85	.081
1.73	1.610	.824	1.181	2.86	.082
1.74	1.614	.829	1.182	2.87	.083
1.75	1.618	.833	1.183	2.89	.084
1.76	1.622	.838	1.184	2.90	.085
1.77	1.626	.842	1.186	2.92	.086
1.78	1.629	.846	1.187	2.93	.088
1.79	1.633	.851	1.188	2.95	.089
1.80	1.637	.855	1.189	2.96	.090
1.81	1.641	.859	1.190	2.97	.091
1.82	1.645	.864	1.191	2.99	.093
1.83	1.649	.868	1.192	3.00	.094
1.84	1.652	.873	1.193	3.02	.095
1.85	1.656	.877	1.195	3.03	.096
1.86	1.660	.881	1.196	3.05	.098
1.87	1.664	.886	1.197	3.06	.099
1.88	1.668	.890	1.198	3.08	.100
1.89	1.672	.894	1.199	3.09	.101
1.90	1.675	.899	1.200	3.11	.102
1.91	1.679	.903	1.202	3.12	.104
1.92	1.683	.907	1.203	3.14	.105
1.93	1.687	.911	1.204	3.15	.107
1.94	1.691	.916	1.205	3.17	.108
1.95	1.695	.920	1.206	3.18	.109
1.96	1.698	.925	1.207	3.20	.111
1.97	1.702	.929	1.208	3.21	.112
1.98	1.706	.933	1.209	3.23	.113
1.99	1.710	.937	1.211	3.24	.115
2.00	1.714	.942	1.212	3.26	.116

TABLE 1a (*Continued*). $\gamma = 1.4$

$\frac{\Delta P}{\mathcal{A}}, \frac{\Delta Q}{\mathcal{A}}$	M_S	$\frac{\lvert\Delta \mathcal{U}\rvert}{\mathcal{A}}$	$\frac{\mathcal{A}'}{\mathcal{A}}$	$\frac{p'}{p}$	ΔS
2.00	1.714	.942	1.212	3.26	.116
2.01	1.718	.946	1.213	3.28	.117
2.02	1.721	.950	1.214	3.29	.119
2.03	1.725	.954	1.215	3.31	.120
2.04	1.729	.959	1.216	3.32	.122
2.05	1.733	.963	1.217	3.34	.123
2.06	1.737	.967	1.219	3.35	.125
2.07	1.740	.971	1.220	3.37	.126
2.08	1.744	.976	1.221	3.38	.127
2.09	1.748	.980	1.222	3.40	.128
2.10	1.752	.984	1.223	3.41	.130
2.11	1.756	.988	1.224	3.43	.131
2.12	1.760	.993	1.226	3.45	.133
2.13	1.763	.997	1.227	3.46	.134
2.14	1.767	1.001	1.228	3.48	.136
2.15	1.771	1.005	1.229	3.49	.137
2.16	1.775	1.010	1.230	3.51	.138
2.17	1.779	1.014	1.231	3.53	.140
2.18	1.783	1.018	1.232	3.54	.141
2.19	1.787	1.023	1.234	3.56	.143
2.20	1.791	1.027	1.235	3.57	.144
2.21	1.794	1.031	1.236	3.59	.146
2.22	1.798	1.035	1.237	3.61	.147
2.23	1.802	1.039	1.238	3.62	.149
2.24	1.806	1.044	1.239	3.64	.150
2.25	1.810	1.048	1.241	3.66	.152
2.26	1.814	1.052	1.242	3.67	.153
2.27	1.817	1.056	1.243	3.69	.155
2.28	1.821	1.060	1.244	3.70	.156
2.29	1.825	1.064	1.245	3.72	.158
2.30	1.829	1.069	1.246	3.74	.159
2.31	1.833	1.073	1.248	3.75	.161
2.32	1.837	1.077	1.249	3.77	.162
2.33	1.840	1.081	1.250	3.79	.164
2.34	1.844	1.085	1.251	3.80	.166
2.35	1.848	1.089	1.252	3.82	.167
2.36	1.852	1.093	1.253	3.83	.169
2.37	1.856	1.097	1.255	3.85	.171
2.38	1.859	1.101	1.256	3.87	.172
2.39	1.863	1.105	1.257	3.88	.174
2.40	1.867	1.110	1.258	3.90	.175
2.41	1.871	1.114	1.259	3.92	.177
2.42	1.875	1.118	1.260	3.93	.178
2.43	1.879	1.122	1.262	3.95	.180
2.44	1.883	1.126	1.263	3.97	.182
2.45	1.887	1.130	1.264	3.99	.183
2.46	1.890	1.134	1.265	4.00	.185
2.47	1.894	1.139	1.266	4.02	.187
2.48	1.898	1.143	1.267	4.04	.188
2.49	1.902	1.147	1.269	4.05	.190
2.50	1.906	1.151	1.270	4.07	.191
2.50	1.906	1.151	1.270	4.07	.191
2.51	1.910	1.155	1.271	4.09	.193
2.52	1.914	1.159	1.272	4.11	.195
2.53	1.918	1.163	1.273	4.12	.196
2.54	1.921	1.167	1.275	4.14	.198
2.55	1.925	1.171	1.276	4.16	.200
2.56	1.929	1.175	1.277	4.17	.201
2.57	1.933	1.179	1.278	4.19	.203
2.58	1.937	1.183	1.279	4.21	.205
2.59	1.940	1.187	1.280	4.23	.207
2.60	1.944	1.191	1.282	4.24	.208
2.61	1.948	1.195	1.283	4.26	.210
2.62	1.952	1.200	1.284	4.28	.212
2.63	1.956	1.204	1.285	4.30	.213
2.64	1.960	1.208	1.287	4.31	.215
2.65	1.963	1.212	1.288	4.33	.217
2.66	1.967	1.216	1.289	4.35	.219
2.67	1.971	1.220	1.290	4.37	.221
2.68	1.975	1.224	1.291	4.38	.223
2.69	1.979	1.228	1.292	4.40	.224
2.70	1.983	1.232	1.294	4.42	.226
2.71	1.986	1.236	1.295	4.44	.227
2.72	1.990	1.240	1.296	4.45	.229
2.73	1.994	1.244	1.297	4.47	.231
2.74	1.998	1.248	1.298	4.49	.233
2.75	2.002	1.252	1.300	4.51	.234
2.76	2.006	1.256	1.301	4.53	.236
2.77	2.010	1.260	1.302	4.55	.238
2.78	2.013	1.264	1.303	4.56	.240
2.79	2.017	1.268	1.304	4.58	.242
2.80	2.021	1.272	1.306	4.60	.244
2.81	2.025	1.276	1.307	4.62	.245
2.82	2.029	1.280	1.308	4.64	.247
2.83	2.033	1.284	1.309	4.65	.249
2.84	2.037	1.288	1.310	4.67	.250
2.85	2.040	1.292	1.312	4.69	.252
2.86	2.044	1.296	1.313	4.71	.254
2.87	2.048	1.300	1.314	4.73	.256
2.88	2.052	1.304	1.315	4.75	.258
2.89	2.056	1.308	1.317	4.76	.260
2.90	2.059	1.312	1.318	4.78	.261
2.91	2.063	1.315	1.319	4.80	.263
2.92	2.067	1.319	1.320	4.82	.265
2.93	2.071	1.323	1.321	4.84	.267
2.94	2.075	1.327	1.323	4.86	.269
2.95	2.078	1.331	1.324	4.87	.271
2.96	2.082	1.335	1.325	4.89	.273
2.97	2.086	1.339	1.326	4.91	.274
2.98	2.090	1.343	1.327	4.93	.276
2.99	2.094	1.347	1.329	4.95	.278
3.00	2.098	1.351	1.330	4.97	.280

TABLE 1a (*Continued*). $\gamma = 1.4$

$\frac{\Delta P}{\mathfrak{a}}, \frac{\Delta Q}{\mathfrak{a}}$	M_S	$\frac{\lvert\Delta \mathfrak{u}\rvert}{\mathfrak{a}}$	$\frac{\mathfrak{a}'}{\mathfrak{a}}$	$\frac{p'}{p}$	ΔS
3.00	2.098	1.351	1.330	4.97	.280
3.01	2.102	1.355	1.331	4.99	.282
3.02	2.105	1.359	1.332	5.00	.284
3.03	2.109	1.363	1.333	5.02	.286
3.04	2.113	1.367	1.335	5.04	.288
3.05	2.117	1.370	1.336	5.06	.290
3.06	2.121	1.374	1.337	5.08	.291
3.07	2.124	1.378	1.338	5.10	.293
3.08	2.128	1.382	1.340	5.12	.295
3.09	2.132	1.386	1.341	5.14	.297
3.10	2.136	1.390	1.342	5.16	.299
3.11	2.140	1.394	1.343	5.18	.301
3.12	2.144	1.398	1.345	5.20	.303
3.13	2.147	1.402	1.346	5.21	.305
3.14	2.151	1.405	1.347	5.23	.307
3.15	2.155	1.409	1.348	5.25	.309
3.16	2.159	1.413	1.349	5.27	.311
3.17	2.163	1.417	1.351	5.29	.313
3.18	2.166	1.421	1.352	5.31	.315
3.19	2.170	1.425	1.353	5.33	.317
3.20	2.174	1.429	1.354	5.35	.319
3.21	2.178	1.432	1.356	5.37	.321
3.22	2.182	1.436	1.357	5.39	.322
3.23	2.186	1.440	1.358	5.41	.324
3.24	2.189	1.444	1.359	5.43	.326
3.25	2.193	1.448	1.360	5.45	.328
3.26	2.197	1.452	1.362	5.47	.330
3.27	2.201	1.455	1.363	5.48	.332
3.28	2.205	1.459	1.364	5.50	.334
3.29	2.208	1.463	1.365	5.52	.336
3.30	2.212	1.467	1.367	5.54	.338
3.31	2.216	1.471	1.368	5.56	.340
3.32	2.220	1.475	1.369	5.58	.342
3.33	2.224	1.478	1.370	5.60	.344
3.34	2.228	1.482	1.372	5.62	.346
3.35	2.231	1.486	1.373	5.64	.348
3.36	2.235	1.490	1.374	5.66	.350
3.37	2.239	1.494	1.375	5.68	.352
3.38	2.243	1.497	1.376	5.70	.354
3.39	2.247	1.501	1.378	5.72	.356
3.40	2.251	1.505	1.379	5.74	.358
3.41	2.254	1.509	1.380	5.76	.360
3.42	2.258	1.513	1.381	5.78	.362
3.43	2.262	1.516	1.383	5.80	.364
3.44	2.266	1.520	1.384	5.82	.366
3.45	2.269	1.524	1.385	5.84	.368
3.46	2.273	1.528	1.387	5.86	.370
3.47	2.277	1.531	1.388	5.88	.372
3.48	2.281	1.535	1.389	5.90	.375
3.49	2.285	1.539	1.390	5.92	.377
3.50	2.289	1.543	1.391	5.94	.379

$\frac{\Delta P}{\mathfrak{a}}, \frac{\Delta Q}{\mathfrak{a}}$	M_S	$\frac{\lvert\Delta \mathfrak{u}\rvert}{\mathfrak{a}}$	$\frac{\mathfrak{a}'}{\mathfrak{a}}$	$\frac{p'}{p}$	ΔS
3.50	2.289	1.543	1.391	5.94	.379
3.51	2.292	1.547	1.393	5.96	.381
3.52	2.296	1.550	1.394	5.98	.383
3.53	2.300	1.554	1.395	6.00	.385
3.54	2.303	1.558	1.396	6.02	.387
3.55	2.307	1.561	1.398	6.04	.389
3.56	2.311	1.565	1.399	6.07	.391
3.57	2.315	1.569	1.400	6.09	.393
3.58	2.319	1.573	1.401	6.11	.395
3.59	2.322	1.576	1.403	6.13	.397
3.60	2.326	1.580	1.404	6.15	.399
3.61	2.330	1.584	1.405	6.17	.401
3.62	2.334	1.588	1.406	6.19	.403
3.63	2.338	1.592	1.408	6.21	.405
3.64	2.342	1.595	1.409	6.23	.407
3.65	2.345	1.599	1.410	6.25	.409
3.66	2.349	1.603	1.411	6.27	.411
3.67	2.353	1.607	1.413	6.29	.414
3.68	2.357	1.610	1.414	6.31	.416
3.69	2.360	1.614	1.415	6.33	.418
3.70	2.364	1.618	1.416	6.35	.420
3.71	2.368	1.621	1.418	6.37	.422
3.72	2.372	1.625	1.419	6.40	.424
3.73	2.376	1.629	1.420	6.42	.426
3.74	2.379	1.633	1.422	6.44	.428
3.75	2.383	1.636	1.423	6.46	.430
3.76	2.387	1.640	1.424	6.48	.432
3.77	2.391	1.644	1.425	6.50	.435
3.78	2.395	1.648	1.427	6.52	.437
3.79	2.398	1.651	1.428	6.54	.439
3.80	2.402	1.655	1.429	6.57	.441
3.81	2.406	1.659	1.430	6.59	.443
3.82	2.410	1.662	1.432	6.61	.445
3.83	2.413	1.666	1.433	6.63	.447
3.84	2.417	1.669	1.434	6.65	.449
3.85	2.421	1.673	1.435	6.67	.451
3.86	2.425	1.677	1.437	6.69	.453
3.87	2.428	1.681	1.438	6.71	.455
3.88	2.432	1.684	1.439	6.73	.457
3.89	2.436	1.688	1.440	6.76	.460
3.90	2.440	1.692	1.442	6.78	.462
3.91	2.444	1.695	1.443	6.80	.464
3.92	2.447	1.699	1.444	6.82	.466
3.93	2.451	1.703	1.446	6.84	.468
3.94	2.455	1.706	1.447	6.86	.470
3.95	2.458	1.710	1.448	6.88	.473
3.96	2.462	1.713	1.449	6.91	.475
3.97	2.466	1.717	1.451	6.93	.477
3.98	2.470	1.721	1.452	6.95	.479
3.99	2.474	1.725	1.453	6.97	.481
4.00	2.477	1.728	1.454	6.99	.483

TABLE 1a (*Continued*). $\gamma = 1.4$

$\frac{\Delta P}{\mathfrak{a}}, \frac{\Delta Q}{\mathfrak{a}}$	M_S	$\frac{\lvert\Delta \mathfrak{u}\rvert}{\mathfrak{a}}$	$\frac{\mathfrak{a}'}{\mathfrak{a}}$	$\frac{p'}{p}$	ΔS
4.00	2.477	1.728	1.454	6.99	.483
4.01	2.481	1.732	1.456	7.02	.486
4.02	2.485	1.735	1.457	7.04	.488
4.03	2.489	1.739	1.458	7.06	.490
4.04	2.493	1.743	1.459	7.08	.492
4.05	2.496	1.746	1.461	7.10	.494
4.06	2.500	1.750	1.462	7.12	.496
4.07	2.504	1.754	1.463	7.15	.499
4.08	2.507	1.757	1.465	7.17	.501
4.09	2.511	1.761	1.466	7.19	.503
4.10	2.515	1.764	1.467	7.21	.505
4.11	2.519	1.768	1.468	7.24	.507
4.12	2.523	1.772	1.470	7.26	.509
4.13	2.526	1.775	1.471	7.28	.511
4.14	2.530	1.779	1.472	7.30	.514
4.15	2.534	1.783	1.474	7.33	.516
4.16	2.538	1.786	1.475	7.35	.518
4.17	2.541	1.790	1.476	7.37	.520
4.18	2.545	1.794	1.477	7.39	.522
4.19	2.549	1.797	1.479	7.41	.524
4.20	2.553	1.801	1.480	7.44	.527
4.21	2.556	1.804	1.481	7.46	.529
4.22	2.560	1.808	1.482	7.48	.531
4.23	2.564	1.811	1.484	7.50	.533
4.24	2.567	1.815	1.485	7.52	.535
4.25	2.571	1.819	1.486	7.55	.537
4.26	2.575	1.822	1.488	7.57	.540
4.27	2.579	1.826	1.489	7.59	.542
4.28	2.583	1.829	1.490	7.61	.544
4.29	2.586	1.833	1.491	7.64	.546
4.30	2.590	1.837	1.493	7.66	.548
4.31	2.594	1.840	1.494	7.68	.551
4.32	2.598	1.844	1.495	7.71	.553
4.33	2.601	1.848	1.497	7.73	.555
4.34	2.605	1.851	1.498	7.75	.557
4.35	2.609	1.855	1.499	7.77	.559
4.36	2.613	1.858	1.500	7.80	.562
4.37	2.616	1.862	1.502	7.82	.564
4.38	2.620	1.865	1.503	7.84	.566
4.39	2.624	1.869	1.504	7.86	.568
4.40	2.627	1.872	1.505	7.89	.570
4.41	2.631	1.876	1.507	7.91	.572
4.42	2.635	1.879	1.508	7.93	.575
4.43	2.639	1.883	1.509	7.96	.577
4.44	2.643	1.887	1.511	7.98	.579
4.45	2.646	1.890	1.512	8.00	.581
4.46	2.650	1.894	1.513	8.03	.584
4.47	2.654	1.897	1.515	8.05	.586
4.48	2.657	1.901	1.516	8.07	.588
4.49	2.661	1.904	1.517	8.09	.590
4.50	2.665	1.908	1.518	8.12	.592

$\frac{\Delta P}{\mathfrak{a}}, \frac{\Delta Q}{\mathfrak{a}}$	M_S	$\frac{\lvert\Delta \mathfrak{u}\rvert}{\mathfrak{a}}$	$\frac{\mathfrak{a}'}{\mathfrak{a}}$	$\frac{p'}{p}$	ΔS
4.50	2.665	1.908	1.518	8.12	.592
4.51	2.668	1.911	1.520	8.14	.594
4.52	2.672	1.915	1.521	8.16	.597
4.53	2.676	1.919	1.522	8.19	.599
4.54	2.680	1.922	1.524	8.21	.601
4.55	2.684	1.926	1.525	8.24	.603
4.56	2.687	1.929	1.526	8.26	.606
4.57	2.691	1.933	1.528	8.28	.608
4.58	2.695	1.936	1.529	8.31	.610
4.59	2.698	1.940	1.530	8.33	.612
4.60	2.702	1.943	1.531	8.35	.615
4.61	2.706	1.947	1.533	8.38	.617
4.62	2.710	1.950	1.534	8.40	.619
4.63	2.713	1.954	1.535	8.42	.621
4.64	2.717	1.957	1.537	8.45	.624
4.65	2.721	1.961	1.538	8.47	.626
4.66	2.724	1.964	1.539	8.49	.628
4.67	2.728	1.968	1.540	8.52	.630
4.68	2.732	1.971	1.542	8.54	.632
4.69	2.735	1.975	1.543	8.56	.635
4.70	2.739	1.978	1.544	8.59	.637
4.71	2.743	1.982	1.546	8.61	.639
4.72	2.747	1.985	1.547	8.63	.641
4.73	2.750	1.989	1.548	8.66	.644
4.74	2.754	1.993	1.550	8.68	.646
4.75	2.758	1.996	1.551	8.71	.648
4.76	2.762	2.000	1.552	8.73	.650
4.77	2.765	2.003	1.553	8.76	.653
4.78	2.769	2.007	1.555	8.78	.655
4.79	2.773	2.010	1.556	8.80	.657
4.80	2.777	2.014	1.557	8.83	.659
4.81	2.780	2.017	1.559	8.85	.661
4.82	2.784	2.021	1.560	8.88	.664
4.83	2.788	2.024	1.561	8.90	.666
4.84	2.791	2.028	1.563	8.92	.668
4.85	2.795	2.031	1.564	8.95	.670
4.86	2.799	2.035	1.565	8.97	.672
4.87	2.802	2.038	1.566	9.00	.675
4.88	2.806	2.042	1.568	9.02	.677
4.89	2.810	2.045	1.569	9.04	.679
4.90	2.814	2.048	1.570	9.07	.681
4.91	2.817	2.052	1.572	9.09	.684
4.92	2.821	2.055	1.573	9.12	.686
4.93	2.825	2.059	1.574	9.14	.688
4.94	2.828	2.062	1.576	9.17	.690
4.95	2.832	2.066	1.577	9.19	.693
4.96	2.836	2.069	1.578	9.22	.695
4.97	2.839	2.073	1.579	9.24	.697
4.98	2.843	2.076	1.581	9.26	.700
4.99	2.847	2.080	1.582	9.29	.702
5.00	2.851	2.083	1.583	9.31	.704

TABLE 1a (*Continued*). $\gamma = 1.4$

$\frac{\Delta P}{\mathfrak{A}}, \frac{\Delta Q}{\mathfrak{A}}$	M_s	$\frac{\lvert\Delta\mathfrak{U}\rvert}{\mathfrak{A}}$	$\frac{\mathfrak{A}'}{\mathfrak{A}}$	$\frac{p'}{p}$	ΔS
5.00	2.851	2.083	1.583	9.31	.704
5.01	2.854	2.087	1.585	9.34	.706
5.02	2.858	2.090	1.586	9.36	.708
5.03	2.862	2.094	1.587	9.39	.711
5.04	2.865	2.097	1.589	9.41	.713
5.05	2.869	2.100	1.590	9.44	.715
5.06	2.873	2.104	1.591	9.46	.717
5.07	2.876	2.107	1.593	9.49	.720
5.08	2.880	2.111	1.594	9.51	.722
5.09	2.884	2.114	1.595	9.54	.724
5.10	2.888	2.118	1.597	9.56	.727
5.11	2.891	2.121	1.598	9.59	.729
5.12	2.895	2.125	1.599	9.61	.731
5.13	2.899	2.128	1.601	9.64	.733
5.14	2.902	2.132	1.602	9.66	.736
5.15	2.906	2.135	1.603	9.69	.738
5.16	2.910	2.138	1.604	9.71	.740
5.17	2.913	2.142	1.606	9.74	.742
5.18	2.917	2.145	1.607	9.76	.744
5.19	2.921	2.149	1.608	9.79	.747
5.20	2.925	2.152	1.610	9.81	.749
5.21	2.928	2.156	1.611	9.84	.751
5.22	2.932	2.159	1.612	9.86	.753
5.23	2.936	2.163	1.614	9.89	.756
5.24	2.939	2.166	1.615	9.91	.758
5.25	2.943	2.169	1.616	9.94	.760
5.26	2.947	2.173	1.618	9.96	.762
5.27	2.950	2.176	1.619	9.99	.765
5.28	2.954	2.180	1.620	10.01	.767
5.29	2.957	2.183	1.622	10.04	.769
5.30	2.961	2.186	1.623	10.06	.771
5.31	2.965	2.190	1.624	10.09	.774
5.32	2.968	2.193	1.625	10.11	.776
5.33	2.972	2.196	1.627	10.14	.778
5.34	2.976	2.200	1.628	10.17	.781
5.35	2.980	2.203	1.629	10.19	.783
5.36	2.983	2.207	1.631	10.22	.785
5.37	2.987	2.210	1.632	10.24	.787
5.38	2.990	2.213	1.633	10.27	.789
5.39	2.994	2.217	1.635	10.29	.792
5.40	2.998	2.220	1.636	10.32	.794
5.41	3.002	2.224	1.637	10.34	.796
5.42	3.005	2.227	1.639	10.37	.798
5.43	3.009	2.231	1.640	10.40	.801
5.44	3.013	2.234	1.641	10.42	.803
5.45	3.016	2.237	1.643	10.45	.805
5.46	3.020	2.241	1.644	10.47	.808
5.47	3.024	2.244	1.645	10.50	.810
5.48	3.027	2.248	1.647	10.53	.812
5.49	3.031	2.251	1.648	10.55	.814
5.50	3.035	2.254	1.649	10.58	.817

$\frac{\Delta P}{\mathfrak{A}}, \frac{\Delta Q}{\mathfrak{A}}$	M_s	$\frac{\lvert\Delta\mathfrak{U}\rvert}{\mathfrak{A}}$	$\frac{\mathfrak{A}'}{\mathfrak{A}}$	$\frac{p'}{p}$	ΔS
5.50	3.035	2.254	1.649	10.58	.81
5.51	3.039	2.258	1.651	10.61	.81
5.52	3.042	2.261	1.652	10.63	.82
5.53	3.046	2.264	1.653	10.65	.82
5.54	3.049	2.268	1.655	10.68	.82
5.55	3.053	2.271	1.656	10.71	.82
5.56	3.057	2.275	1.657	10.74	.83
5.57	3.061	2.278	1.659	10.76	.83
5.58	3.064	2.281	1.660	10.78	.83
5.59	3.067	2.285	1.661	10.81	.83
5.60	3.071	2.288	1.662	10.84	.83
5.61	3.075	2.291	1.664	10.86	.84
5.62	3.079	2.295	1.665	10.89	.84
5.63	3.082	2.298	1.666	10.92	.84
5.64	3.086	2.302	1.668	10.94	.84
5.65	3.090	2.305	1.669	10.97	.85
5.66	3.093	2.308	1.671	11.00	.85
5.67	3.097	2.312	1.672	11.02	.85
5.68	3.101	2.315	1.673	11.05	.85
5.69	3.104	2.319	1.674	11.08	.85
5.70	3.108	2.322	1.676	11.10	.86
5.71	3.111	2.325	1.677	11.13	.86
5.72	3.115	2.328	1.678	11.15	.86
5.73	3.119	2.332	1.680	11.18	.86
5.74	3.123	2.335	1.681	11.21	.87
5.75	3.126	2.339	1.682	11.24	.87
5.76	3.130	2.342	1.684	11.26	.87
5.77	3.134	2.345	1.685	11.29	.87
5.78	3.137	2.349	1.686	11.32	.88
5.79	3.141	2.352	1.688	11.34	.88
5.80	3.144	2.355	1.689	11.37	.88
5.81	3.148	2.359	1.690	11.39	.88
5.82	3.152	2.362	1.692	11.42	.88
5.83	3.155	2.365	1.693	11.45	.89
5.84	3.159	2.369	1.694	11.48	.89
5.85	3.163	2.372	1.696	11.50	.89
5.86	3.166	2.376	1.697	11.53	.89
5.87	3.170	2.379	1.698	11.56	.90
5.88	3.174	2.382	1.700	11.59	.90
5.89	3.177	2.385	1.701	11.61	.90
5.90	3.181	2.389	1.702	11.64	.90
5.91	3.185	2.392	1.704	11.67	.90
5.92	3.188	2.396	1.705	11.69	.91
5.93	3.192	2.399	1.706	11.72	.91
5.94	3.195	2.402	1.708	11.74	.91
5.95	3.199	2.405	1.709	11.77	.91
5.96	3.203	2.409	1.710	11.80	.92
5.97	3.206	2.412	1.712	11.83	.92
5.98	3.210	2.415	1.713	11.85	.92
5.99	3.214	2.419	1.714	11.88	.92
6.00	3.217	2.422	1.716	11.91	.92

TABLE 1a (*Continued*). $\gamma = 1.4$

$\frac{\Delta P}{\mathfrak{A}}, \frac{\Delta Q}{\mathfrak{A}}$	M_s	$\frac{\lvert\Delta \mathfrak{U}\rvert}{\mathfrak{A}}$	$\frac{\mathfrak{A}'}{\mathfrak{A}}$	$\frac{p'}{p}$	ΔS
6.00	3.217	2.422	1.716	11.91	.929
6.01	3.221	2.426	1.717	11.94	.932
6.02	3.225	2.429	1.718	11.97	.934
6.03	3.228	2.432	1.720	11.99	.936
6.04	3.232	2.435	1.721	12.02	.938
6.05	3.235	2.439	1.722	12.05	.941
6.06	3.239	2.442	1.724	12.07	.943
6.07	3.243	2.445	1.725	12.10	.945
6.08	3.246	2.449	1.726	12.13	.948
6.09	3.250	2.452	1.728	12.16	.950
6.10	3.254	2.455	1.729	12.19	.952
6.11	3.258	2.459	1.730	12.21	.954
6.12	3.261	2.462	1.732	12.24	.957
6.13	3.264	2.465	1.733	12.27	.959
6.14	3.268	2.468	1.734	12.29	.961
6.15	3.272	2.472	1.736	12.32	.963
6.16	3.276	2.475	1.737	12.35	.966
6.17	3.279	2.478	1.738	12.38	.968
6.18	3.282	2.482	1.740	12.40	.970
6.19	3.286	2.485	1.741	12.43	.972
6.20	3.290	2.488	1.742	12.46	.975
6.21	3.294	2.492	1.744	12.49	.977
6.22	3.297	2.495	1.745	12.52	.979
6.23	3.301	2.498	1.746	12.55	.981
6.24	3.304	2.501	1.748	12.57	.983
6.25	3.308	2.505	1.749	12.60	.986
6.26	3.312	2.508	1.750	12.63	.988
6.27	3.315	2.511	1.752	12.66	.990
6.28	3.319	2.515	1.753	12.68	.992
6.29	3.323	2.518	1.755	12.71	.995
6.30	3.326	2.521	1.756	12.74	.997
6.31	3.330	2.525	1.757	12.77	.999
6.32	3.333	2.528	1.758	12.80	1.001
6.33	3.337	2.531	1.760	12.82	1.004
6.34	3.341	2.534	1.761	12.85	1.006
6.35	3.344	2.538	1.763	12.88	1.008
6.36	3.348	2.541	1.764	12.91	1.010
6.37	3.351	2.544	1.765	12.94	1.013
6.38	3.355	2.547	1.767	12.97	1.015
6.39	3.359	2.551	1.768	12.99	1.017
6.40	3.362	2.554	1.769	13.02	1.019
6.41	3.366	2.557	1.771	13.05	1.022
6.42	3.370	2.561	1.772	13.08	1.024
6.43	3.373	2.564	1.773	13.11	1.026
6.44	3.377	2.567	1.775	13.14	1.028
6.45	3.380	2.570	1.776	13.16	1.030
6.46	3.384	2.574	1.777	13.19	1.033
6.47	3.388	2.577	1.779	13.22	1.035
6.48	3.391	2.580	1.780	13.25	1.037
6.49	3.395	2.584	1.781	13.28	1.039
6.50	3.398	2.587	1.783	13.31	1.042

$\frac{\Delta P}{\mathfrak{A}}, \frac{\Delta Q}{\mathfrak{A}}$	M_s	$\frac{\lvert\Delta \mathfrak{U}\rvert}{\mathfrak{A}}$	$\frac{\mathfrak{A}'}{\mathfrak{A}}$	$\frac{p'}{p}$	ΔS
6.50	3.398	2.587	1.783	13.31	1.042
6.51	3.402	2.590	1.784	13.34	1.044
6.52	3.406	2.594	1.785	13.37	1.046
6.53	3.410	2.597	1.787	13.40	1.048
6.54	3.413	2.600	1.788	13.42	1.051
6.55	3.416	2.603	1.789	13.45	1.053
6.56	3.420	2.606	1.791	13.48	1.055
6.57	3.424	2.610	1.792	13.51	1.057
6.58	3.428	2.613	1.794	13.54	1.060
6.59	3.431	2.616	1.795	13.56	1.062
6.60	3.434	2.619	1.796	13.59	1.064
6.61	3.438	2.623	1.798	13.62	1.066
6.62	3.442	2.626	1.799	13.65	1.069
6.63	3.445	2.629	1.800	13.68	1.071
6.64	3.449	2.633	1.802	13.71	1.073
6.65	3.452	2.636	1.803	13.74	1.075
6.66	3.456	2.639	1.804	13.77	1.077
6.67	3.460	2.642	1.806	13.80	1.080
6.68	3.464	2.646	1.807	13.83	1.082
6.69	3.467	2.649	1.808	13.86	1.084
6.70	3.471	2.652	1.810	13.89	1.086
6.71	3.474	2.655	1.811	13.91	1.089
6.72	3.478	2.659	1.812	13.94	1.091
6.73	3.482	2.662	1.814	13.97	1.093
6.74	3.485	2.665	1.815	14.00	1.095
6.75	3.488	2.668	1.816	14.03	1.097
6.76	3.492	2.672	1.818	14.06	1.100
6.77	3.496	2.675	1.819	14.09	1.102
6.78	3.500	2.678	1.821	14.12	1.104
6.79	3.503	2.681	1.822	14.15	1.106
6.80	3.507	2.684	1.823	14.18	1.109
6.81	3.510	2.688	1.825	14.21	1.111
6.82	3.514	2.691	1.826	14.24	1.113
6.83	3.518	2.694	1.827	14.27	1.116
6.84	3.521	2.697	1.829	14.30	1.118
6.85	3.524	2.701	1.830	14.33	1.120
6.86	3.528	2.704	1.831	14.36	1.122
6.87	3.532	2.707	1.833	14.39	1.124
6.88	3.536	2.711	1.834	14.42	1.127
6.89	3.539	2.714	1.835	14.45	1.129
6.90	3.542	2.717	1.837	14.47	1.131
6.91	3.546	2.720	1.838	14.50	1.133
6.92	3.550	2.723	1.840	14.53	1.136
6.93	3.554	2.727	1.841	14.56	1.138
6.94	3.557	2.730	1.842	14.59	1.140
6.95	3.560	2.733	1.843	14.62	1.142
6.96	3.564	2.736	1.845	14.65	1.144
6.97	3.568	2.740	1.846	14.68	1.147
6.98	3.572	2.743	1.848	14.71	1.149
6.99	3.575	2.746	1.849	14.74	1.151
7.00	3.578	2.749	1.850	14.77	1.153

TABLE 1a (*Continued*). $\gamma = 1.4$

$\frac{\Delta P}{\mathcal{Q}}, \frac{\Delta Q}{\mathcal{Q}}$	M_s	$\frac{\lvert\Delta\mathcal{U}\rvert}{\mathcal{Q}}$	$\frac{\mathcal{Q}'}{\mathcal{Q}}$	$\frac{p'}{p}$	ΔS
7.00	3.578	2.749	1.850	14.77	1.153
7.01	3.582	2.752	1.852	14.80	1.155
7.02	3.586	2.756	1.853	14.83	1.157
7.03	3.589	2.759	1.854	14.86	1.160
7.04	3.593	2.762	1.856	14.89	1.162
7.05	3.596	2.765	1.857	14.92	1.164
7.06	3.600	2.768	1.858	14.95	1.166
7.07	3.604	2.772	1.860	14.98	1.169
7.08	3.607	2.775	1.861	15.01	1.171
7.09	3.611	2.778	1.863	15.04	1.173
7.10	3.614	2.781	1.864	15.07	1.175
7.11	3.618	2.785	1.865	15.10	1.177
7.12	3.622	2.788	1.867	15.13	1.180
7.13	3.625	2.791	1.868	15.16	1.182
7.14	3.628	2.794	1.869	15.19	1.184
7.15	3.632	2.797	1.871	15.22	1.186
7.16	3.636	2.801	1.872	15.26	1.189
7.17	3.640	2.804	1.873	15.29	1.191
7.18	3.643	2.807	1.875	15.32	1.193
7.19	3.646	2.810	1.876	15.35	1.195
7.20	3.650	2.813	1.877	15.38	1.197
7.21	3.654	2.817	1.879	15.41	1.200
7.22	3.657	2.820	1.880	15.44	1.202
7.23	3.661	2.823	1.882	15.47	1.204
7.24	3.664	2.826	1.883	15.50	1.206
7.25	3.668	2.830	1.884	15.53	1.208
7.26	3.672	2.833	1.886	15.56	1.211
7.27	3.675	2.836	1.887	15.59	1.213
7.28	3.679	2.839	1.888	15.62	1.215
7.29	3.682	2.842	1.890	15.65	1.217
7.30	3.686	2.846	1.891	15.68	1.219
7.31	3.690	2.849	1.892	15.72	1.222
7.32	3.693	2.852	1.894	15.75	1.224
7.33	3.697	2.855	1.895	15.78	1.226
7.34	3.700	2.858	1.897	15.81	1.228
7.35	3.704	2.861	1.898	15.84	1.230
7.36	3.707	2.865	1.899	15.87	1.232
7.37	3.711	2.868	1.900	15.90	1.234
7.38	3.715	2.871	1.902	15.93	1.237
7.39	3.718	2.874	1.903	15.96	1.239
7.40	3.722	2.877	1.905	15.99	1.241
7.41	3.725	2.881	1.906	16.02	1.244
7.42	3.729	2.884	1.907	16.05	1.246
7.43	3.732	2.887	1.909	16.09	1.248
7.44	3.736	2.890	1.910	16.12	1.250
7.45	3.739	2.893	1.911	16.15	1.252
7.46	3.743	2.897	1.913	16.18	1.254
7.47	3.747	2.900	1.914	16.21	1.256
7.48	3.750	2.903	1.916	16.24	1.259
7.49	3.754	2.906	1.917	16.27	1.261
7.50	3.757	2.909	1.918	16.30	1.263

$\frac{\Delta P}{\mathcal{Q}}, \frac{\Delta Q}{\mathcal{Q}}$	M_s	$\frac{\lvert\Delta\mathcal{U}\rvert}{\mathcal{Q}}$	$\frac{\mathcal{Q}'}{\mathcal{Q}}$	$\frac{p'}{p}$	ΔS
7.50	3.757	2.909	1.918	16.30	1.263
7.51	3.761	2.912	1.920	16.33	1.265
7.52	3.765	2.916	1.921	16.37	1.268
7.53	3.768	2.919	1.922	16.40	1.270
7.54	3.771	2.922	1.924	16.43	1.272
7.55	3.775	2.925	1.925	16.46	1.274
7.56	3.779	2.928	1.926	16.49	1.276
7.57	3.782	2.932	1.928	16.52	1.278
7.58	3.786	2.935	1.929	16.55	1.280
7.59	3.789	2.938	1.930	16.59	1.283
7.60	3.793	2.941	1.932	16.62	1.285
7.61	3.796	2.944	1.933	16.65	1.287
7.62	3.800	2.948	1.935	16.68	1.289
7.63	3.804	2.951	1.936	16.71	1.292
7.64	3.807	2.954	1.937	16.74	1.294
7.65	3.811	2.957	1.939	16.78	1.296
7.66	3.814	2.960	1.940	16.81	1.298
7.67	3.818	2.963	1.941	16.84	1.300
7.68	3.821	2.966	1.943	16.87	1.302
7.69	3.825	2.970	1.944	16.90	1.304
7.70	3.828	2.973	1.946	16.93	1.307
7.71	3.832	2.976	1.947	16.97	1.309
7.72	3.836	2.979	1.948	17.00	1.311
7.73	3.839	2.982	1.950	17.03	1.313
7.74	3.843	2.985	1.951	17.06	1.315
7.75	3.846	2.989	1.952	17.09	1.318
7.76	3.850	2.992	1.954	17.12	1.320
7.77	3.853	2.995	1.955	17.16	1.322
7.78	3.857	2.998	1.957	17.19	1.324
7.79	3.860	3.001	1.958	17.22	1.326
7.80	3.864	3.004	1.959	17.25	1.328
7.81	3.868	3.008	1.961	17.29	1.331
7.82	3.871	3.011	1.962	17.32	1.333
7.83	3.875	3.014	1.963	17.35	1.335
7.84	3.879	3.017	1.965	17.38	1.337
7.85	3.882	3.020	1.966	17.41	1.339
7.86	3.885	3.023	1.967	17.45	1.341
7.87	3.889	3.027	1.969	17.48	1.343
7.88	3.893	3.030	1.970	17.51	1.346
7.89	3.896	3.033	1.972	17.54	1.348
7.90	3.900	3.036	1.973	17.58	1.350
7.91	3.904	3.039	1.974	17.61	1.352
7.92	3.907	3.042	1.976	17.64	1.354
7.93	3.910	3.045	1.977	17.67	1.356
7.94	3.914	3.049	1.978	17.71	1.358
7.95	3.917	3.052	1.980	17.74	1.361
7.96	3.921	3.055	1.981	17.77	1.363
7.97	3.925	3.058	1.983	17.80	1.365
7.98	3.928	3.061	1.984	17.83	1.367
7.99	3.932	3.064	1.985	17.87	1.369
8.00	3.935	3.068	1.987	17.90	1.371

TABLE 1a (*Continued*). $\gamma = 1.4$

$\frac{\Delta P}{\mathfrak{A}}, \frac{\Delta Q}{\mathfrak{A}}$	M_S	$\frac{\lvert\Delta \mathfrak{U}\rvert}{\mathfrak{A}}$	$\frac{\mathfrak{A}'}{\mathfrak{A}}$	$\frac{p'}{p}$	ΔS
8.00	3.935	3.068	1.987	17.90	1.371
8.01	3.939	3.071	1.988	17.93	1.373
8.02	3.942	3.074	1.989	17.97	1.376
8.03	3.946	3.077	1.991	18.00	1.378
8.04	3.950	3.080	1.992	18.03	1.380
8.05	3.953	3.083	1.993	18.06	1.382
8.06	3.956	3.086	1.995	18.10	1.384
8.07	3.960	3.090	1.996	18.13	1.386
8.08	3.963	3.093	1.997	18.16	1.388
8.09	3.967	3.096	1.999	18.19	1.390
8.10	3.971	3.099	2.000	18.23	1.392
8.11	3.974	3.102	2.002	18.26	1.395
8.12	3.978	3.105	2.003	18.29	1.397
8.13	3.981	3.109	2.004	18.33	1.399
8.14	3.985	3.112	2.006	18.36	1.401
8.15	3.988	3.115	2.007	18.39	1.403
8.16	3.992	3.118	2.009	18.42	1.406
8.17	3.996	3.121	2.010	18.46	1.408
8.18	3.999	3.124	2.011	18.49	1.410
8.19	4.002	3.127	2.013	18.52	1.412
8.20	4.006	3.130	2.014	18.56	1.414
8.21	4.010	3.134	2.015	18.59	1.416
8.22	4.013	3.137	2.017	18.62	1.419
8.23	4.016	3.140	2.018	18.65	1.421
8.24	4.020	3.143	2.020	18.69	1.423
8.25	4.024	3.146	2.021	18.72	1.425
8.26	4.027	3.149	2.022	18.75	1.427
8.27	4.031	3.152	2.024	18.79	1.429
8.28	4.034	3.155	2.025	18.82	1.431
8.29	4.038	3.159	2.026	18.86	1.434
8.30	4.042	3.162	2.028	18.89	1.436
8.31	4.045	3.165	2.029	18.92	1.438
8.32	4.049	3.168	2.031	18.96	1.440
8.33	4.052	3.171	2.032	18.99	1.442
8.34	4.055	3.174	2.033	19.02	1.444
8.35	4.059	3.177	2.035	19.06	1.446
8.36	4.063	3.180	2.036	19.09	1.448
8.37	4.066	3.183	2.037	19.12	1.450
8.38	4.070	3.187	2.039	19.15	1.452
8.39	4.073	3.190	2.040	19.19	1.455
8.40	4.077	3.193	2.041	19.22	1.457
8.41	4.080	3.196	2.043	19.26	1.459
8.42	4.084	3.199	2.044	19.29	1.461
8.43	4.087	3.202	2.046	19.32	1.463
8.44	4.091	3.205	2.047	19.36	1.465
8.45	4.095	3.209	2.048	19.39	1.467
8.46	4.098	3.212	2.050	19.42	1.469
8.47	4.101	3.215	2.051	19.46	1.471
8.48	4.105	3.218	2.053	19.49	1.474
8.49	4.109	3.221	2.054	19.53	1.476
8.50	4.112	3.224	2.055	19.56	1.478

$\frac{\Delta P}{\mathfrak{A}}, \frac{\Delta Q}{\mathfrak{A}}$	M_S	$\frac{\lvert\Delta \mathfrak{U}\rvert}{\mathfrak{A}}$	$\frac{\mathfrak{A}'}{\mathfrak{A}}$	$\frac{p'}{p}$	ΔS
8.50	4.112	3.224	2.055	19.56	1.478
8.51	4.116	3.227	2.057	19.60	1.480
8.52	4.119	3.230	2.058	19.63	1.482
8.53	4.123	3.233	2.059	19.66	1.484
8.54	4.126	3.237	2.061	19.70	1.486
8.55	4.130	3.240	2.062	19.73	1.489
8.56	4.133	3.243	2.063	19.76	1.491
8.57	4.137	3.246	2.065	19.80	1.493
8.58	4.140	3.249	2.066	19.83	1.495
8.59	4.144	3.252	2.068	19.87	1.497
8.60	4.147	3.255	2.069	19.90	1.499
8.61	4.151	3.258	2.070	19.94	1.501
8.62	4.154	3.261	2.072	19.97	1.503
8.63	4.158	3.265	2.073	20.00	1.505
8.64	4.162	3.268	2.075	20.04	1.508
8.65	4.165	3.271	2.076	20.07	1.510
8.66	4.168	3.274	2.077	20.11	1.512
8.67	4.172	3.277	2.079	20.14	1.514
8.68	4.176	3.280	2.080	20.18	1.516
8.69	4.179	3.283	2.081	20.21	1.518
8.70	4.183	3.286	2.083	20.24	1.520
8.71	4.186	3.289	2.084	20.28	1.522
8.72	4.189	3.292	2.086	20.31	1.524
8.73	4.193	3.296	2.087	20.35	1.526
8.74	4.197	3.299	2.088	20.38	1.529
8.75	4.200	3.302	2.090	20.41	1.531
8.76	4.204	3.305	2.091	20.45	1.533
8.77	4.207	3.308	2.092	20.48	1.535
8.78	4.211	3.311	2.094	20.52	1.537
8.79	4.214	3.314	2.095	20.56	1.539
8.80	4.218	3.317	2.097	20.59	1.541
8.81	4.221	3.320	2.098	20.62	1.543
8.82	4.225	3.324	2.099	20.66	1.545
8.83	4.228	3.327	2.101	20.69	1.547
8.84	4.232	3.330	2.102	20.73	1.549
8.85	4.235	3.333	2.103	20.76	1.551
8.86	4.239	3.336	2.105	20.80	1.553
8.87	4.243	3.339	2.106	20.83	1.556
8.88	4.246	3.342	2.108	20.87	1.558
8.89	4.249	3.345	2.109	20.90	1.560
8.90	4.253	3.348	2.110	20.94	1.562
8.91	4.256	3.351	2.112	20.97	1.564
8.92	4.260	3.354	2.113	21.01	1.566
8.93	4.264	3.358	2.115	21.04	1.568
8.94	4.267	3.361	2.116	21.08	1.570
8.95	4.270	3.364	2.117	21.11	1.572
8.96	4.274	3.367	2.119	21.15	1.574
8.97	4.278	3.370	2.120	21.18	1.576
8.98	4.281	3.373	2.121	21.21	1.578
8.99	4.285	3.376	2.123	21.25	1.581
9.00	4.288	3.379	2.124	21.29	1.583

TABLE 1a (*Continued*). $\gamma = 1.4$

$\frac{\Delta P}{\mathcal{Q}}, \frac{\Delta Q}{\mathcal{Q}}$	M_S	$\frac{\lvert\Delta \mathcal{U}\rvert}{\mathcal{Q}}$	$\frac{\mathcal{Q}'}{\mathcal{Q}}$	$\frac{p'}{p}$	ΔS
9.00	4.288	3.379	2.124	21.29	1.583
9.01	4.292	3.382	2.126	21.32	1.585
9.02	4.295	3.385	2.127	21.36	1.587
9.03	4.299	3.388	2.128	21.39	1.589
9.04	4.302	3.392	2.130	21.43	1.591
9.05	4.306	3.395	2.131	21.46	1.593
9.06	4.309	3.398	2.133	21.50	1.595
9.07	4.313	3.401	2.134	21.54	1.597
9.08	4.316	3.404	2.135	21.57	1.599
9.09	4.320	3.407	2.137	21.60	1.601
9.10	4.324	3.410	2.138	21.64	1.603
9.11	4.327	3.413	2.139	21.67	1.605
9.12	4.330	3.416	2.141	21.71	1.607
9.13	4.334	3.419	2.142	21.75	1.609
9.14	4.337	3.422	2.144	21.78	1.611
9.15	4.341	3.425	2.145	21.82	1.614
9.16	4.344	3.428	2.146	21.85	1.616
9.17	4.348	3.432	2.148	21.89	1.618
9.18	4.352	3.435	2.149	21.93	1.620
9.19	4.355	3.438	2.150	21.96	1.622
9.20	4.359	3.441	2.152	22.00	1.624
9.21	4.362	3.444	2.153	22.03	1.626
9.22	4.365	3.447	2.155	22.07	1.628
9.23	4.369	3.450	2.156	22.10	1.630
9.24	4.372	3.453	2.157	22.14	1.632
9.25	4.376	3.456	2.159	22.17	1.634
9.26	4.380	3.459	2.160	22.21	1.636
9.27	4.383	3.462	2.162	22.24	1.638
9.28	4.387	3.465	2.163	22.28	1.640
9.29	4.390	3.468	2.164	22.31	1.642
9.30	4.393	3.472	2.166	22.35	1.644
9.31	4.397	3.475	2.167	22.39	1.646
9.32	4.400	3.478	2.169	22.42	1.649
9.33	4.404	3.481	2.170	22.46	1.651
9.34	4.407	3.484	2.171	22.50	1.653
9.35	4.411	3.487	2.173	22.53	1.655
9.36	4.415	3.490	2.174	22.57	1.657
9.37	4.418	3.493	2.175	22.60	1.659
9.38	4.422	3.496	2.177	22.64	1.661
9.39	4.425	3.499	2.178	22.68	1.663
9.40	4.428	3.502	2.180	22.71	1.665
9.41	4.432	3.505	2.181	22.75	1.667
9.42	4.436	3.508	2.182	22.79	1.669
9.43	4.439	3.511	2.184	22.82	1.671
9.44	4.443	3.515	2.185	22.86	1.673
9.45	4.446	3.518	2.187	22.90	1.675
9.46	4.450	3.521	2.188	22.93	1.677
9.47	4.453	3.524	2.189	22.97	1.679
9.48	4.456	3.527	2.191	23.00	1.681
9.49	4.460	3.530	2.192	23.04	1.683
9.50	4.463	3.533	2.193	23.08	1.685

$\frac{\Delta P}{\mathcal{Q}}, \frac{\Delta Q}{\mathcal{Q}}$	M_S	$\frac{\lvert\Delta \mathcal{U}\rvert}{\mathcal{Q}}$	$\frac{\mathcal{Q}'}{\mathcal{Q}}$	$\frac{p'}{p}$	ΔS
9.50	4.463	3.533	2.193	23.08	1.685
9.51	4.467	3.536	2.195	23.11	1.687
9.52	4.470	3.539	2.196	23.15	1.689
9.53	4.474	3.542	2.198	23.19	1.691
9.54	4.478	3.545	2.199	23.22	1.693
9.55	4.481	3.548	2.200	23.26	1.695
9.56	4.485	3.551	2.202	23.30	1.698
9.57	4.488	3.554	2.203	23.33	1.700
9.58	4.492	3.557	2.205	23.37	1.702
9.59	4.495	3.560	2.206	23.41	1.704
9.60	4.498	3.563	2.207	23.44	1.706
9.61	4.502	3.567	2.209	23.48	1.708
9.62	4.506	3.570	2.210	23.52	1.710
9.63	4.509	3.573	2.212	23.55	1.712
9.64	4.513	3.576	2.213	23.59	1.714
9.65	4.516	3.579	2.214	23.62	1.716
9.66	4.520	3.582	2.216	23.66	1.718
9.67	4.523	3.585	2.217	23.70	1.720
9.68	4.526	3.588	2.218	23.74	1.722
9.69	4.530	3.591	2.220	23.78	1.724
9.70	4.533	3.594	2.221	23.81	1.726
9.71	4.537	3.597	2.223	23.85	1.728
9.72	4.541	3.600	2.224	23.89	1.730
9.73	4.544	3.603	2.225	23.92	1.732
9.74	4.547	3.606	2.227	23.96	1.734
9.75	4.551	3.609	2.228	24.00	1.736
9.76	4.554	3.612	2.230	24.03	1.738
9.77	4.558	3.615	2.231	24.07	1.740
9.78	4.562	3.619	2.232	24.11	1.742
9.79	4.565	3.622	2.234	24.14	1.744
9.80	4.568	3.625	2.235	24.18	1.746
9.81	4.572	3.628	2.237	24.22	1.748
9.82	4.576	3.631	2.238	24.26	1.750
9.83	4.579	3.634	2.239	24.29	1.752
9.84	4.583	3.637	2.241	24.33	1.754
9.85	4.586	3.640	2.242	24.37	1.756
9.86	4.589	3.643	2.243	24.41	1.758
9.87	4.593	3.646	2.245	24.45	1.760
9.88	4.596	3.649	2.246	24.48	1.762
9.89	4.600	3.652	2.248	24.52	1.764
9.90	4.603	3.655	2.249	24.56	1.766
9.91	4.607	3.658	2.251	24.60	1.768
9.92	4.610	3.661	2.252	24.63	1.770
9.93	4.614	3.664	2.253	24.67	1.772
9.94	4.617	3.667	2.255	24.70	1.774
9.95	4.621	3.670	2.256	24.74	1.776
9.96	4.624	3.673	2.257	24.78	1.778
9.97	4.628	3.676	2.259	24.82	1.780
9.98	4.631	3.680	2.260	24.86	1.782
9.99	4.635	3.683	2.262	24.89	1.784
10.00	4.638	3.686	2.263	24.93	1.786

TABLE 1b. SHOCK WAVE RELATIONS FOR $\gamma = 5/3$

$\frac{\Delta P}{\mathcal{Q}}, \frac{\Delta Q}{\mathcal{Q}}$	M_S	$\frac{\lvert\Delta \mathcal{U}\rvert}{\mathcal{Q}}$	$\frac{\mathcal{Q}'}{\mathcal{Q}}$	$\frac{p'}{p}$	ΔS
.00	1.000	.000	1.000	1.00	.000
.01	1.003	.005	1.002	1.01	.000
.02	1.006	.010	1.003	1.02	.000
.03	1.010	.015	1.005	1.03	.000
.04	1.013	.020	1.007	1.03	.000
.05	1.017	.025	1.008	1.04	.000
.06	1.020	.030	1.010	1.05	.000
.07	1.024	.035	1.012	1.06	.000
.08	1.027	.040	1.013	1.07	.000
.09	1.030	.045	1.015	1.08	.000
.10	1.033	.050	1.017	1.09	.000
.11	1.037	.055	1.018	1.10	.000
.12	1.041	.060	1.020	1.10	.000
.13	1.044	.065	1.022	1.11	.000
.14	1.048	.070	1.023	1.12	.000
.15	1.051	.075	1.025	1.13	.000
.16	1.055	.080	1.027	1.14	.000
.17	1.058	.085	1.028	1.15	.000
.18	1.062	.090	1.030	1.16	.000
.19	1.065	.095	1.032	1.17	.000
.20	1.069	.100	1.034	1.18	.000
.21	1.072	.105	1.035	1.19	.000
.22	1.076	.110	1.037	1.20	.000
.23	1.080	.115	1.039	1.21	.000
.24	1.083	.120	1.040	1.22	.000
.25	1.086	.125	1.042	1.23	.000
.26	1.090	.130	1.043	1.24	.000
.27	1.094	.135	1.045	1.25	.001
.28	1.097	.140	1.047	1.26	.001
.29	1.101	.145	1.049	1.27	.001
.30	1.105	.150	1.050	1.28	.001
.31	1.109	.155	1.052	1.29	.001
.32	1.112	.159	1.053	1.30	.001
.33	1.115	.164	1.055	1.31	.001
.34	1.119	.169	1.057	1.32	.001
.35	1.123	.174	1.059	1.33	.001
.36	1.127	.179	1.060	1.34	.001
.37	1.131	.184	1.062	1.35	.001
.38	1.134	.189	1.063	1.36	.001
.39	1.138	.194	1.065	1.37	.002
.40	1.142	.199	1.067	1.38	.002
.41	1.145	.204	1.069	1.39	.002
.42	1.149	.209	1.070	1.40	.002
.43	1.153	.214	1.072	1.41	.002
.44	1.157	.219	1.074	1.42	.002
.45	1.160	.224	1.075	1.43	.002
.46	1.164	.229	1.077	1.44	.002
.47	1.168	.234	1.079	1.45	.003
.48	1.172	.239	1.081	1.47	.003
.49	1.176	.244	1.082	1.48	.003
.50	1.179	.249	1.084	1.49	.003

$\frac{\Delta P}{\mathcal{Q}}, \frac{\Delta Q}{\mathcal{Q}}$	M_S	$\frac{\lvert\Delta \mathcal{U}\rvert}{\mathcal{Q}}$	$\frac{\mathcal{Q}'}{\mathcal{Q}}$	$\frac{p'}{p}$	ΔS
.50	1.179	.249	1.084	1.49	.003
.51	1.183	.253	1.086	1.50	.003
.52	1.187	.258	1.087	1.51	.003
.53	1.191	.263	1.089	1.52	.004
.54	1.195	.268	1.091	1.53	.004
.55	1.198	.273	1.093	1.55	.004
.56	1.202	.278	1.094	1.56	.004
.57	1.206	.283	1.096	1.57	.004
.58	1.210	.288	1.097	1.58	.005
.59	1.214	.293	1.099	1.59	.005
.60	1.218	.297	1.101	1.60	.005
.61	1.222	.302	1.103	1.62	.005
.62	1.225	.307	1.104	1.63	.005
.63	1.229	.312	1.106	1.64	.006
.64	1.233	.317	1.108	1.65	.006
.65	1.237	.322	1.110	1.66	.006
.66	1.241	.327	1.111	1.68	.006
.67	1.245	.332	1.113	1.69	.007
.68	1.249	.337	1.115	1.70	.007
.69	1.253	.341	1.116	1.71	.007
.70	1.257	.346	1.118	1.73	.008
.71	1.261	.351	1.120	1.74	.008
.72	1.265	.356	1.121	1.75	.008
.73	1.269	.360	1.123	1.76	.009
.74	1.273	.365	1.125	1.78	.009
.75	1.277	.370	1.127	1.79	.009
.76	1.281	.375	1.128	1.80	.010
.77	1.285	.380	1.130	1.81	.010
.78	1.289	.385	1.132	1.83	.010
.79	1.293	.389	1.134	1.84	.011
.80	1.297	.394	1.135	1.85	.011
.81	1.301	.399	1.137	1.87	.011
.82	1.305	.404	1.139	1.88	.012
.83	1.309	.409	1.141	1.89	.012
.84	1.313	.413	1.143	1.91	.013
.85	1.317	.418	1.144	1.92	.013
.86	1.321	.423	1.146	1.93	.013
.87	1.325	.428	1.147	1.94	.014
.88	1.329	.433	1.149	1.96	.014
.89	1.333	.438	1.151	1.97	.014
.90	1.337	.442	1.153	1.99	.015
.91	1.341	.447	1.155	2.00	.015
.92	1.345	.451	1.156	2.01	.016
.93	1.349	.456	1.158	2.03	.016
.94	1.353	.461	1.160	2.04	.017
.95	1.358	.466	1.162	2.05	.017
.96	1.362	.471	1.163	2.07	.018
.97	1.366	.475	1.165	2.08	.018
.98	1.370	.480	1.167	2.10	.019
.99	1.374	.485	1.169	2.11	.019
1.00	1.378	.489	1.170	2.12	.020

TABLE 1b (*Continued*). $\gamma = 5/3$

$\frac{\Delta P}{\mathfrak{a}}, \frac{\Delta Q}{\mathfrak{a}}$	M_S	$\frac{\lvert\Delta\mathfrak{u}\rvert}{\mathfrak{a}}$	$\frac{\mathfrak{a}'}{\mathfrak{a}}$	$\frac{p'}{p}$	ΔS
1.00	1.378	.489	1.170	2.12	.020
1.01	1.382	.494	1.172	2.14	.021
1.02	1.386	.499	1.174	2.15	.021
1.03	1.390	.503	1.176	2.17	.021
1.04	1.395	.508	1.178	2.18	.022
1.05	1.399	.513	1.179	2.20	.023
1.06	1.403	.518	1.181	2.21	.023
1.07	1.407	.522	1.183	2.23	.024
1.08	1.411	.527	1.185	2.24	.024
1.09	1.415	.531	1.186	2.25	.025
1.10	1.419	.536	1.188	2.27	.025
1.11	1.423	.541	1.190	2.28	.026
1.12	1.428	.545	1.191	2.30	.026
1.13	1.432	.550	1.193	2.31	.027
1.14	1.436	.555	1.195	2.33	.028
1.15	1.440	.559	1.197	2.34	.028
1.16	1.444	.564	1.199	2.36	.029
1.17	1.449	.569	1.201	2.37	.030
1.18	1.453	.573	1.202	2.39	.031
1.19	1.457	.578	1.204	2.40	.031
1.20	1.461	.583	1.206	2.42	.032
1.21	1.465	.587	1.208	2.43	.033
1.22	1.470	.592	1.209	2.45	.033
1.23	1.474	.596	1.211	2.47	.034
1.24	1.478	.601	1.213	2.48	.034
1.25	1.482	.606	1.215	2.50	.035
1.26	1.486	.610	1.217	2.51	.036
1.27	1.491	.615	1.219	2.53	.037
1.28	1.495	.619	1.220	2.54	.038
1.29	1.499	.624	1.222	2.56	.038
1.30	1.503	.628	1.224	2.57	.039
1.31	1.507	.633	1.226	2.59	.039
1.32	1.512	.638	1.228	2.61	.040
1.33	1.516	.642	1.229	2.62	.041
1.34	1.520	.647	1.231	2.64	.042
1.35	1.524	.651	1.233	2.65	.043
1.36	1.528	.656	1.235	2.67	.044
1.37	1.533	.660	1.237	2.69	.044
1.38	1.537	.665	1.238	2.70	.045
1.39	1.541	.669	1.240	2.72	.046
1.40	1.545	.673	1.242	2.73	.046
1.41	1.550	.678	1.244	2.75	.047
1.42	1.554	.683	1.246	2.77	.048
1.43	1.558	.687	1.248	2.79	.049
1.44	1.562	.692	1.249	2.80	.050
1.45	1.567	.696	1.251	2.82	.051
1.46	1.571	.701	1.253	2.83	.052
1.47	1.575	.705	1.255	2.85	.052
1.48	1.580	.710	1.257	2.87	.053
1.49	1.584	.714	1.259	2.89	.054
1.50	1.588	.719	1.260	2.90	.055

$\frac{\Delta P}{\mathfrak{a}}, \frac{\Delta Q}{\mathfrak{a}}$	M_S	$\frac{\lvert\Delta\mathfrak{u}\rvert}{\mathfrak{a}}$	$\frac{\mathfrak{a}'}{\mathfrak{a}}$	$\frac{p'}{p}$	ΔS
1.50	1.588	.719	1.260	2.90	.055
1.51	1.592	.723	1.262	2.92	.056
1.52	1.596	.728	1.264	2.94	.057
1.53	1.601	.732	1.266	2.95	.058
1.54	1.605	.737	1.268	2.97	.059
1.55	1.609	.741	1.270	2.99	.060
1.56	1.614	.745	1.272	3.01	.061
1.57	1.618	.750	1.273	3.02	.061
1.58	1.622	.754	1.275	3.04	.062
1.59	1.626	.759	1.277	3.06	.063
1.60	1.631	.763	1.279	3.07	.064
1.61	1.635	.768	1.281	3.09	.065
1.62	1.639	.772	1.283	3.11	.066
1.63	1.643	.776	1.285	3.13	.067
1.64	1.648	.781	1.287	3.15	.068
1.65	1.652	.785	1.288	3.16	.069
1.66	1.657	.790	1.290	3.18	.070
1.67	1.661	.794	1.292	3.20	.071
1.68	1.665	.798	1.294	3.22	.072
1.69	1.669	.803	1.296	3.23	.073
1.70	1.674	.808	1.298	3.25	.074
1.71	1.678	.812	1.300	3.27	.075
1.72	1.682	.816	1.301	3.29	.076
1.73	1.687	.820	1.303	3.31	.077
1.74	1.691	.825	1.305	3.32	.078
1.75	1.696	.829	1.307	3.34	.079
1.76	1.700	.834	1.309	3.36	.080
1.77	1.704	.838	1.311	3.38	.081
1.78	1.708	.842	1.313	3.40	.082
1.79	1.713	.847	1.315	3.42	.083
1.80	1.717	.851	1.316	3.44	.084
1.81	1.721	.855	1.318	3.45	.085
1.82	1.725	.859	1.320	3.47	.086
1.83	1.730	.864	1.322	3.49	.087
1.84	1.734	.868	1.324	3.51	.088
1.85	1.738	.872	1.326	3.53	.089
1.86	1.743	.877	1.328	3.55	.090
1.87	1.747	.881	1.330	3.57	.092
1.88	1.752	.886	1.331	3.59	.093
1.89	1.756	.890	1.333	3.60	.094
1.90	1.760	.894	1.335	3.62	.095
1.91	1.764	.898	1.337	3.64	.096
1.92	1.769	.903	1.339	3.66	.097
1.93	1.773	.907	1.341	3.68	.098
1.94	1.777	.911	1.343	3.70	.099
1.95	1.782	.915	1.345	3.72	.100
1.96	1.786	.920	1.347	3.74	.101
1.97	1.791	.924	1.349	3.76	.103
1.98	1.795	.928	1.350	3.78	.104
1.99	1.799	.932	1.352	3.80	.105
2.00	1.804	.937	1.354	3.82	.106

TABLE 1b (*Continued*). $\gamma = 5/3$

$\frac{\Delta P}{\mathcal{A}}, \frac{\Delta Q}{\mathcal{A}}$	M_s	$\frac{\lvert\Delta \mathfrak{U}\rvert}{\mathcal{A}}$	$\frac{\mathcal{A}'}{\mathcal{A}}$	$\frac{p'}{p}$	ΔS
2.00	1.804	.937	1.354	3.82	.106
2.01	1.808	.941	1.356	3.84	.107
2.02	1.812	.945	1.358	3.86	.108
2.03	1.817	.950	1.360	3.88	.110
2.04	1.821	.954	1.362	3.90	.111
2.05	1.825	.958	1.364	3.92	.112
2.06	1.830	.963	1.366	3.94	.113
2.07	1.834	.967	1.368	3.96	.114
2.08	1.839	.971	1.370	3.98	.115
2.09	1.843	.975	1.371	4.00	.116
2.10	1.847	.979	1.373	4.02	.118
2.11	1.852	.984	1.375	4.04	.119
2.12	1.856	.988	1.377	4.06	.120
2.13	1.860	.992	1.379	4.08	.121
2.14	1.864	.996	1.381	4.10	.123
2.15	1.869	1.001	1.383	4.12	.124
2.16	1.873	1.005	1.385	4.14	.125
2.17	1.878	1.009	1.387	4.16	.126
2.18	1.882	1.013	1.389	4.18	.128
2.19	1.886	1.017	1.391	4.20	.129
2.20	1.891	1.021	1.393	4.22	.130
2.21	1.895	1.026	1.395	4.24	.131
2.22	1.900	1.030	1.397	4.26	.133
2.23	1.904	1.034	1.399	4.28	.134
2.24	1.908	1.038	1.401	4.30	.135
2.25	1.913	1.042	1.402	4.32	.136
2.26	1.917	1.047	1.404	4.34	.137
2.27	1.921	1.051	1.406	4.36	.139
2.28	1.926	1.055	1.408	4.39	.140
2.29	1.930	1.059	1.410	4.41	.141
2.30	1.934	1.063	1.412	4.43	.143
2.31	1.939	1.068	1.414	4.45	.144
2.32	1.943	1.072	1.416	4.47	.145
2.33	1.947	1.076	1.418	4.49	.146
2.34	1.952	1.080	1.420	4.51	.148
2.35	1.956	1.084	1.422	4.53	.149
2.36	1.961	1.088	1.424	4.56	.151
2.37	1.965	1.092	1.426	4.58	.152
2.38	1.969	1.096	1.428	4.60	.153
2.39	1.974	1.100	1.430	4.62	.154
2.40	1.978	1.104	1.432	4.64	.156
2.41	1.983	1.109	1.434	4.66	.157
2.42	1.987	1.113	1.436	4.69	.158
2.43	1.991	1.117	1.438	4.71	.160
2.44	1.996	1.121	1.440	4.73	.161
2.45	2.000	1.125	1.442	4.75	.162
2.46	2.004	1.129	1.444	4.77	.163
2.47	2.009	1.133	1.446	4.79	.165
2.48	2.013	1.137	1.448	4.82	.166
2.49	2.018	1.142	1.450	4.84	.168
2.50	2.022	1.146	1.451	4.86	.169

$\frac{\Delta P}{\mathcal{A}}, \frac{\Delta Q}{\mathcal{A}}$	M_s	$\frac{\lvert\Delta \mathfrak{U}\rvert}{\mathcal{A}}$	$\frac{\mathcal{A}'}{\mathcal{A}}$	$\frac{p'}{p}$	ΔS
2.50	2.022	1.146	1.451	4.86	.169
2.51	2.027	1.150	1.453	4.88	.170
2.52	2.031	1.154	1.455	4.91	.172
2.53	2.035	1.158	1.457	4.93	.173
2.54	2.040	1.162	1.459	4.95	.174
2.55	2.044	1.166	1.461	4.97	.176
2.56	2.048	1.170	1.463	4.99	.177
2.57	2.053	1.174	1.465	5.02	.178
2.58	2.057	1.178	1.467	5.04	.180
2.59	2.061	1.182	1.469	5.06	.181
2.60	2.066	1.186	1.471	5.08	.182
2.61	2.070	1.190	1.473	5.11	.184
2.62	2.075	1.194	1.475	5.13	.185
2.63	2.079	1.198	1.477	5.15	.187
2.64	2.083	1.202	1.479	5.17	.188
2.65	2.087	1.206	1.481	5.20	.189
2.66	2.092	1.210	1.483	5.22	.191
2.67	2.096	1.214	1.485	5.24	.192
2.68	2.101	1.218	1.487	5.27	.194
2.69	2.105	1.222	1.489	5.29	.195
2.70	2.109	1.227	1.491	5.31	.196
2.71	2.114	1.231	1.493	5.33	.198
2.72	2.118	1.235	1.495	5.36	.199
2.73	2.123	1.239	1.497	5.38	.200
2.74	2.127	1.243	1.499	5.40	.202
2.75	2.131	1.247	1.501	5.43	.203
2.76	2.136	1.251	1.503	5.45	.205
2.77	2.140	1.255	1.505	5.48	.206
2.78	2.144	1.259	1.507	5.50	.208
2.79	2.149	1.263	1.509	5.52	.209
2.80	2.153	1.267	1.511	5.55	.211
2.81	2.158	1.271	1.513	5.57	.212
2.82	2.162	1.275	1.515	5.59	.213
2.83	2.166	1.279	1.517	5.62	.215
2.84	2.171	1.283	1.519	5.64	.216
2.85	2.175	1.287	1.521	5.66	.218
2.86	2.180	1.291	1.523	5.69	.219
2.87	2.184	1.294	1.525	5.71	.221
2.88	2.188	1.298	1.527	5.73	.222
2.89	2.193	1.302	1.529	5.76	.223
2.90	2.197	1.306	1.531	5.78	.225
2.91	2.201	1.310	1.533	5.81	.227
2.92	2.206	1.314	1.535	5.83	.228
2.93	2.210	1.318	1.537	5.86	.229
2.94	2.214	1.322	1.539	5.88	.231
2.95	2.219	1.326	1.541	5.90	.232
2.96	2.223	1.330	1.543	5.93	.234
2.97	2.228	1.334	1.545	5.95	.235
2.98	2.232	1.338	1.547	5.98	.237
2.99	2.237	1.342	1.549	6.00	.238
3.00	2.241	1.346	1.551	6.03	.240

TABLE 1b (*Continued*). $\gamma = 5/3$

$\frac{\Delta P}{\mathcal{Q}}, \frac{\Delta Q}{\mathcal{Q}}$	M_s	$\frac{\lvert\Delta \mathcal{U}\rvert}{\mathcal{Q}}$	$\frac{\mathcal{Q}'}{\mathcal{Q}}$	$\frac{p'}{p}$	ΔS
3.00	2.241	1.346	1.551	6.03	.240
3.01	2.245	1.350	1.553	6.05	.241
3.02	2.250	1.354	1.555	6.08	.243
3.03	2.254	1.358	1.557	6.10	.244
3.04	2.258	1.362	1.559	6.13	.246
3.05	2.263	1.366	1.561	6.15	.247
3.06	2.267	1.369	1.563	6.17	.248
3.07	2.271	1.373	1.565	6.20	.250
3.08	2.276	1.377	1.567	6.22	.252
3.09	2.280	1.381	1.569	6.25	.253
3.10	2.285	1.385	1.572	6.27	.254
3.11	2.289	1.389	1.574	6.30	.256
3.12	2.293	1.393	1.576	6.32	.257
3.13	2.298	1.397	1.578	6.35	.259
3.14	2.302	1.401	1.580	6.37	.260
3.15	2.306	1.404	1.582	6.40	.262
3.16	2.311	1.408	1.584	6.42	.263
3.17	2.315	1.412	1.586	6.45	.265
3.18	2.320	1.416	1.588	6.48	.266
3.19	2.324	1.420	1.590	6.50	.268
3.20	2.328	1.424	1.592	6.52	.269
3.21	2.333	1.428	1.594	6.55	.271
3.22	2.337	1.432	1.596	6.58	.272
3.23	2.341	1.436	1.598	6.60	.274
3.24	2.346	1.440	1.600	6.63	.275
3.25	2.350	1.443	1.602	6.65	.277
3.26	2.354	1.447	1.604	6.68	.278
3.27	2.359	1.451	1.606	6.70	.280
3.28	2.363	1.455	1.608	6.73	.281
3.29	2.368	1.459	1.610	6.76	.283
3.30	2.372	1.463	1.612	6.78	.284
3.31	2.376	1.466	1.614	6.81	.286
3.32	2.381	1.470	1.616	6.83	.287
3.33	2.385	1.474	1.619	6.86	.289
3.34	2.389	1.478	1.621	6.89	.291
3.35	2.394	1.482	1.623	6.91	.292
3.36	2.398	1.486	1.625	6.94	.293
3.37	2.402	1.490	1.627	6.96	.295
3.38	2.407	1.493	1.629	6.99	.297
3.39	2.411	1.497	1.631	7.02	.298
3.40	2.415	1.501	1.633	7.04	.300
3.41	2.420	1.505	1.635	7.07	.301
3.42	2.424	1.509	1.637	7.10	.303
3.43	2.429	1.513	1.639	7.12	.304
3.44	2.433	1.517	1.641	7.15	.306
3.45	2.437	1.520	1.643	7.18	.307
3.46	2.441	1.524	1.645	7.20	.309
3.47	2.446	1.528	1.647	7.23	.310
3.48	2.450	1.532	1.649	7.26	.312
3.49	2.455	1.536	1.651	7.28	.313
3.50	2.459	1.540	1.654	7.31	.315

$\frac{\Delta P}{\mathcal{Q}}, \frac{\Delta Q}{\mathcal{Q}}$	M_s	$\frac{\lvert\Delta \mathcal{U}\rvert}{\mathcal{Q}}$	$\frac{\mathcal{Q}'}{\mathcal{Q}}$	$\frac{p'}{p}$	ΔS
3.50	2.459	1.540	1.654	7.31	.315
3.51	2.464	1.544	1.656	7.34	.317
3.52	2.468	1.547	1.658	7.36	.318
3.53	2.472	1.551	1.660	7.39	.320
3.54	2.476	1.555	1.662	7.42	.321
3.55	2.481	1.559	1.664	7.44	.323
3.56	2.486	1.563	1.666	7.47	.324
3.57	2.490	1.566	1.668	7.50	.326
3.58	2.494	1.570	1.670	7.53	.327
3.59	2.499	1.574	1.672	7.55	.329
3.60	2.503	1.578	1.674	7.58	.331
3.61	2.507	1.581	1.676	7.61	.332
3.62	2.511	1.585	1.678	7.63	.334
3.63	2.516	1.589	1.680	7.66	.335
3.64	2.520	1.593	1.682	7.69	.336
3.65	2.524	1.596	1.684	7.72	.338
3.66	2.529	1.600	1.686	7.75	.340
3.67	2.533	1.604	1.688	7.77	.341
3.68	2.538	1.608	1.691	7.80	.343
3.69	2.542	1.612	1.693	7.83	.344
3.70	2.546	1.615	1.695	7.85	.346
3.71	2.551	1.619	1.697	7.88	.347
3.72	2.555	1.623	1.699	7.91	.349
3.73	2.559	1.626	1.701	7.94	.351
3.74	2.564	1.630	1.703	7.97	.352
3.75	2.568	1.634	1.705	7.99	.354
3.76	2.573	1.638	1.707	8.02	.355
3.77	2.577	1.642	1.709	8.05	.357
3.78	2.581	1.645	1.711	8.08	.358
3.79	2.586	1.649	1.714	8.11	.360
3.80	2.590	1.653	1.716	8.14	.362
3.81	2.594	1.657	1.718	8.16	.363
3.82	2.599	1.660	1.720	8.19	.365
3.83	2.603	1.664	1.722	8.22	.366
3.84	2.608	1.668	1.724	8.25	.368
3.85	2.612	1.672	1.726	8.28	.369
3.86	2.616	1.675	1.728	8.30	.371
3.87	2.621	1.679	1.730	8.33	.372
3.88	2.625	1.683	1.732	8.36	.374
3.89	2.629	1.687	1.734	8.39	.376
3.90	2.634	1.690	1.736	8.42	.377
3.91	2.638	1.694	1.738	8.45	.379
3.92	2.642	1.698	1.741	8.48	.380
3.93	2.647	1.702	1.743	8.51	.382
3.94	2.651	1.705	1.745	8.53	.383
3.95	2.655	1.709	1.747	8.56	.385
3.96	2.660	1.713	1.749	8.59	.387
3.97	2.664	1.716	1.751	8.62	.388
3.98	2.668	1.720	1.753	8.65	.390
3.99	2.673	1.724	1.755	8.68	.391
4.00	2.677	1.728	1.757	8.71	.393

TABLE 1b (*Continued*). $\gamma = 5/3$

$\frac{\Delta P}{\mathfrak{a}}, \frac{\Delta Q}{\mathfrak{a}}$	M_S	$\frac{\lvert\Delta \mathfrak{u}\rvert}{\mathfrak{a}}$	$\frac{\mathfrak{a}'}{\mathfrak{a}}$	$\frac{p'}{p}$	ΔS
4.00	2.677	1.728	1.757	8.71	.393
4.01	2.682	1.732	1.760	8.74	.395
4.02	2.686	1.735	1.762	8.77	.396
4.03	2.690	1.739	1.764	8.80	.398
4.04	2.695	1.743	1.766	8.83	.399
4.05	2.699	1.746	1.768	8.85	.401
4.06	2.703	1.750	1.770	8.88	.402
4.07	2.707	1.753	1.772	8.91	.404
4.08	2.712	1.757	1.774	8.94	.405
4.09	2.716	1.761	1.776	8.97	.407
4.10	2.721	1.765	1.778	9.00	.409
4.11	2.725	1.768	1.781	9.03	.410
4.12	2.729	1.772	1.783	9.06	.412
4.13	2.734	1.776	1.785	9.09	.413
4.14	2.738	1.779	1.787	9.12	.415
4.15	2.742	1.783	1.789	9.15	.416
4.16	2.747	1.787	1.791	9.18	.418
4.17	2.751	1.791	1.793	9.21	.419
4.18	2.755	1.794	1.795	9.24	.421
4.19	2.760	1.798	1.797	9.27	.423
4.20	2.764	1.802	1.799	9.30	.424
4.21	2.768	1.805	1.802	9.33	.426
4.22	2.773	1.809	1.804	9.36	.427
4.23	2.777	1.813	1.806	9.39	.429
4.24	2.781	1.816	1.808	9.42	.431
4.25	2.786	1.820	1.810	9.45	.432
4.26	2.790	1.824	1.812	9.48	.434
4.27	2.794	1.827	1.814	9.51	.435
4.28	2.799	1.831	1.816	9.54	.437
4.29	2.803	1.835	1.818	9.57	.438
4.30	2.808	1.839	1.821	9.60	.440
4.31	2.812	1.842	1.823	9.63	.442
4.32	2.816	1.846	1.825	9.66	.443
4.33	2.821	1.850	1.827	9.69	.445
4.34	2.825	1.853	1.829	9.72	.447
4.35	2.829	1.857	1.831	9.75	.448
4.36	2.833	1.860	1.833	9.78	.449
4.37	2.838	1.864	1.835	9.82	.451
4.38	2.842	1.868	1.837	9.85	.453
4.39	2.846	1.871	1.839	9.88	.454
4.40	2.851	1.875	1.842	9.91	.456
4.41	2.855	1.879	1.844	9.94	.457
4.42	2.859	1.882	1.846	9.97	.459
4.43	2.864	1.886	1.848	10.00	.461
4.44	2.868	1.890	1.850	10.03	.462
4.45	2.872	1.893	1.852	10.06	.464
4.46	2.877	1.897	1.854	10.10	.465
4.47	2.881	1.900	1.856	10.13	.467
4.48	2.885	1.904	1.859	10.16	.468
4.49	2.890	1.908	1.861	10.19	.470
4.50	2.894	1.911	1.863	10.22	.472

$\frac{\Delta P}{\mathfrak{a}}, \frac{\Delta Q}{\mathfrak{a}}$	M_S	$\frac{\lvert\Delta \mathfrak{u}\rvert}{\mathfrak{a}}$	$\frac{\mathfrak{a}'}{\mathfrak{a}}$	$\frac{p'}{p}$	ΔS
4.50	2.894	1.911	1.863	10.22	.472
4.51	2.898	1.915	1.865	10.25	.473
4.52	2.903	1.919	1.867	10.28	.475
4.53	2.907	1.922	1.869	10.31	.476
4.54	2.911	1.926	1.871	10.34	.478
4.55	2.915	1.929	1.873	10.38	.479
4.56	2.920	1.933	1.876	10.41	.481
4.57	2.924	1.937	1.878	10.44	.482
4.58	2.929	1.940	1.880	10.47	.484
4.59	2.933	1.944	1.882	10.50	.486
4.60	2.937	1.948	1.884	10.54	.487
4.61	2.942	1.951	1.886	10.57	.489
4.62	2.946	1.955	1.888	10.60	.491
4.63	2.950	1.959	1.890	10.63	.492
4.64	2.955	1.962	1.893	10.66	.494
4.65	2.959	1.966	1.895	10.69	.495
4.66	2.964	1.970	1.897	10.73	.497
4.67	2.968	1.973	1.899	10.76	.498
4.68	2.972	1.977	1.901	10.79	.500
4.69	2.976	1.980	1.903	10.82	.501
4.70	2.981	1.984	1.905	10.86	.503
4.71	2.985	1.987	1.907	10.89	.505
4.72	2.989	1.991	1.909	10.92	.506
4.73	2.994	1.995	1.912	10.95	.508
4.74	2.998	1.998	1.914	10.98	.509
4.75	3.002	2.002	1.916	11.02	.511
4.76	3.007	2.006	1.918	11.05	.513
4.77	3.011	2.009	1.920	11.08	.514
4.78	3.015	2.013	1.922	11.11	.516
4.79	3.020	2.016	1.925	11.15	.517
4.80	3.024	2.020	1.927	11.18	.519
4.81	3.028	2.023	1.929	11.21	.520
4.82	3.032	2.027	1.931	11.24	.522
4.83	3.037	2.031	1.933	11.28	.524
4.84	3.041	2.034	1.935	11.31	.525
4.85	3.046	2.038	1.937	11.35	.527
4.86	3.050	2.042	1.939	11.38	.528
4.87	3.054	2.045	1.942	11.41	.530
4.88	3.058	2.049	1.944	11.44	.531
4.89	3.063	2.052	1.946	11.47	.533
4.90	3.067	2.056	1.948	11.51	.534
4.91	3.071	2.059	1.950	11.54	.536
4.92	3.076	2.063	1.952	11.57	.537
4.93	3.080	2.066	1.954	11.61	.539
4.94	3.084	2.070	1.957	11.64	.541
4.95	3.089	2.074	1.959	11.67	.542
4.96	3.093	2.077	1.961	11.71	.544
4.97	3.097	2.081	1.963	11.74	.546
4.98	3.102	2.084	1.965	11.78	.547
4.99	3.106	2.088	1.967	11.81	.549
5.00	3.110	2.091	1.969	11.84	.550

TABLE 1b (*Continued*). $\gamma = 5/3$

$\frac{\Delta P}{\mathfrak{a}}, \frac{\Delta Q}{\mathfrak{a}}$	M_s	$\frac{\lvert\Delta\mathfrak{u}\rvert}{\mathfrak{a}}$	$\frac{\mathfrak{a}'}{\mathfrak{a}}$	$\frac{p'}{p}$	ΔS
5.00	3.110	2.091	1.969	11.84	.550
5.01	3.115	2.095	1.972	11.88	.552
5.02	3.119	2.099	1.974	11.91	.553
5.03	3.123	2.102	1.976	11.94	.555
5.04	3.128	2.106	1.978	11.98	.556
5.05	3.132	2.109	1.980	12.01	.558
5.06	3.136	2.113	1.982	12.04	.560
5.07	3.140	2.116	1.984	12.08	.561
5.08	3.145	2.120	1.987	12.11	.563
5.09	3.149	2.124	1.989	12.15	.564
5.10	3.153	2.127	1.991	12.18	.566
5.11	3.157	2.131	1.993	12.21	.567
5.12	3.162	2.134	1.995	12.25	.569
5.13	3.166	2.138	1.997	12.28	.570
5.14	3.170	2.141	1.999	12.31	.572
5.15	3.175	2.145	2.002	12.35	.574
5.16	3.179	2.149	2.004	12.39	.575
5.17	3.184	2.152	2.006	12.42	.577
5.18	3.188	2.156	2.008	12.45	.578
5.19	3.192	2.159	2.010	12.49	.580
5.20	3.197	2.163	2.012	12.52	.581
5.21	3.201	2.166	2.015	12.56	.583
5.22	3.205	2.170	2.017	12.59	.584
5.23	3.209	2.173	2.019	12.62	.586
5.24	3.214	2.177	2.021	12.66	.588
5.25	3.218	2.180	2.023	12.69	.589
5.26	3.222	2.184	2.025	12.73	.591
5.27	3.227	2.188	2.028	12.77	.592
5.28	3.231	2.191	2.030	12.80	.594
5.29	3.235	2.195	2.032	12.83	.595
5.30	3.239	2.198	2.034	12.87	.597
5.31	3.244	2.202	2.036	12.90	.599
5.32	3.248	2.205	2.038	12.94	.600
5.33	3.252	2.209	2.040	12.97	.602
5.34	3.257	2.212	2.042	13.01	.603
5.35	3.261	2.216	2.045	13.04	.605
5.36	3.265	2.219	2.047	13.08	.606
5.37	3.270	2.223	2.049	13.11	.608
5.38	3.274	2.226	2.051	13.15	.609
5.39	3.278	2.230	2.053	13.18	.611
5.40	3.283	2.234	2.055	13.22	.612
5.41	3.287	2.237	2.058	13.25	.614
5.42	3.291	2.240	2.060	13.29	.615
5.43	3.295	2.244	2.062	13.32	.617
5.44	3.300	2.248	2.064	13.36	.619
5.45	3.304	2.251	2.066	13.40	.620
5.46	3.308	2.255	2.068	13.43	.622
5.47	3.312	2.258	2.070	13.47	.623
5.48	3.317	2.262	2.073	13.50	.625
5.49	3.321	2.265	2.075	13.54	.626
5.50	3.325	2.269	2.077	13.57	.628

$\frac{\Delta P}{\mathfrak{a}}, \frac{\Delta Q}{\mathfrak{a}}$	M_s	$\frac{\lvert\Delta\mathfrak{u}\rvert}{\mathfrak{a}}$	$\frac{\mathfrak{a}'}{\mathfrak{a}}$	$\frac{p'}{p}$	ΔS
5.50	3.325	2.269	2.077	13.57	.628
5.51	3.330	2.272	2.079	13.61	.629
5.52	3.334	2.276	2.081	13.65	.631
5.53	3.338	2.279	2.083	13.68	.632
5.54	3.343	2.283	2.086	13.72	.634
5.55	3.347	2.286	2.088	13.75	.635
5.56	3.351	2.290	2.090	13.79	.637
5.57	3.356	2.293	2.092	13.83	.639
5.58	3.360	2.297	2.094	13.86	.640
5.59	3.364	2.300	2.097	13.90	.642
5.60	3.368	2.304	2.099	13.93	.643
5.61	3.373	2.307	2.101	13.97	.645
5.62	3.377	2.311	2.103	14.01	.646
5.63	3.381	2.314	2.105	14.04	.648
5.64	3.386	2.318	2.107	14.08	.649
5.65	3.390	2.321	2.109	14.11	.651
5.66	3.394	2.325	2.112	14.15	.653
5.67	3.399	2.328	2.114	14.19	.654
5.68	3.403	2.332	2.116	14.22	.656
5.69	3.407	2.335	2.118	14.26	.657
5.70	3.412	2.339	2.120	14.30	.659
5.71	3.416	2.342	2.123	14.34	.660
5.72	3.420	2.346	2.125	14.37	.662
5.73	3.424	2.349	2.127	14.41	.663
5.74	3.429	2.353	2.129	14.45	.665
5.75	3.433	2.356	2.131	14.48	.666
5.76	3.437	2.360	2.133	14.52	.668
5.77	3.441	2.363	2.135	14.55	.669
5.78	3.446	2.367	2.138	14.59	.671
5.79	3.450	2.370	2.140	14.63	.672
5.80	3.454	2.374	2.142	14.67	.674
5.81	3.459	2.377	2.144	14.70	.675
5.82	3.463	2.381	2.146	14.74	.677
5.83	3.467	2.384	2.148	14.78	.678
5.84	3.471	2.387	2.151	14.81	.680
5.85	3.476	2.391	2.153	14.85	.682
5.86	3.480	2.395	2.155	14.89	.683
5.87	3.484	2.398	2.157	14.93	.684
5.88	3.489	2.401	2.159	14.96	.686
5.89	3.493	2.405	2.162	15.00	.688
5.90	3.497	2.409	2.164	15.04	.689
5.91	3.502	2.412	2.166	15.08	.691
5.92	3.506	2.415	2.168	15.11	.692
5.93	3.510	2.419	2.170	15.15	.693
5.94	3.515	2.423	2.173	15.19	.695
5.95	3.519	2.426	2.175	15.23	.697
5.96	3.523	2.429	2.177	15.26	.698
5.97	3.527	2.433	2.179	15.30	.700
5.98	3.531	2.436	2.181	15.34	.701
5.99	3.536	2.440	2.183	15.38	.703
6.00	3.540	2.443	2.185	15.42	.704

TABLE 1b (*Continued*). $\gamma = 5/3$

$\frac{\Delta P}{\mathcal{Q}}, \frac{\Delta Q}{\mathcal{Q}}$	M_S	$\frac{\vert\Delta \mathcal{U}\vert}{\mathcal{Q}}$	$\frac{\mathcal{Q}'}{\mathcal{Q}}$	$\frac{p'}{p}$	ΔS
6.00	3.540	2.443	2.185	15.42	.704
6.01	3.544	2.447	2.188	15.45	.706
6.02	3.549	2.450	2.190	15.49	.707
6.03	3.553	2.453	2.192	15.53	.709
6.04	3.557	2.457	2.194	15.57	.710
6.05	3.562	2.461	2.196	15.61	.712
6.06	3.566	2.464	2.199	15.64	.713
6.07	3.570	2.467	2.201	15.68	.715
6.08	3.574	2.471	2.203	15.72	.716
6.09	3.579	2.475	2.205	15.76	.718
6.10	3.583	2.478	2.207	15.80	.719
6.11	3.587	2.481	2.209	15.83	.721
6.12	3.591	2.485	2.212	15.87	.722
6.13	3.596	2.488	2.214	15.91	.724
6.14	3.600	2.492	2.216	15.95	.725
6.15	3.604	2.495	2.218	15.99	.727
6.16	3.609	2.499	2.220	16.03	.728
6.17	3.613	2.502	2.223	16.07	.730
6.18	3.617	2.505	2.225	16.10	.731
6.19	3.621	2.509	2.227	16.14	.733
6.20	3.626	2.512	2.229	16.18	.734
6.21	3.630	2.516	2.231	16.22	.736
6.22	3.634	2.519	2.233	16.26	.737
6.23	3.638	2.523	2.236	16.30	.739
6.24	3.643	2.526	2.238	16.34	.740
6.25	3.647	2.530	2.240	16.38	.742
6.26	3.652	2.533	2.242	16.42	.743
6.27	3.656	2.537	2.244	16.46	.745
6.28	3.660	2.540	2.247	16.49	.746
6.29	3.664	2.543	2.249	16.53	.748
6.30	3.668	2.547	2.251	16.57	.749
6.31	3.673	2.551	2.253	16.61	.751
6.32	3.677	2.554	2.255	16.65	.752
6.33	3.681	2.557	2.257	16.69	.754
6.34	3.686	2.561	2.260	16.73	.755
6.35	3.690	2.564	2.262	16.77	.757
6.36	3.694	2.567	2.264	16.81	.758
6.37	3.699	2.571	2.266	16.85	.760
6.38	3.703	2.575	2.269	16.89	.761
6.39	3.707	2.578	2.271	16.93	.763
6.40	3.711	2.581	2.273	16.97	.764
6.41	3.716	2.585	2.275	17.01	.766
6.42	3.720	2.588	2.277	17.05	.767
6.43	3.724	2.592	2.279	17.09	.769
6.44	3.728	2.595	2.282	17.13	.770
6.45	3.733	2.599	2.284	17.17	.772
6.46	3.737	2.602	2.286	17.20	.773
6.47	3.741	2.605	2.288	17.24	.775
6.48	3.746	2.609	2.290	17.29	.776
6.49	3.750	2.612	2.292	17.33	.777
6.50	3.754	2.616	2.295	17.37	.779

$\frac{\Delta P}{\mathcal{Q}}, \frac{\Delta Q}{\mathcal{Q}}$	M_S	$\frac{\vert\Delta \mathcal{U}\vert}{\mathcal{Q}}$	$\frac{\mathcal{Q}'}{\mathcal{Q}}$	$\frac{p'}{p}$	ΔS
6.50	3.754	2.616	2.295	17.37	.779
6.51	3.758	2.619	2.297	17.40	.781
6.52	3.762	2.622	2.299	17.44	.782
6.53	3.767	2.626	2.301	17.48	.783
6.54	3.771	2.630	2.304	17.53	.785
6.55	3.775	2.633	2.306	17.57	.786
6.56	3.780	2.636	2.308	17.61	.788
6.57	3.784	2.640	2.310	17.65	.789
6.58	3.788	2.643	2.312	17.69	.791
6.59	3.793	2.647	2.314	17.73	.792
6.60	3.797	2.650	2.317	17.77	.794
6.61	3.801	2.653	2.319	17.81	.795
6.62	3.805	2.657	2.321	17.85	.797
6.63	3.809	2.660	2.323	17.89	.798
6.64	3.814	2.664	2.325	17.93	.800
6.65	3.818	2.667	2.328	17.97	.801
6.66	3.822	2.671	2.330	18.01	.803
6.67	3.827	2.674	2.332	18.05	.804
6.68	3.831	2.677	2.334	18.09	.806
6.69	3.835	2.681	2.336	18.13	.807
6.70	3.840	2.684	2.339	18.18	.809
6.71	3.844	2.688	2.341	18.22	.810
6.72	3.848	2.691	2.343	18.26	.811
6.73	3.852	2.694	2.345	18.30	.813
6.74	3.856	2.698	2.347	18.34	.814
6.75	3.861	2.701	2.349	18.38	.816
6.76	3.865	2.705	2.352	18.42	.817
6.77	3.869	2.708	2.354	18.46	.819
6.78	3.874	2.712	2.356	18.51	.820
6.79	3.878	2.715	2.358	18.55	.822
6.80	3.882	2.718	2.360	18.59	.823
6.81	3.886	2.722	2.363	18.63	.824
6.82	3.890	2.725	2.365	18.67	.826
6.83	3.895	2.729	2.367	18.71	.827
6.84	3.899	2.732	2.369	18.75	.829
6.85	3.903	2.735	2.371	18.80	.830
6.86	3.908	2.739	2.374	18.84	.832
6.87	3.912	2.742	2.376	18.88	.833
6.88	3.916	2.745	2.378	18.92	.835
6.89	3.920	2.749	2.380	18.96	.836
6.90	3.925	2.753	2.383	19.00	.838
6.91	3.929	2.756	2.385	19.05	.839
6.92	3.933	2.759	2.387	19.09	.841
6.93	3.937	2.763	2.389	19.13	.842
6.94	3.942	2.766	2.391	19.17	.843
6.95	3.946	2.769	2.393	19.21	.845
6.96	3.950	2.773	2.396	19.25	.846
6.97	3.955	2.776	2.398	19.30	.848
6.98	3.959	2.780	2.400	19.34	.849
6.99	3.963	2.783	2.402	19.38	.851
7.00	3.967	2.786	2.404	19.42	.852

TABLE 1b (*Continued*). $\gamma = 5/3$

$\frac{\Delta P}{\mathcal{Q}}, \frac{\Delta Q}{\mathcal{Q}}$	M_s	$\frac{\lvert\Delta\mathcal{U}\rvert}{\mathcal{Q}}$	$\frac{\mathcal{Q}'}{\mathcal{Q}}$	$\frac{p'}{p}$	ΔS
7.00	3.967	2.786	2.404	19.42	.852
7.01	3.971	2.790	2.407	19.46	.853
7.02	3.976	2.793	2.409	19.51	.855
7.03	3.980	2.796	2.411	19.55	.856
7.04	3.984	2.800	2.413	19.59	.858
7.05	3.989	2.803	2.416	19.64	.859
7.06	3.993	2.807	2.418	19.68	.861
7.07	3.997	2.810	2.420	19.72	.862
7.08	4.001	2.814	2.422	19.76	.863
7.09	4.005	2.817	2.424	19.80	.865
7.10	4.010	2.820	2.426	19.85	.866
7.11	4.014	2.824	2.429	19.89	.868
7.12	4.018	2.827	2.431	19.93	.869
7.13	4.023	2.831	2.433	19.98	.871
7.14	4.027	2.834	2.435	20.02	.872
7.15	4.031	2.837	2.437	20.06	.873
7.16	4.035	2.841	2.440	20.10	.875
7.17	4.039	2.844	2.442	20.15	.876
7.18	4.044	2.848	2.444	20.19	.878
7.19	4.048	2.851	2.446	20.24	.879
7.20	4.052	2.854	2.449	20.28	.881
7.21	4.057	2.858	2.451	20.32	.882
7.22	4.061	2.861	2.453	20.36	.884
7.23	4.065	2.864	2.455	20.41	.885
7.24	4.069	2.868	2.457	20.45	.886
7.25	4.073	2.871	2.459	20.49	.888
7.26	4.078	2.874	2.462	20.53	.889
7.27	4.082	2.878	2.464	20.58	.891
7.28	4.087	2.881	2.466	20.62	.892
7.29	4.091	2.885	2.468	20.67	.893
7.30	4.095	2.888	2.471	20.71	.895
7.31	4.099	2.891	2.473	20.75	.896
7.32	4.103	2.895	2.475	20.80	.898
7.33	4.108	2.898	2.477	20.84	.899
7.34	4.112	2.901	2.479	20.88	.901
7.35	4.116	2.905	2.482	20.93	.902
7.36	4.120	2.908	2.484	20.97	.903
7.37	4.125	2.912	2.486	21.02	.905
7.38	4.129	2.915	2.488	21.06	.906
7.39	4.133	2.918	2.490	21.10	.908
7.40	4.137	2.922	2.493	21.15	.909
7.41	4.142	2.925	2.495	21.19	.910
7.42	4.146	2.928	2.497	21.23	.912
7.43	4.150	2.932	2.499	21.28	.913
7.44	4.155	2.935	2.502	21.33	.915
7.45	4.159	2.939	2.504	21.37	.916
7.46	4.163	2.942	2.506	21.41	.918
7.47	4.167	2.945	2.508	21.46	.919
7.48	4.171	2.949	2.510	21.50	.920
7.49	4.176	2.952	2.513	21.54	.922
7.50	4.180	2.955	2.515	21.59	.923

$\frac{\Delta P}{\mathcal{Q}}, \frac{\Delta Q}{\mathcal{Q}}$	M_s	$\frac{\lvert\Delta\mathcal{U}\rvert}{\mathcal{Q}}$	$\frac{\mathcal{Q}'}{\mathcal{Q}}$	$\frac{p'}{p}$	ΔS
7.50	4.180	2.955	2.515	21.59	.923
7.51	4.184	2.959	2.517	21.63	.924
7.52	4.189	2.962	2.519	21.68	.926
7.53	4.193	2.966	2.521	21.72	.927
7.54	4.197	2.969	2.524	21.77	.929
7.55	4.201	2.972	2.526	21.81	.930
7.56	4.205	2.976	2.528	21.86	.932
7.57	4.210	2.979	2.530	21.90	.933
7.58	4.214	2.982	2.532	21.94	.935
7.59	4.218	2.986	2.535	21.99	.936
7.60	4.222	2.989	2.537	22.03	.937
7.61	4.227	2.993	2.539	22.08	.939
7.62	4.231	2.996	2.541	22.13	.940
7.63	4.235	2.999	2.544	22.17	.941
7.64	4.239	3.003	2.546	22.22	.943
7.65	4.244	3.006	2.548	22.26	.944
7.66	4.248	3.009	2.550	22.30	.945
7.67	4.252	3.013	2.552	22.35	.947
7.68	4.256	3.016	2.555	22.39	.948
7.69	4.260	3.019	2.557	22.44	.950
7.70	4.265	3.023	2.559	22.48	.951
7.71	4.269	3.026	2.561	22.53	.953
7.72	4.273	3.030	2.563	22.58	.954
7.73	4.278	3.033	2.566	22.62	.955
7.74	4.282	3.036	2.568	22.67	.957
7.75	4.286	3.040	2.570	22.71	.958
7.76	4.290	3.043	2.572	22.76	.959
7.77	4.295	3.046	2.575	22.81	.961
7.78	4.299	3.050	2.577	22.85	.962
7.79	4.303	3.053	2.579	22.90	.964
7.80	4.307	3.056	2.581	22.94	.965
7.81	4.312	3.060	2.583	22.99	.966
7.82	4.316	3.063	2.586	23.03	.968
7.83	4.320	3.066	2.588	23.08	.969
7.84	4.324	3.070	2.590	23.12	.970
7.85	4.328	3.073	2.592	23.17	.972
7.86	4.333	3.076	2.594	23.21	.973
7.87	4.337	3.080	2.597	23.26	.975
7.88	4.341	3.083	2.599	23.31	.976
7.89	4.346	3.087	2.601	23.35	.977
7.90	4.350	3.090	2.603	23.40	.979
7.91	4.354	3.093	2.606	23.45	.980
7.92	4.358	3.097	2.608	23.49	.981
7.93	4.362	3.100	2.610	23.54	.983
7.94	4.367	3.103	2.612	23.58	.984
7.95	4.371	3.107	2.614	23.63	.986
7.96	4.375	3.110	2.617	23.68	.987
7.97	4.379	3.113	2.619	23.72	.988
7.98	4.383	3.116	2.621	23.77	.990
7.99	4.388	3.120	2.623	23.82	.991
8.00	4.392	3.123	2.626	23.86	.992

TABLE 1b (*Continued*). $\gamma = 5/3$

$\frac{\Delta P}{\mathfrak{a}}, \frac{\Delta Q}{\mathfrak{a}}$	M_S	$\frac{\lvert\Delta \mathfrak{U}\rvert}{\mathfrak{a}}$	$\frac{\mathfrak{a}'}{\mathfrak{a}}$	$\frac{p'}{p}$	ΔS
8.00	4.392	3.123	2.626	23.86	.992
8.01	4.396	3.127	2.628	23.91	.994
8.02	4.401	3.130	2.630	23.96	.995
8.03	4.405	3.133	2.632	24.00	.996
8.04	4.409	3.137	2.634	24.05	.998
8.05	4.413	3.140	2.637	24.09	.999
8.06	4.417	3.143	2.639	24.14	1.001
8.07	4.422	3.147	2.641	24.19	1.002
8.08	4.426	3.150	2.643	24.23	1.003
8.09	4.430	3.154	2.646	24.29	1.005
8.10	4.435	3.157	2.648	24.33	1.006
8.11	4.439	3.160	2.650	24.38	1.007
8.12	4.443	3.163	2.652	24.42	1.009
8.13	4.447	3.167	2.654	24.47	1.010
8.14	4.451	3.170	2.657	24.52	1.011
8.15	4.456	3.173	2.659	24.56	1.013
8.16	4.460	3.177	2.661	24.61	1.014
8.17	4.464	3.180	2.663	24.66	1.015
8.18	4.468	3.183	2.665	24.71	1.017
8.19	4.472	3.187	2.668	24.75	1.018
8.20	4.477	3.190	2.670	24.80	1.019
8.21	4.481	3.193	2.672	24.85	1.021
8.22	4.485	3.197	2.674	24.89	1.022
8.23	4.490	3.200	2.677	24.94	1.024
8.24	4.494	3.203	2.679	24.99	1.025
8.25	4.498	3.207	2.681	25.04	1.026
8.26	4.502	3.210	2.683	25.09	1.028
8.27	4.506	3.213	2.685	25.13	1.029
8.28	4.511	3.217	2.688	25.18	1.030
8.29	4.515	3.220	2.690	25.23	1.032
8.30	4.519	3.223	2.692	25.28	1.033
8.31	4.523	3.227	2.694	25.32	1.034
8.32	4.527	3.230	2.697	25.37	1.036
8.33	4.532	3.234	2.699	25.42	1.037
8.34	4.536	3.237	2.701	25.47	1.038
8.35	4.540	3.240	2.703	25.52	1.040
8.36	4.545	3.243	2.706	25.57	1.041
8.37	4.549	3.247	2.708	25.61	1.042
8.38	4.553	3.250	2.710	25.66	1.044
8.39	4.557	3.253	2.712	25.71	1.045
8.40	4.561	3.257	2.714	25.76	1.046
8.41	4.566	3.260	2.717	25.81	1.048
8.42	4.570	3.263	2.719	25.85	1.049
8.43	4.574	3.267	2.721	25.90	1.050
8.44	4.578	3.270	2.723	25.95	1.052
8.45	4.582	3.273	2.725	26.00	1.053
8.46	4.587	3.276	2.728	26.05	1.054
8.47	4.591	3.280	2.730	26.10	1.056
8.48	4.595	3.283	2.732	26.15	1.057
8.49	4.600	3.287	2.734	26.19	1.058
8.50	4.604	3.290	2.737	26.24	1.060
8.50	4.604	3.290	2.737	26.24	1.060
8.51	4.608	3.293	2.739	26.29	1.061
8.52	4.612	3.297	2.741	26.34	1.062
8.53	4.616	3.300	2.743	26.39	1.064
8.54	4.621	3.303	2.745	26.44	1.065
8.55	4.625	3.306	2.748	26.49	1.066
8.56	4.629	3.310	2.750	26.53	1.068
8.57	4.633	3.313	2.752	26.58	1.069
8.58	4.637	3.316	2.754	26.63	1.070
8.59	4.642	3.320	2.757	26.68	1.072
8.60	4.646	3.323	2.759	26.73	1.073
8.61	4.650	3.327	2.761	26.78	1.074
8.62	4.655	3.330	2.763	26.83	1.076
8.63	4.659	3.333	2.766	26.88	1.077
8.64	4.663	3.336	2.768	26.93	1.078
8.65	4.667	3.340	2.770	26.98	1.080
8.66	4.671	3.343	2.772	27.03	1.081
8.67	4.676	3.346	2.774	27.08	1.082
8.68	4.680	3.350	2.777	27.13	1.084
8.69	4.684	3.353	2.779	27.17	1.085
8.70	4.688	3.356	2.781	27.22	1.086
8.71	4.692	3.359	2.783	27.27	1.087
8.72	4.697	3.363	2.786	27.32	1.089
8.73	4.701	3.366	2.788	27.37	1.090
8.74	4.705	3.369	2.790	27.42	1.091
8.75	4.709	3.373	2.792	27.47	1.093
8.76	4.713	3.376	2.794	27.52	1.094
8.77	4.718	3.380	2.797	27.57	1.095
8.78	4.722	3.383	2.799	27.62	1.097
8.79	4.726	3.386	2.801	27.67	1.098
8.80	4.731	3.389	2.803	27.72	1.099
8.81	4.735	3.393	2.806	27.77	1.101
8.82	4.739	3.396	2.808	27.82	1.102
8.83	4.743	3.399	2.810	27.87	1.103
8.84	4.747	3.403	2.812	27.92	1.104
8.85	4.752	3.406	2.815	27.97	1.106
8.86	4.756	3.409	2.817	28.02	1.107
8.87	4.760	3.412	2.819	28.07	1.108
8.88	4.764	3.416	2.821	28.12	1.110
8.89	4.768	3.419	2.823	28.17	1.111
8.90	4.773	3.423	2.826	28.23	1.112
8.91	4.777	3.426	2.828	28.28	1.114
8.92	4.781	3.429	2.830	28.33	1.115
8.93	4.786	3.432	2.833	28.38	1.116
8.94	4.790	3.436	2.835	28.43	1.117
8.95	4.794	3.439	2.837	28.48	1.119
8.96	4.798	3.442	2.839	28.53	1.120
8.97	4.802	3.446	2.841	28.58	1.121
8.98	4.807	3.449	2.844	28.63	1.123
8.99	4.811	3.452	2.846	28.68	1.124
9.00	4.815	3.455	2.848	28.73	1.125

TABLE 1b (*Continued*). $\gamma = 5/3$

$\frac{\Delta P}{\mathfrak{A}}, \frac{\Delta Q}{\mathfrak{A}}$	M_S	$\frac{\lvert\Delta \mathfrak{U}\rvert}{\mathfrak{A}}$	$\frac{\mathfrak{A}'}{\mathfrak{A}}$	$\frac{p'}{p}$	ΔS
9.00	4.815	3.455	2.848	28.73	1.125
9.01	4.819	3.459	2.850	28.78	1.126
9.02	4.823	3.462	2.853	28.83	1.128
9.03	4.828	3.465	2.855	28.88	1.129
9.04	4.832	3.469	2.857	28.93	1.130
9.05	4.836	3.472	2.859	28.98	1.131
9.06	4.840	3.475	2.861	29.03	1.133
9.07	4.844	3.478	2.864	29.08	1.134
9.08	4.849	3.482	2.866	29.14	1.135
9.09	4.853	3.485	2.868	29.19	1.137
9.10	4.857	3.488	2.870	29.24	1.138
9.11	4.861	3.492	2.873	29.29	1.139
9.12	4.866	3.495	2.875	29.34	1.141
9.13	4.870	3.498	2.877	29.40	1.142
9.14	4.874	3.502	2.879	29.45	1.143
9.15	4.878	3.505	2.882	29.50	1.144
9.16	4.883	3.508	2.884	29.55	1.146
9.17	4.887	3.512	2.886	29.60	1.147
9.18	4.891	3.515	2.888	29.65	1.148
9.19	4.895	3.518	2.891	29.70	1.150
9.20	4.899	3.521	2.893	29.75	1.151
9.21	4.904	3.525	2.895	29.81	1.152
9.22	4.908	3.528	2.897	29.86	1.153
9.23	4.912	3.531	2.899	29.91	1.155
9.24	4.916	3.535	2.902	29.96	1.156
9.25	4.920	3.538	2.904	30.01	1.157
9.26	4.925	3.541	2.906	30.06	1.158
9.27	4.929	3.544	2.908	30.12	1.160
9.28	4.933	3.548	2.911	30.17	1.161
9.29	4.937	3.551	2.913	30.22	1.162
9.30	4.941	3.554	2.915	30.27	1.164
9.31	4.946	3.558	2.917	30.32	1.165
9.32	4.950	3.561	2.920	30.38	1.166
9.33	4.954	3.564	2.922	30.43	1.167
9.34	4.959	3.568	2.924	30.48	1.169
9.35	4.963	3.571	2.926	30.54	1.170
9.36	4.967	3.574	2.929	30.59	1.171
9.37	4.971	3.578	2.931	30.64	1.172
9.38	4.975	3.581	2.933	30.69	1.174
9.39	4.980	3.584	2.935	30.74	1.175
9.40	4.984	3.587	2.938	30.80	1.176
9.41	4.988	3.591	2.940	30.85	1.177
9.42	4.992	3.594	2.942	30.90	1.179
9.43	4.996	3.597	2.944	30.95	1.180
9.44	5.001	3.600	2.946	31.01	1.181
9.45	5.005	3.604	2.949	31.06	1.182
9.46	5.009	3.607	2.951	31.11	1.184
9.47	5.013	3.610	2.953	31.16	1.185
9.48	5.017	3.614	2.955	31.22	1.186
9.49	5.022	3.617	2.958	31.27	1.187
9.50	5.026	3.620	2.960	31.32	1.189

$\frac{\Delta P}{\mathfrak{A}}, \frac{\Delta Q}{\mathfrak{A}}$	M_S	$\frac{\lvert\Delta \mathfrak{U}\rvert}{\mathfrak{A}}$	$\frac{\mathfrak{A}'}{\mathfrak{A}}$	$\frac{p'}{p}$	ΔS
9.50	5.026	3.620	2.960	31.32	1.189
9.51	5.030	3.623	2.962	31.38	1.190
9.52	5.034	3.627	2.964	31.43	1.191
9.53	5.038	3.630	2.967	31.48	1.193
9.54	5.043	3.633	2.969	31.53	1.194
9.55	5.047	3.636	2.971	31.59	1.195
9.56	5.051	3.640	2.973	31.64	1.196
9.57	5.055	3.643	2.975	31.69	1.197
9.58	5.059	3.646	2.978	31.75	1.199
9.59	5.064	3.650	2.980	31.80	1.200
9.60	5.068	3.653	2.982	31.85	1.201
9.61	5.072	3.656	2.985	31.91	1.202
9.62	5.077	3.660	2.987	31.96	1.204
9.63	5.081	3.663	2.989	32.02	1.205
9.64	5.085	3.666	2.991	32.07	1.206
9.65	5.089	3.670	2.994	32.12	1.208
9.66	5.093	3.673	2.996	32.18	1.209
9.67	5.098	3.676	2.998	32.23	1.210
9.68	5.102	3.679	3.000	32.28	1.211
9.69	5.106	3.683	3.002	32.34	1.212
9.70	5.110	3.686	3.005	32.39	1.214
9.71	5.114	3.689	3.007	32.45	1.215
9.72	5.119	3.692	3.009	32.50	1.216
9.73	5.123	3.696	3.011	32.55	1.217
9.74	5.127	3.699	3.014	32.61	1.219
9.75	5.131	3.702	3.016	32.66	1.220
9.76	5.135	3.706	3.018	32.71	1.221
9.77	5.140	3.709	3.020	32.77	1.222
9.78	5.144	3.712	3.023	32.82	1.224
9.79	5.148	3.715	3.025	32.88	1.225
9.80	5.152	3.719	3.027	32.93	1.226
9.81	5.156	3.722	3.029	32.98	1.227
9.82	5.161	3.725	3.032	33.04	1.229
9.83	5.165	3.728	3.034	33.09	1.230
9.84	5.169	3.732	3.036	33.15	1.231
9.85	5.173	3.735	3.038	33.20	1.232
9.86	5.177	3.738	3.040	33.26	1.233
9.87	5.182	3.741	3.043	33.31	1.235
9.88	5.186	3.745	3.045	33.36	1.236
9.89	5.190	3.748	3.047	33.42	1.237
9.90	5.194	3.751	3.049	33.47	1.238
9.91	5.198	3.755	3.052	33.53	1.240
9.92	5.203	3.758	3.054	33.58	1.241
9.93	5.207	3.761	3.056	33.64	1.242
9.94	5.211	3.765	3.059	33.70	1.243
9.95	5.216	3.768	3.061	33.75	1.245
9.96	5.220	3.771	3.063	33.81	1.246
9.97	5.224	3.774	3.065	33.86	1.247
9.98	5.228	3.778	3.067	33.92	1.248
9.99	5.232	3.781	3.070	33.97	1.249
10.00	5.237	3.784	3.072	34.03	1.251

TABLE 1c. SHOCK WAVE RELATIONS FOR $\gamma = 4/3$

$\frac{\Delta P}{\mathcal{Q}}, \frac{\Delta Q}{\mathcal{Q}}$	M_s	$\frac{\lvert\Delta \mathfrak{u}\rvert}{\mathcal{Q}}$	$\frac{\mathcal{Q}'}{\mathcal{Q}}$	$\frac{p'}{p}$	ΔS
.00	1.000	.000	1.000	1.00	.000
.01	1.003	.005	1.001	1.01	.000
.02	1.006	.010	1.002	1.01	.000
.03	1.009	.015	1.003	1.02	.000
.04	1.012	.020	1.003	1.03	.000
.05	1.015	.025	1.004	1.03	.000
.06	1.018	.030	1.005	1.04	.000
.07	1.021	.035	1.006	1.05	.000
.08	1.024	.040	1.007	1.06	.000
.09	1.027	.045	1.008	1.06	.000
.10	1.030	.050	1.008	1.07	.000
.11	1.033	.055	1.009	1.08	.000
.12	1.036	.060	1.010	1.08	.000
.13	1.039	.065	1.011	1.09	.000
.14	1.042	.070	1.012	1.10	.000
.15	1.045	.075	1.013	1.10	.000
.16	1.048	.080	1.013	1.11	.000
.17	1.051	.085	1.014	1.12	.000
.18	1.054	.090	1.015	1.13	.000
.19	1.057	.095	1.016	1.13	.000
.20	1.060	.100	1.017	1.14	.000
.21	1.063	.105	1.018	1.15	.000
.22	1.066	.110	1.018	1.16	.000
.23	1.069	.115	1.019	1.16	.000
.24	1.072	.120	1.020	1.17	.000
.25	1.075	.125	1.021	1.18	.000
.26	1.078	.130	1.022	1.19	.001
.27	1.082	.135	1.023	1.20	.001
.28	1.085	.140	1.023	1.20	.001
.29	1.088	.144	1.024	1.21	.001
.30	1.091	.149	1.025	1.22	.001
.31	1.094	.154	1.026	1.23	.001
.32	1.098	.159	1.027	1.23	.001
.33	1.101	.164	1.028	1.24	.001
.34	1.104	.169	1.028	1.25	.001
.35	1.107	.174	1.029	1.26	.001
.36	1.110	.179	1.030	1.27	.001
.37	1.114	.184	1.031	1.27	.001
.38	1.117	.189	1.032	1.28	.002
.39	1.120	.194	1.033	1.29	.002
.40	1.123	.199	1.033	1.30	.002
.41	1.126	.204	1.034	1.31	.002
.42	1.129	.209	1.035	1.31	.002
.43	1.133	.214	1.036	1.32	.002
.44	1.136	.219	1.037	1.33	.002
.45	1.139	.224	1.038	1.34	.002
.46	1.143	.229	1.038	1.35	.003
.47	1.146	.234	1.039	1.36	.003
.48	1.149	.239	1.040	1.37	.003
.49	1.152	.244	1.041	1.38	.003
.50	1.155	.249	1.042	1.38	.003

$\frac{\Delta P}{\mathcal{Q}}, \frac{\Delta Q}{\mathcal{Q}}$	M_s	$\frac{\lvert\Delta \mathfrak{u}\rvert}{\mathcal{Q}}$	$\frac{\mathcal{Q}'}{\mathcal{Q}}$	$\frac{p'}{p}$	ΔS
.50	1.155	.249	1.042	1.38	.003
.51	1.159	.254	1.043	1.39	.003
.52	1.162	.259	1.044	1.40	.004
.53	1.166	.264	1.044	1.41	.004
.54	1.169	.269	1.045	1.42	.004
.55	1.172	.274	1.046	1.43	.004
.56	1.175	.278	1.047	1.44	.004
.57	1.179	.283	1.048	1.45	.005
.58	1.182	.288	1.049	1.45	.005
.59	1.185	.293	1.050	1.46	.005
.60	1.188	.298	1.050	1.47	.005
.61	1.192	.303	1.051	1.48	.005
.62	1.195	.307	1.052	1.49	.006
.63	1.199	.312	1.053	1.50	.006
.64	1.202	.317	1.054	1.51	.006
.65	1.206	.322	1.055	1.52	.006
.66	1.209	.327	1.056	1.53	.007
.67	1.212	.332	1.056	1.54	.007
.68	1.215	.337	1.057	1.55	.007
.69	1.219	.342	1.058	1.56	.007
.70	1.222	.347	1.059	1.57	.008
.71	1.226	.352	1.060	1.58	.008
.72	1.229	.356	1.061	1.58	.008
.73	1.233	.361	1.062	1.59	.009
.74	1.236	.366	1.062	1.60	.009
.75	1.240	.371	1.063	1.61	.009
.76	1.243	.376	1.064	1.62	.010
.77	1.247	.381	1.065	1.63	.010
.78	1.250	.386	1.066	1.64	.010
.79	1.253	.390	1.067	1.65	.011
.80	1.256	.395	1.068	1.66	.011
.81	1.260	.400	1.068	1.67	.011
.82	1.264	.405	1.069	1.68	.012
.83	1.267	.410	1.070	1.69	.012
.84	1.271	.415	1.071	1.70	.012
.85	1.274	.419	1.072	1.71	.013
.86	1.278	.424	1.073	1.72	.013
.87	1.281	.429	1.074	1.73	.013
.88	1.285	.434	1.074	1.74	.014
.89	1.288	.439	1.075	1.75	.014
.90	1.291	.443	1.076	1.76	.015
.91	1.295	.448	1.077	1.77	.015
.92	1.298	.453	1.078	1.78	.016
.93	1.302	.457	1.079	1.79	.016
.94	1.305	.462	1.080	1.81	.017
.95	1.309	.467	1.081	1.82	.017
.96	1.312	.472	1.081	1.83	.018
.97	1.316	.476	1.082	1.84	.019
.98	1.319	.481	1.083	1.85	.019
.99	1.323	.486	1.084	1.86	.020
1.00	1.326	.491	1.085	1.87	.020

TABLE 1c (*Continued*). $\gamma = 4/3$

$\frac{\Delta P}{\mathfrak{A}}, \frac{\Delta Q}{\mathfrak{A}}$	M_S	$\frac{\lvert\Delta \mathfrak{U}\rvert}{\mathfrak{A}}$	$\frac{\mathfrak{A}'}{\mathfrak{A}}$	$\frac{p'}{p}$	ΔS	$\frac{\Delta P}{\mathfrak{A}}, \frac{\Delta Q}{\mathfrak{A}}$	M_S	$\frac{\lvert\Delta \mathfrak{U}\rvert}{\mathfrak{A}}$	$\frac{\mathfrak{A}'}{\mathfrak{A}}$	$\frac{p'}{p}$	ΔS
1.00	1.326	.491	1.085	1.87	.020	1.50	1.507	.722	1.130	2.45	.058
1.01	1.330	.496	1.086	1.88	.021	1.51	1.511	.727	1.131	2.47	.059
1.02	1.334	.500	1.087	1.89	.021	1.52	1.514	.732	1.131	2.48	.060
1.03	1.337	.505	1.088	1.90	.022	1.53	1.518	.736	1.132	2.49	.061
1.04	1.341	.510	1.088	1.91	.022	1.54	1.521	.740	1.133	2.50	.062
1.05	1.345	.515	1.089	1.92	.023	1.55	1.525	.745	1.134	2.52	.063
1.06	1.348	.520	1.090	1.93	.023	1.56	1.529	.750	1.135	2.53	.064
1.07	1.352	.525	1.091	1.95	.024	1.57	1.533	.755	1.136	2.54	.065
1.08	1.355	.529	1.092	1.96	.024	1.58	1.537	.759	1.137	2.56	.066
1.09	1.359	.534	1.093	1.97	.025	1.59	1.540	.764	1.138	2.57	.068
1.10	1.362	.538	1.094	1.98	.026	1.60	1.544	.768	1.139	2.58	.069
1.11	1.365	.543	1.095	1.99	.027	1.61	1.547	.772	1.140	2.59	.070
1.12	1.369	.547	1.095	2.00	.027	1.62	1.551	.776	1.141	2.61	.071
1.13	1.373	.552	1.096	2.01	.028	1.63	1.555	.781	1.141	2.62	.072
1.14	1.376	.557	1.097	2.02	.028	1.64	1.558	.786	1.142	2.63	.073
1.15	1.380	.562	1.098	2.03	.029	1.65	1.562	.790	1.143	2.65	.074
1.16	1.384	.566	1.099	2.05	.030	1.66	1.566	.794	1.144	2.66	.075
1.17	1.387	.571	1.100	2.06	.030	1.67	1.569	.799	1.145	2.67	.077
1.18	1.391	.576	1.101	2.07	.031	1.68	1.573	.803	1.146	2.68	.078
1.19	1.394	.581	1.102	2.08	.031	1.69	1.576	.807	1.147	2.70	.079
1.20	1.398	.585	1.103	2.09	.032	1.70	1.580	.812	1.148	2.71	.080
1.21	1.402	.590	1.103	2.10	.033	1.71	1.584	.817	1.149	2.72	.081
1.22	1.405	.595	1.104	2.11	.034	1.72	1.588	.821	1.150	2.74	.082
1.23	1.409	.599	1.105	2.13	.035	1.73	1.592	.826	1.151	2.75	.083
1.24	1.412	.604	1.106	2.14	.036	1.74	1.595	.830	1.152	2.77	.085
1.25	1.416	.609	1.107	2.15	.036	1.75	1.599	.834	1.153	2.78	.086
1.26	1.419	.613	1.108	2.16	.037	1.76	1.602	.838	1.154	2.79	.087
1.27	1.423	.617	1.109	2.17	.038	1.77	1.606	.843	1.155	2.81	.088
1.28	1.427	.622	1.110	2.18	.039	1.78	1.610	.847	1.155	2.82	.089
1.29	1.431	.627	1.111	2.20	.039	1.79	1.613	.852	1.156	2.83	.090
1.30	1.434	.632	1.111	2.21	.040	1.80	1.617	.856	1.157	2.85	.092
1.31	1.438	.636	1.112	2.22	.041	1.81	1.621	.861	1.158	2.86	.093
1.32	1.441	.641	1.113	2.23	.042	1.82	1.625	.865	1.159	2.87	.094
1.33	1.445	.645	1.114	2.24	.043	1.83	1.628	.869	1.160	2.89	.096
1.34	1.449	.650	1.115	2.26	.044	1.84	1.632	.874	1.161	2.90	.097
1.35	1.452	.654	1.116	2.27	.044	1.85	1.636	.879	1.162	2.92	.098
1.36	1.456	.659	1.117	2.28	.045	1.86	1.640	.883	1.163	2.93	.100
1.37	1.459	.664	1.118	2.29	.046	1.87	1.643	.887	1.164	2.94	.101
1.38	1.463	.668	1.119	2.30	.047	1.88	1.647	.891	1.165	2.96	.102
1.39	1.467	.673	1.120	2.32	.048	1.89	1.651	.896	1.166	2.97	.104
1.40	1.470	.677	1.120	2.33	.049	1.90	1.655	.900	1.167	2.99	.105
1.41	1.474	.682	1.121	2.34	.050	1.91	1.658	.904	1.168	3.00	.106
1.42	1.478	.687	1.122	2.35	.051	1.92	1.662	.909	1.169	3.01	.108
1.43	1.481	.691	1.123	2.37	.052	1.93	1.665	.913	1.170	3.03	.109
1.44	1.485	.696	1.124	2.38	.052	1.94	1.669	.918	1.171	3.04	.110
1.45	1.489	.700	1.125	2.39	.053	1.95	1.673	.922	1.172	3.06	.112
1.46	1.492	.705	1.126	2.40	.054	1.96	1.677	.926	1.172	3.07	.113
1.47	1.496	.709	1.127	2.41	.055	1.97	1.680	.930	1.173	3.08	.114
1.48	1.499	.714	1.128	2.43	.056	1.98	1.684	.934	1.174	3.10	.116
1.49	1.503	.718	1.129	2.44	.057	1.99	1.687	.938	1.175	3.11	.117
1.50	1.507	.722	1.130	2.45	.058	2.00	1.691	.943	1.176	3.13	.118

TABLE 1c (*Continued*). $\gamma = 4/3$

$\frac{\Delta P}{\mathcal{Q}}, \frac{\Delta Q}{\mathcal{Q}}$	M_s	$\frac{\lvert\Delta \mathcal{U}\rvert}{\mathcal{Q}}$	$\frac{\mathcal{Q}'}{\mathcal{Q}}$	$\frac{p'}{p}$	ΔS
2.00	1.691	.943	1.176	3.13	.118
2.01	1.695	.947	1.177	3.14	.120
2.02	1.699	.952	1.178	3.16	.121
2.03	1.703	.956	1.179	3.17	.123
2.04	1.706	.960	1.180	3.18	.125
2.05	1.710	.965	1.181	3.20	.126
2.06	1.714	.969	1.182	3.21	.127
2.07	1.717	.973	1.183	3.23	.129
2.08	1.721	.977	1.184	3.24	.130
2.09	1.725	.981	1.185	3.26	.132
2.10	1.728	.986	1.186	3.27	.133
2.11	1.732	.990	1.187	3.29	.135
2.12	1.736	.994	1.188	3.30	.136
2.13	1.740	.998	1.189	3.32	.138
2.14	1.743	1.003	1.190	3.33	.139
2.15	1.747	1.007	1.191	3.35	.140
2.16	1.751	1.011	1.191	3.36	.142
2.17	1.754	1.015	1.192	3.38	.144
2.18	1.758	1.019	1.193	3.39	.145
2.19	1.762	1.023	1.194	3.40	.147
2.20	1.766	1.028	1.195	3.42	.148
2.21	1.769	1.032	1.196	3.44	.150
2.22	1.773	1.037	1.197	3.45	.151
2.23	1.777	1.041	1.198	3.47	.153
2.24	1.781	1.045	1.199	3.48	.154
2.25	1.784	1.049	1.200	3.50	.156
2.26	1.788	1.053	1.201	3.51	.158
2.27	1.792	1.058	1.202	3.53	.160
2.28	1.796	1.062	1.203	3.54	.161
2.29	1.799	1.066	1.204	3.56	.162
2.30	1.803	1.070	1.205	3.57	.164
2.31	1.806	1.074	1.206	3.59	.166
2.32	1.810	1.078	1.207	3.60	.167
2.33	1.814	1.082	1.208	3.62	.169
2.34	1.818	1.086	1.209	3.63	.171
2.35	1.821	1.090	1.210	3.65	.173
2.36	1.825	1.094	1.211	3.66	.174
2.37	1.828	1.098	1.212	3.68	.176
2.38	1.832	1.103	1.213	3.70	.177
2.39	1.836	1.107	1.214	3.71	.179
2.40	1.840	1.111	1.215	3.73	.181
2.41	1.844	1.116	1.216	3.74	.182
2.42	1.847	1.120	1.217	3.76	.184
2.43	1.851	1.124	1.218	3.77	.186
2.44	1.855	1.128	1.219	3.79	.187
2.45	1.859	1.132	1.220	3.81	.189
2.46	1.862	1.136	1.221	3.82	.191
2.47	1.866	1.140	1.222	3.84	.193
2.48	1.870	1.144	1.223	3.85	.194
2.49	1.874	1.148	1.224	3.87	.196
2.50	1.877	1.152	1.225	3.88	.198
2.50	1.877	1.152	1.225	3.88	.198
2.51	1.881	1.156	1.226	3.90	.200
2.52	1.885	1.161	1.227	3.92	.201
2.53	1.889	1.165	1.228	3.93	.203
2.54	1.892	1.169	1.229	3.95	.205
2.55	1.896	1.173	1.230	3.97	.207
2.56	1.900	1.177	1.231	3.98	.208
2.57	1.903	1.181	1.232	4.00	.210
2.58	1.907	1.185	1.232	4.01	.212
2.59	1.910	1.189	1.233	4.03	.214
2.60	1.914	1.193	1.234	4.04	.215
2.61	1.918	1.197	1.235	4.06	.217
2.62	1.922	1.201	1.236	4.08	.219
2.63	1.925	1.205	1.237	4.09	.221
2.64	1.929	1.209	1.238	4.11	.223
2.65	1.933	1.213	1.239	4.13	.225
2.66	1.937	1.217	1.240	4.14	.227
2.67	1.940	1.221	1.241	4.16	.228
2.68	1.944	1.225	1.242	4.18	.230
2.69	1.948	1.229	1.243	4.19	.232
2.70	1.952	1.233	1.244	4.21	.234
2.71	1.955	1.237	1.245	4.23	.236
2.72	1.959	1.241	1.246	4.24	.238
2.73	1.962	1.245	1.247	4.26	.239
2.74	1.966	1.249	1.248	4.28	.241
2.75	1.970	1.253	1.249	4.29	.243
2.76	1.974	1.257	1.250	4.31	.245
2.77	1.977	1.261	1.251	4.33	.247
2.78	1.981	1.265	1.252	4.34	.249
2.79	1.985	1.269	1.253	4.36	.251
2.80	1.988	1.273	1.254	4.38	.253
2.81	1.992	1.277	1.255	4.39	.255
2.82	1.996	1.281	1.256	4.41	.257
2.83	2.000	1.285	1.257	4.43	.259
2.84	2.003	1.289	1.258	4.44	.261
2.85	2.007	1.293	1.259	4.46	.263
2.86	2.010	1.297	1.260	4.48	.264
2.87	2.014	1.301	1.261	4.49	.266
2.88	2.018	1.305	1.262	4.51	.268
2.89	2.022	1.309	1.263	4.53	.270
2.90	2.025	1.313	1.264	4.55	.272
2.91	2.029	1.317	1.266	4.56	.274
2.92	2.033	1.321	1.267	4.58	.276
2.93	2.036	1.325	1.268	4.60	.278
2.94	2.040	1.328	1.269	4.61	.280
2.95	2.044	1.332	1.270	4.63	.282
2.96	2.048	1.336	1.271	4.65	.284
2.97	2.051	1.340	1.272	4.67	.286
2.98	2.055	1.344	1.273	4.68	.288
2.99	2.059	1.348	1.274	4.70	.290
3.00	2.062	1.352	1.275	4.72	.292

TABLE 1c (*Continued*). $\gamma = 4/3$

$\frac{\Delta P}{\mathfrak{a}}, \frac{\Delta Q}{\mathfrak{a}}$	M_s	$\frac{\lvert\Delta \mathfrak{u}\rvert}{\mathfrak{a}}$	$\frac{\mathfrak{a}'}{\mathfrak{a}}$	$\frac{p'}{p}$	ΔS
3.00	2.062	1.352	1.275	4.72	.292
3.01	2.066	1.356	1.276	4.73	.294
3.02	2.070	1.360	1.277	4.75	.296
3.03	2.074	1.364	1.278	4.77	.298
3.04	2.077	1.368	1.279	4.79	.300
3.05	2.081	1.372	1.280	4.81	.302
3.06	2.084	1.375	1.281	4.82	.304
3.07	2.088	1.379	1.282	4.84	.306
3.08	2.092	1.383	1.283	4.86	.308
3.09	2.096	1.387	1.284	4.88	.311
3.10	2.099	1.391	1.285	4.89	.313
3.11	2.103	1.395	1.286	4.91	.315
3.12	2.106	1.399	1.287	4.93	.317
3.13	2.110	1.402	1.288	4.95	.319
3.14	2.114	1.406	1.289	4.96	.321
3.15	2.118	1.410	1.290	4.98	.323
3.16	2.121	1.414	1.291	5.00	.325
3.17	2.125	1.418	1.292	5.02	.327
3.18	2.129	1.422	1.293	5.04	.329
3.19	2.133	1.426	1.294	5.05	.331
3.20	2.136	1.430	1.295	5.07	.334
3.21	2.140	1.433	1.296	5.09	.336
3.22	2.143	1.437	1.297	5.11	.338
3.23	2.147	1.441	1.298	5.13	.340
3.24	2.151	1.445	1.299	5.15	.342
3.25	2.155	1.449	1.300	5.16	.344
3.26	2.158	1.453	1.301	5.18	.346
3.27	2.162	1.457	1.302	5.20	.348
3.28	2.165	1.460	1.303	5.22	.351
3.29	2.169	1.464	1.304	5.23	.353
3.30	2.173	1.468	1.305	5.25	.355
3.31	2.177	1.472	1.306	5.27	.357
3.32	2.180	1.476	1.308	5.29	.359
3.33	2.184	1.479	1.309	5.31	.361
3.34	2.188	1.483	1.310	5.33	.363
3.35	2.191	1.487	1.311	5.34	.365
3.36	2.195	1.491	1.312	5.36	.368
3.37	2.198	1.494	1.313	5.38	.370
3.38	2.202	1.498	1.314	5.40	.372
3.39	2.206	1.502	1.315	5.42	.374
3.40	2.210	1.506	1.316	5.44	.376
3.41	2.213	1.510	1.317	5.46	.378
3.42	2.217	1.513	1.318	5.47	.380
3.43	2.220	1.517	1.319	5.49	.383
3.44	2.224	1.521	1.320	5.51	.385
3.45	2.228	1.525	1.321	5.53	.387
3.46	2.232	1.529	1.322	5.55	.389
3.47	2.235	1.532	1.323	5.57	.391
3.48	2.239	1.536	1.324	5.59	.394
3.49	2.242	1.540	1.325	5.60	.396
3.50	2.246	1.543	1.326	5.62	.398

$\frac{\Delta P}{\mathfrak{a}}, \frac{\Delta Q}{\mathfrak{a}}$	M_s	$\frac{\lvert\Delta \mathfrak{u}\rvert}{\mathfrak{a}}$	$\frac{\mathfrak{a}'}{\mathfrak{a}}$	$\frac{p'}{p}$	ΔS
3.50	2.246	1.543	1.326	5.62	.398
3.51	2.250	1.547	1.327	5.64	.400
3.52	2.253	1.551	1.328	5.66	.402
3.53	2.257	1.555	1.329	5.68	.405
3.54	2.261	1.559	1.330	5.70	.407
3.55	2.265	1.563	1.331	5.72	.409
3.56	2.268	1.566	1.332	5.74	.411
3.57	2.272	1.570	1.333	5.76	.414
3.58	2.275	1.574	1.334	5.77	.416
3.59	2.279	1.577	1.335	5.79	.418
3.60	2.283	1.581	1.336	5.81	.420
3.61	2.286	1.585	1.338	5.83	.423
3.62	2.290	1.588	1.339	5.85	.425
3.63	2.293	1.592	1.340	5.87	.427
3.64	2.297	1.596	1.341	5.89	.429
3.65	2.301	1.600	1.342	5.91	.432
3.66	2.305	1.604	1.343	5.93	.434
3.67	2.308	1.607	1.344	5.95	.436
3.68	2.312	1.611	1.345	5.97	.439
3.69	2.315	1.614	1.346	5.98	.441
3.70	2.319	1.618	1.347	6.00	.443
3.71	2.323	1.622	1.348	6.02	.445
3.72	2.326	1.625	1.349	6.04	.447
3.73	2.330	1.629	1.350	6.06	.450
3.74	2.333	1.633	1.351	6.08	.452
3.75	2.337	1.636	1.352	6.10	.454
3.76	2.341	1.640	1.353	6.12	.456
3.77	2.345	1.644	1.354	6.14	.458
3.78	2.348	1.648	1.355	6.16	.461
3.79	2.352	1.651	1.356	6.18	.463
3.80	2.355	1.655	1.358	6.20	.466
3.81	2.359	1.659	1.359	6.22	.468
3.82	2.363	1.662	1.360	6.24	.470
3.83	2.366	1.666	1.361	6.26	.472
3.84	2.370	1.670	1.362	6.28	.475
3.85	2.373	1.673	1.363	6.30	.477
3.86	2.377	1.677	1.364	6.32	.480
3.87	2.381	1.681	1.365	6.34	.482
3.88	2.385	1.685	1.366	6.36	.484
3.89	2.388	1.688	1.367	6.38	.486
3.90	2.392	1.692	1.368	6.40	.489
3.91	2.395	1.695	1.369	6.42	.491
3.92	2.399	1.699	1.370	6.43	.493
3.93	2.403	1.703	1.371	6.45	.496
3.94	2.406	1.706	1.372	6.47	.498
3.95	2.410	1.710	1.373	6.49	.500
3.96	2.413	1.714	1.374	6.51	.503
3.97	2.417	1.717	1.375	6.53	.505
3.98	2.421	1.721	1.376	6.55	.507
3.99	2.424	1.724	1.378	6.57	.509
4.00	2.428	1.728	1.379	6.59	.512

TABLE 1c (*Continued*). $\gamma = 4/3$

$\frac{\Delta P}{\mathfrak{a}}, \frac{\Delta Q}{\mathfrak{a}}$	M_s	$\frac{\lvert\Delta \mathfrak{u}\rvert}{\mathfrak{a}}$	$\frac{\mathfrak{a}'}{\mathfrak{a}}$	$\frac{p'}{p}$	ΔS	$\frac{\Delta P}{\mathfrak{a}}, \frac{\Delta Q}{\mathfrak{a}}$	M_s	$\frac{\lvert\Delta \mathfrak{u}\rvert}{\mathfrak{a}}$	$\frac{\mathfrak{a}'}{\mathfrak{a}}$	$\frac{p'}{p}$	ΔS
4.00	2.428	1.728	1.379	6.59	.512	4.50	2.608	1.907	1.432	7.63	.631
4.01	2.431	1.732	1.380	6.61	.514	4.51	2.612	1.910	1.433	7.65	.634
4.02	2.435	1.735	1.381	6.63	.516	4.52	2.615	1.914	1.434	7.67	.636
4.03	2.439	1.739	1.382	6.66	.519	4.53	2.618	1.917	1.436	7.69	.639
4.04	2.443	1.743	1.383	6.68	.522	4.54	2.622	1.921	1.437	7.71	.641
4.05	2.446	1.746	1.384	6.70	.524	4.55	2.626	1.924	1.438	7.74	.644
4.06	2.450	1.750	1.385	6.72	.526	4.56	2.629	1.928	1.439	7.76	.646
4.07	2.453	1.754	1.386	6.74	.528	4.57	2.633	1.931	1.440	7.78	.648
4.08	2.457	1.757	1.387	6.76	.530	4.58	2.636	1.935	1.441	7.80	.651
4.09	2.461	1.761	1.388	6.78	.533	4.59	2.640	1.938	1.442	7.82	.653
4.10	2.464	1.764	1.389	6.80	.536	4.60	2.644	1.942	1.443	7.84	.656
4.11	2.468	1.768	1.390	6.82	.538	4.61	2.647	1.945	1.444	7.87	.658
4.12	2.471	1.772	1.391	6.84	.540	4.62	2.651	1.949	1.445	7.89	.661
4.13	2.475	1.775	1.393	6.86	.543	4.63	2.654	1.952	1.446	7.91	.663
4.14	2.479	1.779	1.394	6.88	.545	4.64	2.658	1.956	1.447	7.93	.666
4.15	2.482	1.782	1.395	6.90	.548	4.65	2.662	1.959	1.449	7.95	.668
4.16	2.486	1.786	1.396	6.92	.550	4.66	2.665	1.963	1.450	7.98	.670
4.17	2.489	1.790	1.397	6.94	.552	4.67	2.669	1.966	1.451	8.00	.673
4.18	2.493	1.793	1.398	6.96	.555	4.68	2.672	1.970	1.452	8.02	.675
4.19	2.497	1.797	1.399	6.98	.557	4.69	2.676	1.973	1.453	8.04	.678
4.20	2.500	1.800	1.400	7.00	.559	4.70	2.680	1.977	1.454	8.06	.680
4.21	2.504	1.804	1.401	7.02	.562	4.71	2.683	1.980	1.455	8.08	.683
4.22	2.507	1.807	1.402	7.04	.564	4.72	2.686	1.984	1.456	8.11	.685
4.23	2.511	1.811	1.403	7.06	.566	4.73	2.690	1.987	1.457	8.13	.688
4.24	2.515	1.815	1.404	7.08	.569	4.74	2.694	1.991	1.458	8.15	.690
4.25	2.518	1.818	1.405	7.10	.571	4.75	2.697	1.994	1.459	8.17	.693
4.26	2.522	1.822	1.406	7.13	.574	4.76	2.701	1.998	1.460	8.19	.695
4.27	2.525	1.825	1.408	7.15	.576	4.77	2.704	2.001	1.462	8.21	.697
4.28	2.529	1.829	1.409	7.17	.578	4.78	2.708	2.004	1.463	8.24	.700
4.29	2.533	1.832	1.410	7.19	.581	4.79	2.711	2.008	1.464	8.26	.702
4.30	2.536	1.836	1.411	7.21	.583	4.80	2.715	2.011	1.465	8.28	.705
4.31	2.540	1.840	1.412	7.23	.586	4.81	2.719	2.015	1.466	8.30	.708
4.32	2.543	1.843	1.413	7.25	.588	4.82	2.722	2.018	1.467	8.32	.710
4.33	2.547	1.847	1.414	7.27	.591	4.83	2.725	2.022	1.468	8.35	.712
4.34	2.551	1.850	1.415	7.29	.593	4.84	2.729	2.025	1.469	8.37	.715
4.35	2.554	1.854	1.416	7.31	.595	4.85	2.733	2.029	1.470	8.39	.717
4.36	2.558	1.857	1.417	7.33	.598	4.86	2.736	2.032	1.471	8.41	.720
4.37	2.561	1.861	1.418	7.36	.600	4.87	2.740	2.036	1.472	8.44	.722
4.38	2.565	1.864	1.419	7.38	.602	4.88	2.743	2.039	1.474	8.46	.725
4.39	2.569	1.868	1.420	7.40	.605	4.89	2.747	2.043	1.475	8.48	.727
4.40	2.572	1.872	1.422	7.42	.607	4.90	2.751	2.046	1.476	8.50	.730
4.41	2.576	1.875	1.423	7.44	.609	4.91	2.754	2.049	1.477	8.53	.732
4.42	2.579	1.878	1.424	7.46	.612	4.92	2.758	2.053	1.478	8.55	.734
4.43	2.583	1.882	1.425	7.48	.615	4.93	2.761	2.056	1.479	8.57	.737
4.44	2.586	1.886	1.426	7.50	.617	4.94	2.765	2.060	1.480	8.59	.740
4.45	2.590	1.889	1.427	7.52	.619	4.95	2.768	2.063	1.481	8.61	.742
4.46	2.594	1.893	1.428	7.55	.622	4.96	2.772	2.067	1.482	8.64	.744
4.47	2.597	1.896	1.429	7.57	.624	4.97	2.775	2.070	1.483	8.66	.747
4.48	2.601	1.900	1.430	7.59	.627	4.98	2.779	2.073	1.484	8.68	.749
4.49	2.604	1.903	1.431	7.61	.629	4.99	2.782	2.077	1.486	8.70	.752
4.50	2.608	1.907	1.432	7.63	.631	5.00	2.786	2.080	1.487	8.73	.755

TABLE 1c (*Continued*). $\gamma = 4/3$

$\frac{\Delta P}{\mathfrak{A}}, \frac{\Delta Q}{\mathfrak{A}}$	M_S	$\frac{\lvert\Delta\mathfrak{U}\rvert}{\mathfrak{A}}$	$\frac{\mathfrak{A}'}{\mathfrak{A}}$	$\frac{p'}{p}$	ΔS
5.00	2.786	2.080	1.487	8.73	.755
5.01	2.789	2.084	1.488	8.75	.757
5.02	2.793	2.087	1.489	8.77	.759
5.03	2.796	2.090	1.490	8.79	.762
5.04	2.800	2.094	1.491	8.82	.765
5.05	2.804	2.097	1.492	8.84	.767
5.06	2.807	2.101	1.493	8.86	.769
5.07	2.811	2.104	1.494	8.89	.772
5.08	2.814	2.108	1.496	8.91	.774
5.09	2.817	2.111	1.497	8.93	.777
5.10	2.821	2.114	1.498	8.95	.779
5.11	2.825	2.118	1.499	8.98	.782
5.12	2.828	2.121	1.500	9.00	.784
5.13	2.832	2.125	1.501	9.02	.787
5.14	2.835	2.128	1.502	9.05	.789
5.15	2.839	2.132	1.503	9.07	.792
5.16	2.842	2.135	1.504	9.09	.794
5.17	2.846	2.138	1.505	9.11	.797
5.18	2.850	2.142	1.507	9.14	.799
5.19	2.853	2.145	1.508	9.16	.802
5.20	2.856	2.148	1.509	9.18	.804
5.21	2.860	2.152	1.510	9.20	.807
5.22	2.863	2.155	1.511	9.23	.809
5.23	2.867	2.159	1.512	9.25	.812
5.24	2.871	2.162	1.513	9.27	.815
5.25	2.874	2.165	1.514	9.30	.817
5.26	2.878	2.169	1.515	9.32	.819
5.27	2.881	2.172	1.516	9.34	.822
5.28	2.884	2.175	1.517	9.37	.824
5.29	2.888	2.179	1.519	9.39	.827
5.30	2.892	2.182	1.520	9.41	.829
5.31	2.895	2.186	1.521	9.44	.832
5.32	2.898	2.189	1.522	9.46	.835
5.33	2.902	2.192	1.523	9.48	.837
5.34	2.906	2.196	1.524	9.51	.840
5.35	2.909	2.199	1.525	9.53	.842
5.36	2.913	2.202	1.526	9.55	.845
5.37	2.916	2.205	1.527	9.57	.847
5.38	2.919	2.209	1.529	9.60	.849
5.39	2.923	2.212	1.530	9.62	.852
5.40	2.927	2.216	1.531	9.65	.854
5.41	2.930	2.219	1.532	9.67	.857
5.42	2.934	2.223	1.533	9.69	.860
5.43	2.937	2.226	1.534	9.72	.862
5.44	2.941	2.229	1.535	9.74	.864
5.45	2.945	2.233	1.536	9.77	.867
5.46	2.948	2.236	1.537	9.79	.870
5.47	2.951	2.239	1.539	9.81	.872
5.48	2.955	2.243	1.540	9.84	.875
5.49	2.958	2.246	1.541	9.86	.877
5.50	2.962	2.250	1.542	9.88	.880

$\frac{\Delta P}{\mathfrak{A}}, \frac{\Delta Q}{\mathfrak{A}}$	M_S	$\frac{\lvert\Delta\mathfrak{U}\rvert}{\mathfrak{A}}$	$\frac{\mathfrak{A}'}{\mathfrak{A}}$	$\frac{p'}{p}$	ΔS
5.50	2.962	2.250	1.542	9.88	.880
5.51	2.966	2.253	1.543	9.91	.882
5.52	2.969	2.256	1.544	9.93	.884
5.53	2.972	2.259	1.545	9.95	.887
5.54	2.976	2.263	1.546	9.98	.890
5.55	2.979	2.266	1.547	10.00	.892
5.56	2.983	2.269	1.549	10.03	.895
5.57	2.987	2.273	1.550	10.05	.897
5.58	2.990	2.276	1.551	10.07	.900
5.59	2.993	2.279	1.552	10.10	.902
5.60	2.997	2.283	1.553	10.12	.905
5.61	3.000	2.286	1.554	10.15	.908
5.62	3.004	2.289	1.555	10.17	.910
5.63	3.007	2.292	1.556	10.19	.912
5.64	3.011	2.296	1.557	10.22	.915
5.65	3.014	2.299	1.558	10.24	.917
5.66	3.018	2.303	1.560	10.27	.920
5.67	3.021	2.306	1.561	10.29	.923
5.68	3.025	2.309	1.562	10.32	.925
5.69	3.028	2.312	1.563	10.34	.928
5.70	3.032	2.316	1.564	10.36	.930
5.71	3.035	2.319	1.565	10.39	.933
5.72	3.038	2.322	1.566	10.41	.935
5.73	3.042	2.326	1.567	10.43	.938
5.74	3.046	2.329	1.569	10.46	.940
5.75	3.049	2.333	1.570	10.48	.942
5.76	3.053	2.336	1.571	10.51	.945
5.77	3.056	2.339	1.572	10.53	.948
5.78	3.059	2.342	1.573	10.55	.950
5.79	3.063	2.346	1.574	10.58	.953
5.80	3.067	2.349	1.575	10.61	.956
5.81	3.070	2.352	1.576	10.63	.958
5.82	3.074	2.356	1.578	10.66	.960
5.83	3.077	2.359	1.579	10.68	.963
5.84	3.081	2.362	1.580	10.70	.965
5.85	3.084	2.365	1.581	10.73	.968
5.86	3.087	2.369	1.582	10.75	.970
5.87	3.091	2.372	1.583	10.78	.973
5.88	3.095	2.375	1.584	10.80	.976
5.89	3.098	2.379	1.585	10.83	.978
5.90	3.101	2.382	1.586	10.85	.980
5.91	3.105	2.385	1.587	10.87	.983
5.92	3.108	2.389	1.589	10.90	.986
5.93	3.112	2.392	1.590	10.92	.988
5.94	3.116	2.395	1.591	10.95	.991
5.95	3.119	2.399	1.592	10.97	.993
5.96	3.122	2.402	1.593	11.00	.995
5.97	3.126	2.405	1.594	11.02	.998
5.98	3.129	2.408	1.595	11.05	1.001
5.99	3.133	2.412	1.597	11.07	1.003
6.00	3.136	2.415	1.598	11.10	1.006

TABLE 1c (*Continued*). $\gamma = 4/3$

$\frac{\Delta P}{\mathcal{Q}}, \frac{\Delta Q}{\mathcal{Q}}$	M_s	$\frac{\lvert\Delta\mathcal{U}\rvert}{\mathcal{Q}}$	$\frac{\mathcal{Q}'}{\mathcal{Q}}$	$\frac{p'}{p}$	ΔS
6.00	3.136	2.415	1.598	11.10	1.006
6.01	3.139	2.418	1.599	11.12	1.008
6.02	3.143	2.421	1.600	11.15	1.011
6.03	3.147	2.425	1.601	11.17	1.014
6.04	3.150	2.428	1.602	11.20	1.016
6.05	3.154	2.431	1.603	11.22	1.019
6.06	3.157	2.434	1.604	11.25	1.021
6.07	3.160	2.438	1.605	11.27	1.023
6.08	3.164	2.441	1.607	11.30	1.026
6.09	3.168	2.444	1.608	11.32	1.029
6.10	3.171	2.447	1.609	11.35	1.031
6.11	3.174	2.450	1.610	11.37	1.033
6.12	3.178	2.454	1.611	11.40	1.036
6.13	3.181	2.457	1.612	11.42	1.039
6.14	3.185	2.461	1.613	11.45	1.041
6.15	3.188	2.464	1.614	11.47	1.044
6.16	3.192	2.467	1.616	11.50	1.047
6.17	3.195	2.470	1.617	11.52	1.049
6.18	3.199	2.474	1.618	11.55	1.051
6.19	3.202	2.477	1.619	11.57	1.054
6.20	3.205	2.480	1.620	11.60	1.056
6.21	3.209	2.483	1.621	11.62	1.059
6.22	3.212	2.487	1.622	11.65	1.062
6.23	3.216	2.490	1.623	11.67	1.064
6.24	3.219	2.493	1.625	11.70	1.066
6.25	3.223	2.496	1.626	11.73	1.069
6.26	3.226	2.500	1.627	11.75	1.072
6.27	3.229	2.503	1.628	11.78	1.074
6.28	3.233	2.506	1.629	11.80	1.077
6.29	3.237	2.509	1.630	11.83	1.079
6.30	3.240	2.512	1.631	11.85	1.081
6.31	3.243	2.516	1.632	11.88	1.084
6.32	3.247	2.519	1.634	11.91	1.087
6.33	3.251	2.522	1.635	11.93	1.089
6.34	3.254	2.525	1.636	11.96	1.092
6.35	3.257	2.529	1.637	11.98	1.094
6.36	3.261	2.532	1.638	12.01	1.097
6.37	3.264	2.535	1.639	12.03	1.099
6.38	3.267	2.538	1.640	12.06	1.102
6.39	3.271	2.542	1.641	12.08	1.104
6.40	3.275	2.545	1.643	12.11	1.107
6.41	3.278	2.548	1.644	12.13	1.109
6.42	3.281	2.551	1.645	12.16	1.112
6.43	3.285	2.555	1.646	12.19	1.114
6.44	3.288	2.558	1.647	12.22	1.117
6.45	3.291	2.561	1.648	12.24	1.119
6.46	3.295	2.564	1.649	12.27	1.122
6.47	3.299	2.568	1.651	12.29	1.125
6.48	3.302	2.571	1.652	12.32	1.127
6.49	3.305	2.574	1.653	12.34	1.129
6.50	3.309	2.577	1.654	12.37	1.132

$\frac{\Delta P}{\mathcal{Q}}, \frac{\Delta Q}{\mathcal{Q}}$	M_s	$\frac{\lvert\Delta\mathcal{U}\rvert}{\mathcal{Q}}$	$\frac{\mathcal{Q}'}{\mathcal{Q}}$	$\frac{p'}{p}$	ΔS
6.50	3.309	2.577	1.654	12.37	1.132
6.51	3.312	2.580	1.655	12.40	1.135
6.52	3.316	2.584	1.656	12.42	1.137
6.53	3.319	2.587	1.657	12.45	1.140
6.54	3.323	2.590	1.658	12.47	1.142
6.55	3.326	2.593	1.660	12.50	1.145
6.56	3.329	2.596	1.661	12.52	1.147
6.57	3.333	2.599	1.662	12.55	1.150
6.58	3.336	2.603	1.663	12.58	1.152
6.59	3.340	2.606	1.664	12.61	1.155
6.60	3.344	2.610	1.665	12.63	1.157
6.61	3.347	2.613	1.666	12.66	1.160
6.62	3.350	2.616	1.667	12.68	1.162
6.63	3.353	2.619	1.669	12.71	1.165
6.64	3.357	2.622	1.670	12.74	1.168
6.65	3.361	2.625	1.671	12.76	1.170
6.66	3.364	2.628	1.672	12.79	1.172
6.67	3.367	2.632	1.673	12.82	1.175
6.68	3.371	2.635	1.674	12.84	1.178
6.69	3.374	2.638	1.675	12.87	1.180
6.70	3.377	2.641	1.676	12.89	1.182
6.71	3.381	2.644	1.678	12.92	1.185
6.72	3.384	2.647	1.679	12.95	1.188
6.73	3.387	2.651	1.680	12.97	1.190
6.74	3.391	2.654	1.681	13.00	1.193
6.75	3.395	2.657	1.682	13.03	1.195
6.76	3.398	2.661	1.683	13.06	1.198
6.77	3.402	2.664	1.685	13.08	1.200
6.78	3.405	2.667	1.686	13.11	1.203
6.79	3.408	2.670	1.687	13.13	1.206
6.80	3.412	2.673	1.688	13.16	1.208
6.81	3.415	2.676	1.689	13.19	1.210
6.82	3.419	2.680	1.690	13.21	1.213
6.83	3.422	2.683	1.691	13.24	1.215
6.84	3.426	2.686	1.692	13.27	1.218
6.85	3.429	2.689	1.693	13.29	1.220
6.86	3.432	2.692	1.695	13.32	1.223
6.87	3.436	2.695	1.696	13.35	1.225
6.88	3.439	2.699	1.697	13.38	1.228
6.89	3.442	2.702	1.698	13.40	1.230
6.90	3.446	2.705	1.699	13.43	1.233
6.91	3.450	2.708	1.700	13.46	1.236
6.92	3.453	2.711	1.702	13.48	1.238
6.93	3.456	2.714	1.703	13.51	1.240
6.94	3.460	2.718	1.704	13.54	1.243
6.95	3.463	2.721	1.705	13.56	1.245
6.96	3.466	2.724	1.706	13.59	1.248
6.97	3.470	2.727	1.707	13.62	1.251
6.98	3.473	2.730	1.708	13.65	1.253
6.99	3.476	2.733	1.709	13.67	1.256
7.00	3.480	2.737	1.711	13.70	1.258

TABLE 1c (*Continued*). $\gamma = 4/3$

$\frac{\Delta P}{\mathfrak{a}}, \frac{\Delta Q}{\mathfrak{a}}$	M_s	$\frac{\lvert\Delta\mathfrak{u}\rvert}{\mathfrak{a}}$	$\frac{\mathfrak{a}'}{\mathfrak{a}}$	$\frac{p'}{p}$	ΔS	$\frac{\Delta P}{\mathfrak{a}}, \frac{\Delta Q}{\mathfrak{a}}$	M_s	$\frac{\lvert\Delta\mathfrak{u}\rvert}{\mathfrak{a}}$	$\frac{\mathfrak{a}'}{\mathfrak{a}}$	$\frac{p'}{p}$	ΔS
7.00	3.480	2.737	1.711	13.70	1.258	7.50	3.650	2.894	1.768	15.08	1.38
7.01	3.484	2.740	1.712	13.73	1.260	7.51	3.654	2.897	1.769	15.11	1.38
7.02	3.487	2.743	1.713	13.75	1.263	7.52	3.657	2.900	1.770	15.14	1.38
7.03	3.490	2.746	1.714	13.78	1.266	7.53	3.660	2.903	1.771	15.17	1.39
7.04	3.494	2.749	1.715	13.81	1.268	7.54	3.664	2.906	1.772	15.20	1.39
7.05	3.497	2.752	1.716	13.83	1.270	7.55	3.667	2.910	1.774	15.23	1.39
7.06	3.501	2.756	1.718	13.86	1.273	7.56	3.670	2.913	1.775	15.25	1.39
7.07	3.504	2.759	1.719	13.89	1.275	7.57	3.674	2.916	1.776	15.28	1.40
7.08	3.507	2.762	1.720	13.92	1.278	7.58	3.677	2.919	1.777	15.31	1.40
7.09	3.511	2.765	1.721	13.94	1.281	7.59	3.680	2.922	1.778	15.34	1.40
7.10	3.514	2.768	1.722	13.97	1.283	7.60	3.684	2.925	1.779	15.37	1.40
7.11	3.517	2.771	1.723	14.00	1.286	7.61	3.687	2.928	1.780	15.39	1.41
7.12	3.521	2.775	1.724	14.03	1.288	7.62	3.690	2.931	1.781	15.42	1.41
7.13	3.525	2.778	1.726	14.06	1.291	7.63	3.694	2.934	1.783	15.45	1.41
7.14	3.528	2.781	1.727	14.08	1.293	7.64	3.698	2.938	1.784	15.48	1.41
7.15	3.531	2.784	1.728	14.11	1.296	7.65	3.701	2.941	1.785	15.51	1.42
7.16	3.535	2.787	1.729	14.14	1.298	7.66	3.704	2.944	1.786	15.54	1.42
7.17	3.538	2.790	1.730	14.16	1.301	7.67	3.708	2.947	1.787	15.57	1.42
7.18	3.541	2.794	1.731	14.19	1.303	7.68	3.711	2.950	1.788	15.60	1.42
7.19	3.545	2.797	1.732	14.22	1.306	7.69	3.714	2.953	1.790	15.63	1.43
7.20	3.548	2.800	1.733	14.24	1.308	7.70	3.718	2.956	1.791	15.66	1.43
7.21	3.552	2.803	1.735	14.27	1.311	7.71	3.721	2.959	1.792	15.68	1.43
7.22	3.555	2.806	1.736	14.30	1.313	7.72	3.725	2.962	1.793	15.71	1.43
7.23	3.558	2.809	1.737	14.33	1.316	7.73	3.728	2.965	1.794	15.74	1.44
7.24	3.562	2.812	1.738	14.36	1.318	7.74	3.731	2.969	1.795	15.77	1.44
7.25	3.565	2.816	1.739	14.38	1.321	7.75	3.735	2.972	1.797	15.80	1.44
7.26	3.568	2.819	1.740	14.41	1.323	7.76	3.738	2.975	1.798	15.83	1.44
7.27	3.572	2.822	1.741	14.44	1.325	7.77	3.741	2.978	1.799	15.86	1.45
7.28	3.576	2.825	1.743	14.47	1.328	7.78	3.745	2.981	1.800	15.88	1.45
7.29	3.579	2.828	1.744	14.49	1.330	7.79	3.748	2.984	1.801	15.91	1.45
7.30	3.582	2.831	1.745	14.52	1.333	7.80	3.751	2.987	1.802	15.94	1.45
7.31	3.586	2.835	1.746	14.55	1.336	7.81	3.755	2.990	1.803	15.97	1.46
7.32	3.589	2.838	1.747	14.58	1.338	7.82	3.758	2.993	1.804	16.00	1.46
7.33	3.592	2.841	1.748	14.60	1.341	7.83	3.762	2.996	1.806	16.03	1.46
7.34	3.596	2.844	1.749	14.63	1.343	7.84	3.765	2.999	1.807	16.06	1.46
7.35	3.599	2.847	1.751	14.66	1.346	7.85	3.768	3.003	1.808	16.09	1.47
7.36	3.602	2.850	1.752	14.69	1.348	7.86	3.772	3.006	1.809	16.11	1.47
7.37	3.606	2.853	1.753	14.72	1.351	7.87	3.775	3.009	1.810	16.14	1.47
7.38	3.610	2.856	1.754	14.75	1.353	7.88	3.778	3.012	1.811	16.17	1.47
7.39	3.613	2.859	1.755	14.77	1.356	7.89	3.782	3.015	1.813	16.20	1.47
7.40	3.616	2.863	1.756	14.80	1.358	7.90	3.785	3.018	1.814	16.23	1.48
7.41	3.620	2.866	1.758	14.83	1.361	7.91	3.789	3.021	1.815	16.26	1.48
7.42	3.623	2.869	1.759	14.86	1.363	7.92	3.792	3.024	1.816	16.29	1.48
7.43	3.626	2.872	1.760	14.89	1.366	7.93	3.795	3.027	1.817	16.32	1.49
7.44	3.630	2.875	1.761	14.92	1.368	7.94	3.799	3.030	1.818	16.35	1.49
7.45	3.633	2.878	1.762	14.94	1.370	7.95	3.802	3.033	1.820	16.38	1.49
7.46	3.637	2.881	1.763	14.97	1.373	7.96	3.805	3.036	1.821	16.41	1.49
7.47	3.640	2.884	1.764	15.00	1.375	7.97	3.809	3.040	1.822	16.44	1.49
7.48	3.643	2.888	1.766	15.03	1.378	7.98	3.812	3.043	1.823	16.47	1.50
7.49	3.646	2.891	1.767	15.05	1.380	7.99	3.815	3.046	1.824	16.49	1.50
7.50	3.650	2.894	1.768	15.08	1.383	8.00	3.819	3.049	1.825	16.52	1.50

TABLE 1c (*Continued*). $\gamma = 4/3$

$\frac{\Delta P}{\mathcal{Q}}, \frac{\Delta Q}{\mathcal{Q}}$	M_s	$\frac{\lvert\Delta\mathfrak{U}\rvert}{\mathcal{Q}}$	$\frac{\mathcal{Q}'}{\mathcal{Q}}$	$\frac{p'}{p}$	ΔS
8.00	3.819	3.049	1.825	16.52	1.507
8.01	3.822	3.052	1.826	16.55	1.509
8.02	3.826	3.055	1.828	16.58	1.512
8.03	3.829	3.058	1.829	16.61	1.514
8.04	3.832	3.061	1.830	16.64	1.517
8.05	3.835	3.064	1.831	16.67	1.519
8.06	3.839	3.067	1.832	16.70	1.522
8.07	3.842	3.070	1.833	16.73	1.524
8.08	3.846	3.073	1.835	16.76	1.526
8.09	3.849	3.076	1.836	16.79	1.529
8.10	3.852	3.080	1.837	16.82	1.531
8.11	3.856	3.083	1.838	16.85	1.534
8.12	3.859	3.086	1.839	16.88	1.536
8.13	3.862	3.089	1.840	16.91	1.538
8.14	3.866	3.092	1.841	16.94	1.541
8.15	3.869	3.095	1.843	16.97	1.543
8.16	3.873	3.098	1.844	17.00	1.546
8.17	3.876	3.101	1.845	17.03	1.548
8.18	3.879	3.104	1.846	17.05	1.551
8.19	3.883	3.107	1.847	17.08	1.553
8.20	3.886	3.110	1.848	17.11	1.556
8.21	3.889	3.113	1.849	17.14	1.558
8.22	3.893	3.116	1.851	17.17	1.560
8.23	3.896	3.119	1.852	17.20	1.563
8.24	3.899	3.123	1.853	17.23	1.566
8.25	3.903	3.126	1.854	17.26	1.568
8.26	3.906	3.129	1.855	17.29	1.570
8.27	3.909	3.132	1.856	17.32	1.573
8.28	3.913	3.135	1.858	17.35	1.575
8.29	3.916	3.138	1.859	17.38	1.578
8.30	3.920	3.141	1.860	17.41	1.580
8.31	3.923	3.144	1.861	17.44	1.583
8.32	3.926	3.147	1.862	17.47	1.585
8.33	3.930	3.150	1.863	17.50	1.587
8.34	3.933	3.153	1.865	17.53	1.590
8.35	3.936	3.156	1.866	17.56	1.592
8.36	3.940	3.159	1.867	17.59	1.595
8.37	3.943	3.162	1.868	17.62	1.597
8.38	3.946	3.165	1.869	17.65	1.600
8.39	3.949	3.168	1.870	17.68	1.602
8.40	3.953	3.171	1.871	17.72	1.604
8.41	3.957	3.174	1.873	17.75	1.607
8.42	3.960	3.177	1.874	17.78	1.609
8.43	3.963	3.181	1.875	17.81	1.612
8.44	3.966	3.184	1.876	17.84	1.614
8.45	3.970	3.187	1.877	17.87	1.617
8.46	3.973	3.190	1.878	17.90	1.619
8.47	3.977	3.193	1.880	17.93	1.622
8.48	3.980	3.196	1.881	17.96	1.624
8.49	3.983	3.199	1.882	17.99	1.626
8.50	3.986	3.202	1.883	18.02	1.629

$\frac{\Delta P}{\mathcal{Q}}, \frac{\Delta Q}{\mathcal{Q}}$	M_s	$\frac{\lvert\Delta\mathfrak{U}\rvert}{\mathcal{Q}}$	$\frac{\mathcal{Q}'}{\mathcal{Q}}$	$\frac{p'}{p}$	ΔS
8.50	3.986	3.202	1.883	18.02	1.629
8.51	3.990	3.205	1.884	18.05	1.631
8.52	3.993	3.208	1.885	18.08	1.634
8.53	3.996	3.211	1.886	18.11	1.636
8.54	4.000	3.214	1.888	18.14	1.638
8.55	4.003	3.217	1.889	18.17	1.641
8.56	4.006	3.220	1.890	18.20	1.643
8.57	4.010	3.223	1.891	18.23	1.646
8.58	4.013	3.226	1.892	18.26	1.648
8.59	4.016	3.229	1.893	18.29	1.650
8.60	4.020	3.232	1.895	18.32	1.653
8.61	4.023	3.236	1.896	18.36	1.655
8.62	4.027	3.239	1.897	18.39	1.658
8.63	4.030	3.242	1.898	18.42	1.660
8.64	4.033	3.245	1.899	18.45	1.663
8.65	4.037	3.248	1.901	18.48	1.665
8.66	4.040	3.251	1.902	18.51	1.667
8.67	4.043	3.254	1.903	18.54	1.670
8.68	4.047	3.257	1.904	18.57	1.672
8.69	4.050	3.260	1.905	18.60	1.674
8.70	4.053	3.263	1.906	18.63	1.677
8.71	4.056	3.266	1.907	18.66	1.679
8.72	4.060	3.269	1.909	18.69	1.682
8.73	4.063	3.272	1.910	18.73	1.684
8.74	4.067	3.275	1.911	18.76	1.687
8.75	4.070	3.278	1.912	18.79	1.689
8.76	4.073	3.281	1.913	18.82	1.691
8.77	4.076	3.284	1.914	18.85	1.694
8.78	4.080	3.287	1.916	18.88	1.696
8.79	4.083	3.290	1.917	18.91	1.699
8.80	4.087	3.293	1.918	18.94	1.701
8.81	4.090	3.296	1.919	18.97	1.703
8.82	4.093	3.299	1.920	19.01	1.706
8.83	4.097	3.302	1.921	19.04	1.709
8.84	4.100	3.305	1.923	19.07	1.711
8.85	4.103	3.308	1.924	19.10	1.713
8.86	4.106	3.311	1.925	19.13	1.716
8.87	4.110	3.314	1.926	19.16	1.718
8.88	4.113	3.317	1.927	19.19	1.720
8.89	4.117	3.320	1.928	19.22	1.723
8.90	4.120	3.323	1.930	19.26	1.725
8.91	4.123	3.326	1.931	19.29	1.727
8.92	4.127	3.329	1.932	19.32	1.730
8.93	4.130	3.332	1.933	19.35	1.732
8.94	4.133	3.335	1.934	19.38	1.735
8.95	4.137	3.338	1.935	19.41	1.737
8.96	4.140	3.341	1.936	19.44	1.739
8.97	4.143	3.344	1.938	19.48	1.742
8.98	4.146	3.347	1.939	19.51	1.744
8.99	4.150	3.350	1.940	19.54	1.747
9.00	4.153	3.353	1.941	19.57	1.749

TABLE 1c (*Continued*). $\gamma = 4/3$

$\frac{\Delta P}{\mathfrak{a}}, \frac{\Delta Q}{\mathfrak{a}}$	M_S	$\frac{\lvert\Delta\mathfrak{U}\rvert}{\mathfrak{a}}$	$\frac{\mathfrak{a}'}{\mathfrak{a}}$	$\frac{p'}{p}$	ΔS
9.00	4.153	3.353	1.941	19.57	1.749
9.01	4.156	3.356	1.942	19.60	1.751
9.02	4.160	3.359	1.944	19.63	1.754
9.03	4.163	3.362	1.945	19.66	1.756
9.04	4.166	3.366	1.946	19.70	1.759
9.05	4.170	3.369	1.947	19.73	1.761
9.06	4.173	3.372	1.948	19.76	1.763
9.07	4.176	3.375	1.949	19.79	1.766
9.08	4.179	3.378	1.950	19.82	1.768
9.09	4.183	3.381	1.952	19.85	1.771
9.10	4.186	3.384	1.953	19.88	1.773
9.11	4.190	3.387	1.954	19.92	1.775
9.12	4.193	3.390	1.955	19.95	1.777
9.13	4.196	3.393	1.956	19.98	1.780
9.14	4.199	3.396	1.957	20.01	1.782
9.15	4.203	3.399	1.959	20.05	1.785
9.16	4.206	3.402	1.960	20.08	1.787
9.17	4.210	3.405	1.961	20.11	1.790
9.18	4.213	3.408	1.962	20.14	1.792
9.19	4.216	3.411	1.963	20.17	1.794
9.20	4.219	3.414	1.964	20.20	1.797
9.21	4.223	3.417	1.966	20.24	1.799
9.22	4.226	3.420	1.967	20.27	1.801
9.23	4.229	3.423	1.968	20.30	1.803
9.24	4.233	3.426	1.969	20.33	1.806
9.25	4.236	3.429	1.970	20.36	1.808
9.26	4.240	3.432	1.971	20.40	1.811
9.27	4.243	3.435	1.973	20.43	1.813
9.28	4.246	3.438	1.974	20.46	1.816
9.29	4.249	3.441	1.975	20.49	1.818
9.30	4.253	3.444	1.976	20.53	1.820
9.31	4.256	3.447	1.977	20.56	1.823
9.32	4.259	3.450	1.978	20.59	1.825
9.33	4.262	3.453	1.980	20.62	1.827
9.34	4.266	3.456	1.981	20.66	1.830
9.35	4.269	3.459	1.982	20.69	1.832
9.36	4.272	3.462	1.983	20.72	1.834
9.37	4.276	3.465	1.984	20.75	1.837
9.38	4.279	3.468	1.985	20.78	1.839
9.39	4.283	3.471	1.987	20.82	1.842
9.40	4.286	3.474	1.988	20.85	1.844
9.41	4.289	3.477	1.989	20.88	1.846
9.42	4.292	3.480	1.990	20.91	1.849
9.43	4.296	3.483	1.991	20.95	1.851
9.44	4.299	3.486	1.993	20.98	1.854
9.45	4.302	3.489	1.994	21.01	1.856
9.46	4.305	3.492	1.995	21.04	1.858
9.47	4.309	3.495	1.996	21.08	1.860
9.48	4.312	3.497	1.997	21.11	1.863
9.49	4.316	3.500	1.998	21.14	1.865
9.50	4.319	3.503	2.000	21.18	1.868

$\frac{\Delta P}{\mathfrak{a}}, \frac{\Delta Q}{\mathfrak{a}}$	M_S	$\frac{\lvert\Delta\mathfrak{U}\rvert}{\mathfrak{a}}$	$\frac{\mathfrak{a}'}{\mathfrak{a}}$	$\frac{p'}{p}$	ΔS
9.50	4.319	3.503	2.000	21.18	1.86
9.51	4.322	3.506	2.001	21.21	1.87
9.52	4.326	3.509	2.002	21.24	1.87
9.53	4.329	3.512	2.003	21.27	1.87
9.54	4.332	3.515	2.004	21.31	1.87
9.55	4.335	3.518	2.005	21.34	1.87
9.56	4.339	3.521	2.007	21.37	1.88
9.57	4.342	3.524	2.008	21.40	1.88
9.58	4.345	3.527	2.009	21.44	1.88
9.59	4.348	3.530	2.010	21.47	1.88
9.60	4.352	3.533	2.011	21.50	1.89
9.61	4.356	3.536	2.012	21.54	1.89
9.62	4.359	3.539	2.014	21.57	1.89
9.63	4.362	3.542	2.015	21.60	1.89
9.64	4.365	3.545	2.016	21.64	1.90
9.65	4.368	3.548	2.017	21.67	1.90
9.66	4.372	3.551	2.018	21.70	1.90
9.67	4.375	3.554	2.019	21.73	1.90
9.68	4.379	3.557	2.021	21.77	1.91
9.69	4.382	3.560	2.022	21.80	1.91
9.70	4.385	3.563	2.023	21.83	1.91
9.71	4.388	3.566	2.024	21.87	1.91
9.72	4.391	3.569	2.025	21.90	1.91
9.73	4.395	3.572	2.026	21.93	1.92
9.74	4.398	3.575	2.027	21.96	1.92
9.75	4.402	3.578	2.029	22.00	1.92
9.76	4.405	3.581	2.030	22.03	1.92
9.77	4.408	3.584	2.031	22.07	1.93
9.78	4.411	3.587	2.032	22.10	1.93
9.79	4.415	3.590	2.033	22.13	1.93
9.80	4.418	3.593	2.035	22.17	1.93
9.81	4.421	3.596	2.036	22.20	1.94
9.82	4.425	3.599	2.037	22.23	1.94
9.83	4.428	3.602	2.038	22.27	1.94
9.84	4.431	3.605	2.039	22.30	1.94
9.85	4.434	3.608	2.040	22.33	1.94
9.86	4.438	3.610	2.042	22.36	1.95
9.87	4.441	3.613	2.043	22.40	1.95
9.88	4.444	3.616	2.044	22.43	1.95
9.89	4.448	3.619	2.045	22.47	1.95
9.90	4.451	3.622	2.046	22.50	1.96
9.91	4.454	3.625	2.048	22.53	1.96
9.92	4.457	3.628	2.049	22.56	1.96
9.93	4.461	3.631	2.050	22.60	1.96
9.94	4.464	3.634	2.051	22.63	1.97
9.95	4.467	3.637	2.052	22.67	1.97
9.96	4.470	3.640	2.053	22.70	1.97
9.97	4.474	3.643	2.054	22.73	1.97
9.98	4.477	3.646	2.056	22.76	1.97
9.99	4.481	3.649	2.057	22.80	1.98
10.00	4.484	3.652	2.058	22.83	1.98

APPENDIX I: STEADY-FLOW RELATIONS

A number of steady-flow relations that may be of interest for wave diagram procedures are collected here for convenient reference. Relations for a stationary shock wave are given in Chapter VI.e.

The *continuity equation* can be expressed by the constancy of any of the following terms

$$\text{Mass flow} = \rho A u = \gamma p A M / a = \left(\frac{\gamma}{RT}\right)^{1/2} p A M$$

$$= \left(\frac{\gamma}{RT_s}\right)^{1/2} p A M \left(1 + \frac{\gamma - 1}{2} M^2\right)^{1/2}$$

$$= \left(\frac{\gamma}{RT_s}\right)^{1/2} p_s A D = \text{const.}$$

where

$$D = \frac{M}{\left(1 + \frac{\gamma - 1}{2} M^2\right)^{\frac{\gamma+1}{2(\gamma-1)}}}$$

In isentropic flow, a simple form of the continuity equation becomes

$$AD = A^* D_{M=1.0} = \text{const.}$$

so that

$$\frac{A}{A^*} = \frac{D_{M=1.0}}{D}$$

The *momentum equation* for a duct of constant cross section has the form

$$p_1 - p_2 = \rho_1 u_1 (u_2 - u_1)$$

and, as a conservation relation, this can also be expressed by

$$p + \rho u^2 = p(1 + \gamma M^2) = \text{const.}$$

The *energy equation* for an adiabatic flow may be required in one of the following forms

$$gJc_pT + \frac{u^2}{2} = gJc_pT_s = \text{const.}$$

$$\frac{2}{\gamma - 1} a^2 + u^2 = a^2 \left(\frac{2}{\gamma - 1} + M^2 \right) = \frac{2}{\gamma - 1} a_s^2$$

$$= \frac{\gamma + 1}{\gamma - 1} a^{*2} = \frac{2\gamma}{\gamma - 1} \frac{p}{\rho} + u^2 = \text{const.}$$

Frequently, the relations between static and stagnation conditions are required; these are expressed by

$$\frac{\mathfrak{a}_s}{\mathfrak{a}} = \left(1 + \frac{\gamma - 1}{2} M^2 \right)^{1/2}$$

$$\frac{T_s}{T} = \left(\frac{\mathfrak{a}_s}{\mathfrak{a}} \right)^2$$

$$\frac{\rho_s}{\rho} = \left(\frac{\mathfrak{a}_s}{\mathfrak{a}} \right)^{\frac{2}{\gamma - 1}}$$

$$\frac{p_s}{p} = \left(\frac{\mathfrak{a}_s}{\mathfrak{a}} \right)^{\frac{2\gamma}{\gamma - 1}}$$

APPENDIX II

A. Improved Boundary Conditions for Open-Ended Duct Configurations

1. *Wave Reflection from an Open End.* As pointed out in Chapters VI.d and VII.e, it is usually satisfactory to assume that the boundary conditions for wave reflection from an open end are the same as if the flow were steady. Occasionally, it may become necessary to account for the lag in establishing the steady-flow boundary conditions to obtain more accurate results or to explain certain experimental observations. Improved boundary conditions for shock reflection from an open end of a duct are given in Chapter VII.e in the form $I(\tau) = (p_e - p_1)/(p_2 - p_1)$ as a function of a dimensionless time $\tau = a_0 t/d$ measured from the arrival of the wave at the exit. For weak waves, this pressure ratio may be replaced by the corresponding speed-of-sound ratio (see page 151) so that the instantaneous boundary conditions are given by

$$\mathfrak{a}(\tau) = \mathfrak{a}_1 + \Delta\mathfrak{a} I(\tau) \tag{A.1}$$

where $\Delta\mathfrak{a}$ is the change of the speed of sound across the shock. The function $I(\tau)$ is plotted in Fig. VIIe.5, and numerical values are given in the table below.

A subsequent study[103] extended these results to arbitrary incident waves based on the idea that the function $I(\tau)$ represents the "response" of the exit pressure to an incident unit pressure step; the response to an arbitrary incident wave can then be derived with the help of Duhamel's integral (see, for instance, reference 98). To compute the instantaneous boundary conditions, the incident wave is divided into small elements by selected waves P_0, P_1, P_2, etc., as shown in Fig. A.1, and

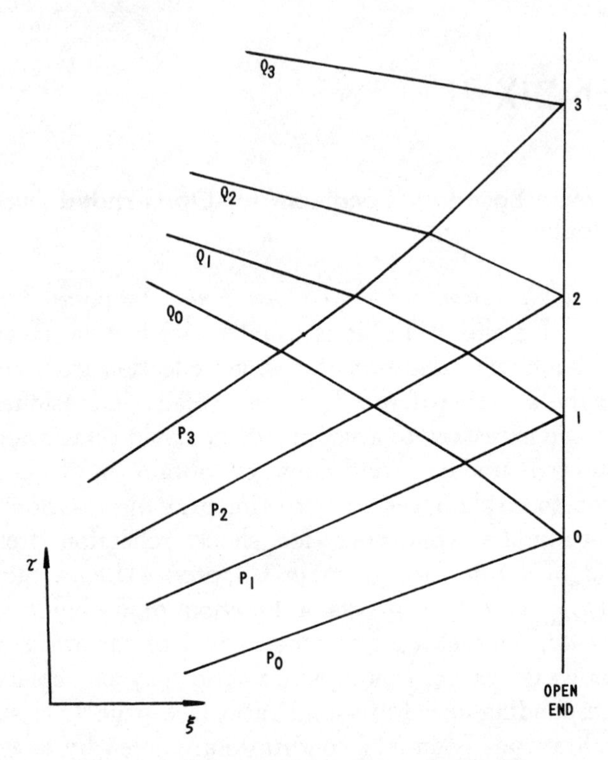

FIG. A.1. Wave reflection from an open end

characterized by increments of the speed of sound

$$\Delta_{-}\mathcal{A}_n = \tfrac{1}{4}(\gamma - 1)(P_n - P_{n-1}) \qquad \text{(A.2)}$$

The incident waves, after interacting with reflections of earlier waves, reach the end of the duct at the times τ_0, τ_1, τ_2, etc., which are obtained as construction of the wave diagram proceeds

(Fig. A.1). Each wave element produces a disturbance of the exit pressure which decays gradually but is superposed on similar disturbances from earlier wave elements. For details of this analysis, reference should be made to the original publication [103], where the boundary condition for any point is derived as

$$\mathcal{A}_n = \mathcal{A}_0 + \sum_{j=1}^{n} \frac{\Delta_- \mathcal{A}_j}{\tau_j - \tau_{j-1}} [\phi(\tau_n - \tau_{j-1}) - \phi(\tau_n - \tau_j)] \quad \text{(A.3)}$$

with the function ϕ being defined as

$$\phi(\tau) = \int_0^\tau I(\theta) d\theta$$

Numerical values of this function are listed in the following table.

THE FUNCTIONS $I(\tau)$ AND $\phi(\tau)$

τ	$I(\tau)$	$\phi(\tau)$	τ	$I(\tau)$	$\phi(\tau)$
0	1.000	0	1.6	0.126	0.774
0.1	0.938	0.097	1.8	0.094	0.795
0.2	0.870	0.187	2.0	0.070	0.811
0.3	0.798	0.271	2.2	0.052	0.823
0.4	0.721	0.346	2.4	0.039	0.832
0.5	0.642	0.414	2.6	0.029	0.839
0.6	0.570	0.475	2.8	0.021	0.844
0.8	0.431	0.574	3.0	0.014	0.847
1.0	0.318	0.648	3.5	0.008	0.853
1.2	0.231	0.705	4.0	0.004	0.856
1.4	0.169	0.745	∞	0	0.858

The terms of the sum in Eq. (A.3) decay quite rapidly, so that only three or four terms need to be computed at each point. The values of $\mathcal{U}_n$ and Q_n then are obtained from $\mathcal{A}_n$ and P_n, so that wave-diagram construction can proceed. The foregoing procedure applies to isentropic outflow from the duct. For extension to isentropic and nonisentropic inflow, as well as experimental verification, reference is made to the original publication.

One interesting consequence of the lag in establishing the steady-flow boundary conditions is that certain discontinuities in the incident wave may not appear in the reflected wave. For example, the derivative $\partial p/\partial \tau$ may be discontinuous at the head of the incident wave (see page 30), but a corresponding discontinuity appears at the head of the reflected wave only if the steady-flow boundary conditions are used and not if the lag phenomena are taken into account.[103]

Another consequence of the lag is that the amount of gas discharged from a duct during the transition from one steady flow to another differs from that computed on the basis of the steady-flow boundary conditions.[117] This error may be quite large (more than 20%), and varies only slowly over a considerable range of the transition time. The reason for this behavior is that the cumulative effects of the instantaneous errors introduced in the computation of a rapid transition are about the same as those of the smaller errors made for a slower transition that persists over a longer time.

2. *Shock Reflection from an Orifice Plate.* If a shock wave impinges on an orifice plate mounted at the end of a duct, the reflected wave represents the superposition of a shock wave that is reflected from the plate and an expansion wave from the orifice itself. If the assumption is made that the steady-flow boundary conditions are instantaneously established, these two waves are immediately combined to a reflected wave of such strength that the finally established flow is compatible with a steady discharge through the orifice. The computed reflected wave is then either a shock wave or a centered expansion wave depending on the strength of the incident shock and the size of the orifice, as discussed in Chapter VII.k for a short nozzle.

This result does not always agree with experimental observations [106] which, under some conditions, indicate a reflected wave that consists of a leading shock front *followed* by an expansion wave regardless of whether the final pressure is higher or lower than that behind the incident shock. The reflected wave thus may exhibit an "overshoot" as qualitatively indicated in Fig. A.2 where the broken lines show the wave form that would be obtained on the basis of the steady-flow boundary conditions.

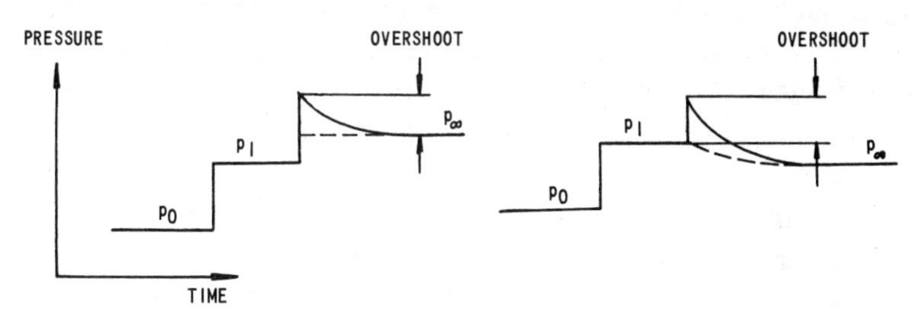

FIG. A.2. Shock reflection from an orifice. Pressure variations at some distance from the orifice. The pressures ahead and behind the incident shock are p_0 and p_1; the pressure finally established behind the reflected wave is p_∞. Broken lines indicate the reflected wave for instantaneous establishment of the steady-flow boundary conditions

This overshoot may reach an appreciable magnitude. Its analysis is outlined here, but reference to the original publication[106] is made for a more detailed discussion of the problem.

When the incident shock reaches the orifice, formation of the wave that is reflected from the low-pressure external region is delayed. The flow therefore behaves at first as if it had entered a sudden contraction of the duct. A reflected shock wave is thus formed that may be computed as outlined in Chapter VII.h and is considered as the initial front of the reflected wave. The pressure behind it is denoted by p_i, while the pressure in the orifice is p_e (corresponding to the pressure in regions 5 and 6 of Fig. VII.h.1).

The final pressure behind the entire reflected wave, p_∞, may be computed as indicated in Chapter VII.k; the corresponding pressure in the orifice, $p_{e,\infty}$, is equal to the initial pressure in the duct, p_0, as long as the discharge is subsonic but may be greater if the flow becomes sonic.

In the final step of the analysis, it is assumed that the transition of the orifice pressure from p_e to $p_{e,\infty}$ is given by the same function $I(\tau)$ that describes the pressure variations in an open end following shock reflection (see Fig. VII.e.5 and the table in the preceding chapter), where the dimensionless time must now be based on the orifice diameter as reference length. Since a jet discharge from an orifice has a smaller diameter than the orifice, the foregoing calculations should be based on an effective, rather than the actual, orifice diameter.

The flow conditions in the orifice and at the end of the duct are related at any instant through the steady-flow equations in the usual manner. Thus one obtains a reflected shock that is gradually being overtaken by an expansion wave, so that the overshoot decreases with increasing distance from the orifice. The rate of this decrease is quite slow because the velocity of the shock differs only little from that of the overtaking characteristics. In the immediate vicinity of the orifice, the results cannot be expected to hold because of the three-dimensional nature of the flow, but good agreement with experimental observations was found less than two duct diameters away.[106]

The initial magniture of the overshoot, $p_i - p_1$, depends on the orifice size and the pressure ratio of the incident shock wave, p_1/p_0. A detailed analysis shows that the overshoot is most pronounced for a "nonreflecting orifice" (see the example in Chapter VII.k) for which $p_\infty = p_1$. Both the absolute and relative magnitudes of the initial overshoot are plotted in Fig. A.3, which shows that the effect is by no means trivial. The ratio of the effective orifice area to that of the duct for a nonreflecting orifice, A_e/A, is also indicated in the figure. The absolute magnitude of the overshoot, $p_i - p_1$, reaches a maximum of about 0.4 p_0 for a shock pressure ratio of about 2.3; this value corresponds to about one-third of the pressure rise across the

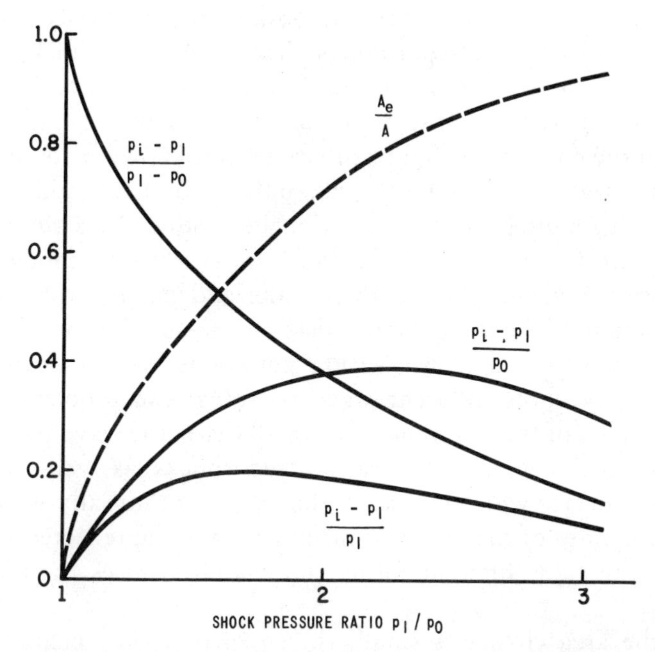

FIG. A.3. Shock reflection from an orifice. Absolute and relative magnitudes of the initial overshoot for a nonreflecting orifice ($\gamma = 1.4$). The broken line represents the ratio of the effective orifice area to the duct area

incident shock, $p_1 - p_0$. For weaker shocks the absolute magnitude of the overshoot decreases, but its magnitude in terms of the pressure rise across the incident shock, $(p_i - p_1)/(p_1 - p_0)$, increases until, for an infinitesimal shock strength, the overshoot becomes equal to the pressure rise across the shock. In terms of the pressure behind the shock front, p_1, the overshoot reaches a maximum of about 0.2 p_1, at shock pressure ratios in the neighborhood of 1.6. For shock pressure ratios that are much larger than 3, both the absolute and relative magnitudes of the overshoot become insignificant.

B. Passage of Strong Shock Waves Through a Discontinuous Contraction

It is frequently convenient to approximate a gradually changing duct area by a discontinuous change. When a shock wave passes through such a discontinuity, the resulting wave pattern may take various forms, as discussed in Chapter VII.h. Although the correct wave pattern often is not known beforehand, it can always be found by trial, but, to be acceptable, this approach requires that the solution be unique. It is therefore important to realize that for shock waves strong enough to produce supersonic flow in their wake and passing through a contraction of the duct, there exists a range of the contraction ratio for which all conservation equations can be formally satisfied by three different wave patterns; the problem thus arises which of these is to be used. In all cases, the wave patterns include the incident and transmitted shocks as well as the contact surface which separates the gases that initially were on different sides of the area discontinuity. It is the reflected wave which can take different forms and thereby causes the uncertainty.

If the area change is small, the supersonic flow behind the incident shock can pass through the contraction without becoming sonic. The reflected wave is then an upstream-facing expansion wave that is swept downstream as illustrated in Fig. A.4(a). With increasing contraction, this solution is possible until the flow in the narrower duct becomes just sonic. With further contraction, the flow can no longer pass into the narrower duct, and an upstream-travelling shock wave is formed of such strength that the flow enters the narrower duct at sonic velocity; the flow is then further accelerated by an upstream-facing expansion wave that is swept downstream, as shown in Fig. A.4(c).

If one starts with the second type of solution, the reflected shock becomes weaker with decreasing contraction until it can no longer advance upstream into the supersonic flow behind the incident shock and becomes stationary at the inlet of the convergence. Any further decrease of the contraction ratio then yields the solution without a reflected shock.

It was noted[109] that the limits for the described two types of wave pattern do not coincide, but that there exists a range of ambiguity where both solutions satisfy all necessary conditions. In fact, even a third solution was found in which a shock wave remains stationary in the convergent section. In this pattern, shown in Fig. A.4(b), the flow enters the narrower duct also at sonic velocity and is further accelerated by a downstream-swept expansion wave.

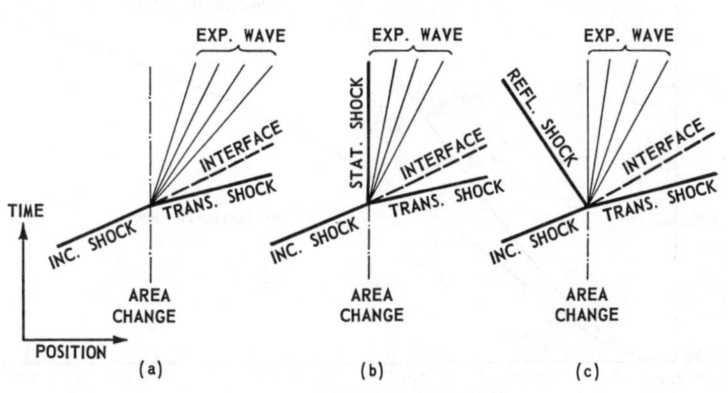

FIG. A.4. Possible wave patterns for shocks passing through a discontinuous contraction in the range of ambiguity

The relationship between shock Mach number, contraction ratio and wave pattern for a diatomic gas is shown in Fig. A.5. If a shock wave is to produce supersonic flow in its wake, its Mach number must exceed about 2.07, for which value the flow become sonic (see Table 1a). Above the lower curve, the solution with a reflected shock, and below the upper curve, the solution without a reflected shock satisfy all necessary conditions. The range of ambiguity is thus represented by the region between these two curves where both solutions are possible.

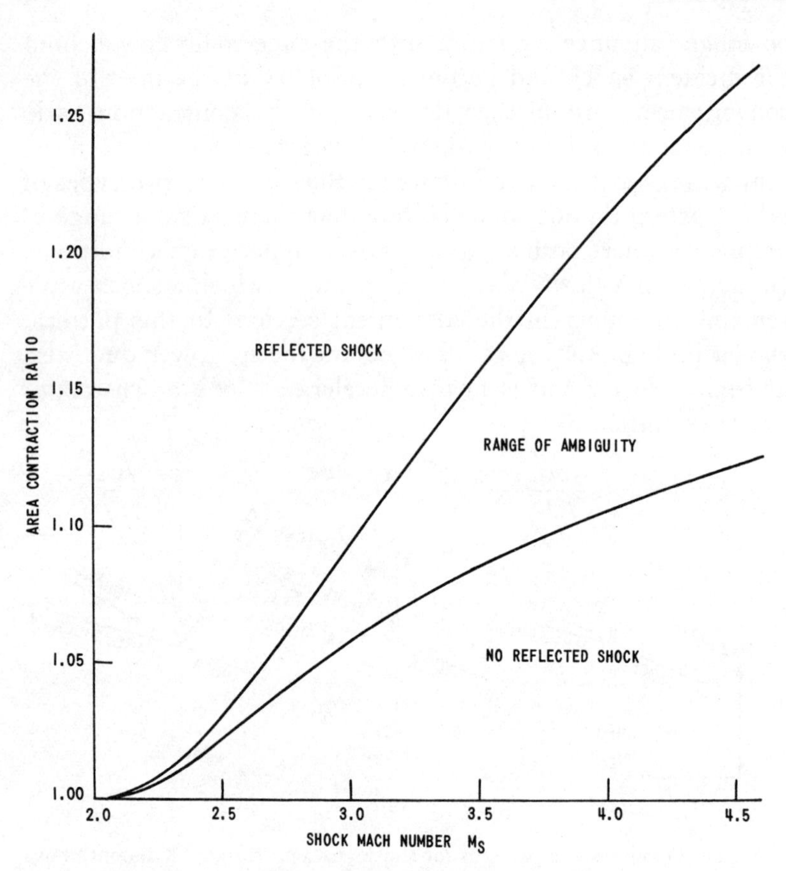

FIG. A.5. Range of ambiguity for shock passage through a discontinuous contraction ($\gamma = 1.4$)

The solution with a stationary shock wave can be ruled out because stationary shocks are unstable in a convergent duct.[97] A detailed investigation[115] to choose between the remaining two possibilities showed that a complete wave diagram for any monotonically convergent duct would always lead to the solution without the reflected shock wave. (If the convergence is not monotonic, the wave pattern is not defined by the over-all contraction ratio alone, and a complete wave diagram must be prepared.) Unfortunately, this discussion does not provide the complete answer to the problem. In the region of ambiguity the wave pattern can be changed back and forth between the two types by sufficiently large disturbances, and it could be shown[116] that the pattern with a reflected shock wave is more stable than the one without a reflected shock. Such disturbances could originate during the initial shock passage as a result of boundary-layer growth. It does not seem possible therefore to state unequivocally which of the two possible wave patterns would actually be established in a particular case, even if a complete wave diagram were prepared, because the assumption of one-dimensional flow may not be adequate for such processes. Flows that are close to the limit of the range of ambiguity are quite sensitive even to small disturbances and may therefore easily develop violent oscillations.

In the foregoing discussion the gas always was at rest prior to the arrival of the incident shock wave, but similar ambiguities arise if there is an initial flow. The shock strength required to produce a supersonic flow into the contraction then depends also on the velocity and direction of the initial flow, and the resulting limits of the region of ambiguity are different from those for an initially quiescent gas. Apparently, this problem has not yet been studied further.

C. Use of Digital Computers

The preparation of wave diagrams clearly involves a great deal of computational labor, and, with the ever wider availability of digital computers, the possibility of using such equipment should be kept in mind. Difficulties arising in the use of digital computers are associated mainly with the variety of procedures that may have to be performed, the appearance of discontinuities, such as shock waves, at points of the wave diagram that cannot be predicted beforehand, and the irregularity of the grid formed by the characteristics.

Any one of the computing procedures described in the preceding chapters could be programmed for a particular computer without difficulty, but if many of them should be needed in a wave diagram a complicated logic would be required to decide at every step which procedure is to be used. This decision is almost trivial if the wave diagram is constructed at the same time as manual computations proceed. If the qualitative appearance of a wave diagram is known beforehand, then it may be possible to program a systematic progression of the computations. The coordinates of each point of the characteristic net must be computed from the slope of the characteristics and the coordinates of preceding points. Some iteration will often be needed but should not introduce significant difficulties. The results of the computations are then printed in tabular form with columns added for the time and position coordinates of each point (see the examples of Chapter IX). This method is practical only if it is known that no discontinuities will appear or if their location in the wave diagram can be specified beforehand. For example, sudden opening or closing of the duct at a specified time would produce a centered expansion wave or a shock wave at well defined points of the wave diagram.

Although a wave diagram gives a clear insight into the wave processes, the irregular grid of the intersection points of the characteristics does not lend itself too well for machine computations except in special cases, such as those outlined in the foregoing discussion. Usually, it seems preferable to choose a predetermined rectangular coordinate grid and to compute the flow conditions at those points. Since the characteristics, in general, do not pass through these points, the appropriate P- and Q- waves and particle paths must be found by suitable forward or backward interpolation.[107, 112, 114, 119]

This procedure does not yet permit the handling of shock waves. It is possible to test for coalescence of characteristics and to include shock points at additional locations in the grid, but shocks usually are handled in a different manner. Shocks may be considered as discontinuities only because the transition distance, being of the order of a few mean free paths of molecular collisions, can normally be neglected. By introducing artificial dissipative terms into the basic differential equations of the flow, corresponding to viscosity, heat transfer and diffusion, shock transition can be stretched out over several grid points. Several such techniques, each having its own advantages and disadvantages, have been devised.[114,119] The merit of these procedures is that the same computations are performed at all grid points and that shocks are automatically accounted for regardless of where they may appear in the wave diagram. The mesh size must be chosen small enough to provide adequate resolution, but the spacing of the grid points need not be uniform. Results of the computations are again obtained in tabular form. The flow conditions at selected positions or times are thus directly obtained, but the physical significance of the characteristics as lines of propagating disturbances is not directly evident from such tabulated results. To plot a wave diagram from these data would require further graphical or numerical interpolations.

In all these techniques, integration proceeds essentially along the characteristic directions. It is possible to integrate the original partial differential equations directly by setting up some finite-difference scheme without recourse to the characteristics, but this approach seems useful only if no discontinuities appear in the solution of a problem.[119]

The foregoing discussion briefly indicates how problems of nonsteady flow may be solved with the aid of digital computers. More detailed discussions of the techniques may be found in the literature (for instance, references 107, 112, 114, and 119), but the search for improved procedures is still going on.

INDEX

SOME DOVER SCIENCE BOOKS

SOME DOVER SCIENCE BOOKS

WHAT IS SCIENCE?,
Norman Campbell
This excellent introduction explains scientific method, role of mathematics, types of scientific laws. Contents: 2 aspects of science, science & nature, laws of science, discovery of laws, explanation of laws, measurement & numerical laws, applications of science. 192pp. 5⅜ x 8. Paperbound $1.25

FADS AND FALLACIES IN THE NAME OF SCIENCE,
Martin Gardner
Examines various cults, quack systems, frauds, delusions which at various times have masqueraded as science. Accounts of hollow-earth fanatics like Symmes; Velikovsky and wandering planets; Hoerbiger; Bellamy and the theory of multiple moons; Charles Fort; dowsing, pseudoscientific methods for finding water, ores, oil. Sections on naturopathy, iridiagnosis, zone therapy, food fads, etc. Analytical accounts of Wilhelm Reich and orgone sex energy; L. Ron Hubbard and Dianetics; A. Korzybski and General Semantics; many others. Brought up to date to include Bridey Murphy, others. Not just a collection of anecdotes, but a fair, reasoned appraisal of eccentric theory. Formerly titled *In the Name of Science*. Preface. Index. x + 384pp. 5⅜ x 8.
Paperbound $1.85

PHYSICS, THE PIONEER SCIENCE,
L. W. Taylor
First thorough text to place all important physical phenomena in cultural-historical framework; remains best work of its kind. Exposition of physical laws, theories developed chronologically, with great historical, illustrative experiments diagrammed, described, worked out mathematically. Excellent physics text for self-study as well as class work. Vol. 1: Heat, Sound: motion, acceleration, gravitation, conservation of energy, heat engines, rotation, heat, mechanical energy, etc. 211 illus. 407pp. 5⅜ x 8. Vol. 2: Light, Electricity: images, lenses, prisms, magnetism, Ohm's law, dynamos, telegraph, quantum theory, decline of mechanical view of nature, etc. Bibliography. 13 table appendix. Index. 551 illus. 2 color plates. 508pp. 5⅜ x 8.
Vol. 1 Paperbound $2.25, Vol. 2 Paperbound $2.25,
The set $4.50

THE EVOLUTION OF SCIENTIFIC THOUGHT FROM NEWTON TO EINSTEIN,
A. d'Abro
Einstein's special and general theories of relativity, with their historical implications, are analyzed in non-technical terms. Excellent accounts of the contributions of Newton, Riemann, Weyl, Planck, Eddington, Maxwell, Lorentz and others are treated in terms of space and time, equations of electromagnetics, finiteness of the universe, methodology of science. 21 diagrams. 482pp. 5⅜ x 8.
Paperbound $2.50

CHANCE, LUCK AND STATISTICS: THE SCIENCE OF CHANCE,
Horace C. Levinson

Theory of probability and science of statistics in simple, non-technical language. Part I deals with theory of probability, covering odd superstitions in regard to "luck," the meaning of betting odds, the law of mathematical expectation, gambling, and applications in poker, roulette, lotteries, dice, bridge, and other games of chance. Part II discusses the misuse of statistics, the concept of statistical probabilities, normal and skew frequency distributions, and statistics applied to various fields—birth rates, stock speculation, insurance rates, advertising, etc. "Presented in an easy humorous style which I consider the best kind of expository writing," Prof. A. C. Cohen, Industry Quality Control. Enlarged revised edition. Formerly titled *The Science of Chance*. Preface and two new appendices by the author. Index. xiv + 365pp. 5⅜ x 8. Paperbound $2.00

BASIC ELECTRONICS,
prepared by the U.S. Navy Training Publications Center

A thorough and comprehensive manual on the fundamentals of electronics. Written clearly, it is equally useful for self-study or course work for those with a knowledge of the principles of basic electricity. Partial contents: Operating Principles of the Electron Tube; Introduction to Transistors; Power Supplies for Electronic Equipment; Tuned Circuits; Electron-Tube Amplifiers; Audio Power Amplifiers; Oscillators; Transmitters; Transmission Lines; Antennas and Propagation; Introduction to Computers; and related topics. Appendix. Index. Hundreds of illustrations and diagrams. vi + 471pp. 6½ x 9¼.
Paperbound $2.75

BASIC THEORY AND APPLICATION OF TRANSISTORS,
prepared by the U.S. Department of the Army

An introductory manual prepared for an army training program. One of the finest available surveys of theory and application of transistor design and operation. Minimal knowledge of physics and theory of electron tubes required. Suitable for textbook use, course supplement, or home study. Chapters: Introduction; fundamental theory of transistors; transistor amplifier fundamentals; parameters, equivalent circuits, and characteristic curves; bias stabilization; transistor analysis and comparison using characteristic curves and charts; audio amplifiers; tuned amplifiers; wide-band amplifiers; oscillators; pulse and switching circuits; modulation, mixing, and demodulation; and additional semiconductor devices. Unabridged, corrected edition. 240 schematic drawings, photographs, wiring diagrams, etc. 2 Appendices. Glossary. Index. 263pp. 6½ x 9¼. Paperbound $1.25

GUIDE TO THE LITERATURE OF MATHEMATICS AND PHYSICS,
N. G. Parke III

Over 5000 entries included under approximately 120 major subject headings of selected most important books, monographs, periodicals, articles in English, plus important works in German, French, Italian, Spanish, Russian (many recently available works). Covers every branch of physics, math, related engineering. Includes author, title, edition, publisher, place, date, number of volumes, number of pages. A 40-page introduction on the basic problems of research and study provides useful information on the organization and use of libraries, the psychology of learning, etc. This reference work will save you hours of time. 2nd revised edition. Indices of authors, subjects, 464pp. 5⅜ x 8.
Paperbound $2.75

THE RISE OF THE NEW PHYSICS (formerly THE DECLINE OF MECHANISM), *A. d'Abro*

This authoritative and comprehensive 2-volume exposition is unique in scientific publishing. Written for intelligent readers not familiar with higher mathematics, it is the only thorough explanation in non-technical language of modern mathematical-physical theory. Combining both history and exposition, it ranges from classical Newtonian concepts up through the electronic theories of Dirac and Heisenberg, the statistical mechanics of Fermi, and Einstein's relativity theories. "A must for anyone doing serious study in the physical sciences," *J. of Franklin Inst.* 97 illustrations. 991pp. 2 volumes.

Vol. 1 Paperbound $2.25, Vol. 2 Paperbound $2.25,
The set $4.50

THE STRANGE STORY OF THE QUANTUM, AN ACCOUNT FOR THE GENERAL READER OF THE GROWTH OF IDEAS UNDERLYING OUR PRESENT ATOMIC KNOWLEDGE, *B. Hoffmann*

Presents lucidly and expertly, with barest amount of mathematics, the problems and theories which led to modern quantum physics. Dr. Hoffmann begins with the closing years of the 19th century, when certain trifling discrepancies were noticed, and with illuminating analogies and examples takes you through the brilliant concepts of Planck, Einstein, Pauli, de Broglie, Bohr, Schroedinger, Heisenberg, Dirac, Sommerfeld, Feynman, etc. This edition includes a new, long postscript carrying the story through 1958. "Of the books attempting an account of the history and contents of our modern atomic physics which have come to my attention, this is the best," H. Margenau, Yale University, in *American Journal of Physics.* 32 tables and line illustrations. Index. 275pp. 5⅜ x 8.

Paperbound $1.75

GREAT IDEAS AND THEORIES OF MODERN COSMOLOGY, *Jagjit Singh*

The theories of Jeans, Eddington, Milne, Kant, Bondi, Gold, Newton, Einstein, Gamow, Hoyle, Dirac, Kuiper, Hubble, Weizsäcker and many others on such cosmological questions as the origin of the universe, space and time, planet formation, "continuous creation," the birth, life, and death of the stars, the origin of the galaxies, etc. By the author of the popular *Great Ideas of Modern Mathematics.* A gifted popularizer of science, he makes the most difficult abstractions crystal-clear even to the most non-mathematical reader. Index. xii + 276pp. 5⅜ x 8½. Paperbound $2.00

GREAT IDEAS OF MODERN MATHEMATICS: THEIR NATURE AND USE, *Jagjit Singh*

Reader with only high school math will understand main mathematical ideas of modern physics, astronomy, genetics, psychology, evolution, etc., better than many who use them as tools, but comprehend little of their basic structure. Author uses his wide knowledge of non-mathematical fields in brilliant exposition of differential equations, matrices, group theory, logic, statistics, problems of mathematical foundations, imaginary numbers, vectors, etc. Original publications, 2 appendices. 2 indexes. 65 illustr. 322pp. 5⅜ x 8. Paperbound $2.00

THE MATHEMATICS OF GREAT AMATEURS, *Julian L. Coolidge*

Great discoveries made by poets, theologians, philosophers, artists and other non-mathematicians: Omar Khayyam, Leonardo da Vinci, Albrecht Dürer, John Napier, Pascal, Diderot, Bolzano, etc. Surprising accounts of what can result from a non-professional preoccupation with the oldest of sciences. 56 figures. viii + 211pp. 5⅜ x 8½. Paperbound $1.50

COLLEGE ALGEBRA, *H. B. Fine*
Standard college text that gives a systematic and deductive structure to algebra; comprehensive, connected, with emphasis on theory. Discusses the commutative, associative, and distributive laws of number in unusual detail, and goes on with undetermined coefficients, quadratic equations, progressions, logarithms, permutations, probability, power series, and much more. Still most valuable elementary-intermediate text on the science and structure of algebra. Index. 1560 problems, all with answers. x + 631pp. 5⅜ x 8. Paperbound $2.75

HIGHER MATHEMATICS FOR STUDENTS OF CHEMISTRY AND PHYSICS, *J. W. Mellor*
Not abstract, but practical, building its problems out of familiar laboratory material, this covers differential calculus, coordinate, analytical geometry, functions, integral calculus, infinite series, numerical equations, differential equations, Fourier's theorem, probability, theory of errors, calculus of variations, determinants. "If the reader is not familiar with this book, it will repay him to examine it," *Chem. & Engineering News.* 800 problems. 189 figures. Bibliography. xxi + 641pp. 5⅜ x 8. Paperbound $2.50

TRIGONOMETRY REFRESHER FOR TECHNICAL MEN, *A. A. Klaf*
A modern question and answer text on plane and spherical trigonometry. Part I covers plane trigonometry: angles, quadrants, trigonometrical functions, graphical representation, interpolation, equations, logarithms, solution of triangles, slide rules, etc. Part II discusses applications to navigation, surveying, elasticity, architecture, and engineering. Small angles, periodic functions, vectors, polar coordinates, De Moivre's theorem, fully covered. Part III is devoted to spherical trigonometry and the solution of spherical triangles, with applications to terrestrial and astronomical problems. Special time-savers for numerical calculation. 913 questions answered for you! 1738 problems; answers to odd numbers. 494 figures. 14 pages of functions, formulae. Index. x + 629pp. 5⅜ x 8. Paperbound $2.00

CALCULUS REFRESHER FOR TECHNICAL MEN, *A. A. Klaf*
Not an ordinary textbook but a unique refresher for engineers, technicians, and students. An examination of the most important aspects of differential and integral calculus by means of 756 key questions. Part I covers simple differential calculus: constants, variables, functions, increments, derivatives, logarithms, curvature, etc. Part II treats fundamental concepts of integration: inspection, substitution, transformation, reduction, areas and volumes, mean value, successive and partial integration, double and triple integration. Stresses practical aspects! A 50 page section gives applications to civil and nautical engineering, electricity, stress and strain, elasticity, industrial engineering, and similar fields. 756 questions answered. 556 problems; solutions to odd numbers. 36 pages of constants, formulae. Index. v + 431pp. 5⅜ x 8. Paperbound $2.00

INTRODUCTION TO THE THEORY OF GROUPS OF FINITE ORDER, *R. Carmichael*
Examines fundamental theorems and their application. Beginning with sets, systems, permutations, etc., it progresses in easy stages through important types of groups: Abelian, prime power, permutation, etc. Except 1 chapter where matrices are desirable, no higher math needed. 783 exercises, problems. Index. xvi + 447pp. 5⅜ x 8. Paperbound $3.00

FIVE VOLUME "THEORY OF FUNCTIONS" SET BY KONRAD KNOPP

This five-volume set, prepared by Konrad Knopp, provides a complete and readily followed account of theory of functions. Proofs are given concisely, yet without sacrifice of completeness or rigor. These volumes are used as texts by such universities as M.I.T., University of Chicago, N. Y. City College, and many others. "Excellent introduction . . . remarkably readable, concise, clear, rigorous," *Journal of the American Statistical Association.*

ELEMENTS OF THE THEORY OF FUNCTIONS,
Konrad Knopp
This book provides the student with background for further volumes in this set, or texts on a similar level. Partial contents: foundations, system of complex numbers and the Gaussian plane of numbers, Riemann sphere of numbers, mapping by linear functions, normal forms, the logarithm, the cyclometric functions and binomial series. "Not only for the young student, but also for the student who knows all about what is in it," *Mathematical Journal.* Bibliography. Index. 140pp. 5⅜ x 8. Paperbound $1.50

THEORY OF FUNCTIONS, PART I,
Konrad Knopp
With volume II, this book provides coverage of basic concepts and theorems. Partial contents: numbers and points, functions of a complex variable, integral of a continuous function, Cauchy's integral theorem, Cauchy's integral formulae, series with variable terms, expansion of analytic functions in power series, analytic continuation and complete definition of analytic functions, entire transcendental functions, Laurent expansion, types of singularities. Bibliography. Index. vii + 146pp. 5⅜ x 8. Paperbound $1.35

THEORY OF FUNCTIONS, PART II,
Konrad Knopp
Application and further development of general theory, special topics. Single valued functions. Entire, Weierstrass, Meromorphic functions. Riemann surfaces. Algebraic functions. Analytical configuration, Riemann surface. Bibliography. Index. x + 150pp. 5⅜ x 8. Paperbound $1.35

PROBLEM BOOK IN THE THEORY OF FUNCTIONS, VOLUME 1.
Konrad Knopp
Problems in elementary theory, for use with Knopp's *Theory of Functions,* or any other text, arranged according to increasing difficulty. Fundamental concepts, sequences of numbers and infinite series, complex variable, integral theorems, development in series, conformal mapping. 182 problems. Answers. viii + 126pp. 5⅜ x 8. Paperbound $1.35

PROBLEM BOOK IN THE THEORY OF FUNCTIONS, VOLUME 2,
Konrad Knopp
Advanced theory of functions, to be used either with Knopp's *Theory of Functions,* or any other comparable text. Singularities, entire & meromorphic functions, periodic, analytic, continuation, multiple-valued functions, Riemann surfaces, conformal mapping. Includes a section of additional elementary problems. "The difficult task of selecting from the immense material of the modern theory of functions the problems just within the reach of the beginner is here masterfully accomplished," *Am. Math. Soc.* Answers. 138pp. 5⅜ x 8.
Paperbound $1.50

NUMERICAL SOLUTIONS OF DIFFERENTIAL EQUATIONS,
H. Levy & E. A. Baggott

Comprehensive collection of methods for solving ordinary differential equations of first and higher order. All must pass 2 requirements: easy to grasp and practical, more rapid than school methods. Partial contents: graphical integration of differential equations, graphical methods for detailed solution. Numerical solution. Simultaneous equations and equations of 2nd and higher orders. "Should be in the hands of all in research in applied mathematics, teaching," *Nature*. 21 figures. viii + 238pp. 5⅜ x 8. Paperbound $1.85

ELEMENTARY STATISTICS, WITH APPLICATIONS IN MEDICINE AND THE BIOLOGICAL SCIENCES, *F. E. Croxton*

A sound introduction to statistics for anyone in the physical sciences, assuming no prior acquaintance and requiring only a modest knowledge of math. All basic formulas carefully explained and illustrated; all necessary reference tables included. From basic terms and concepts, the study proceeds to frequency distribution, linear, non-linear, and multiple correlation, skewness, kurtosis, etc. A large section deals with reliability and significance of statistical methods. Containing concrete examples from medicine and biology, this book will prove unusually helpful to workers in those fields who increasingly must evaluate, check, and interpret statistics. Formerly titled "Elementary Statistics with Applications in Medicine." 101 charts. 57 tables. 14 appendices. Index. vi + 376pp. 5⅜ x 8. Paperbound $2.00

INTRODUCTION TO SYMBOLIC LOGIC,
S. Langer

No special knowledge of math required — probably the clearest book ever written on symbolic logic, suitable for the layman, general scientist, and philosopher. You start with simple symbols and advance to a knowledge of the Boole-Schroeder and Russell-Whitehead systems. Forms, logical structure, classes, the calculus of propositions, logic of the syllogism, etc. are all covered. "One of the clearest and simplest introductions," *Mathematics Gazette*. Second enlarged, revised edition. 368pp. 5⅜ x 8. Paperbound $2.00

A SHORT ACCOUNT OF THE HISTORY OF MATHEMATICS,
W. W. R. Ball

Most readable non-technical history of mathematics treats lives, discoveries of every important figure from Egyptian, Phoenician, mathematicians to late 19th century. Discusses schools of Ionia, Pythagoras, Athens, Cyzicus, Alexandria, Byzantium, systems of numeration; primitive arithmetic; Middle Ages, Renaissance, including Arabs, Bacon, Regiomontanus, Tartaglia, Cardan, Stevinus, Galileo, Kepler; modern mathematics of Descartes, Pascal, Wallis, Huygens, Newton, Leibnitz, d'Alembert, Euler, Lambert, Laplace, Legendre, Gauss, Hermite, Weierstrass, scores more. Index. 25 figures. 546pp. 5⅜ x 8. Paperbound $2.25

INTRODUCTION TO NONLINEAR DIFFERENTIAL AND INTEGRAL EQUATIONS,
Harold T. Davis

Aspects of the problem of nonlinear equations, transformations that lead to equations solvable by classical means, results in special cases, and useful generalizations. Thorough, but easily followed by mathematically sophisticated reader who knows little about non-linear equations. 137 problems for student to solve. xv + 566pp. 5⅜ x 8½. Paperbound $2.00

AN INTRODUCTION TO THE GEOMETRY OF N DIMENSIONS, *D. H. Y. Sommerville*

An introduction presupposing no prior knowledge of the field, the only book in English devoted exclusively to higher dimensional geometry. Discusses fundamental ideas of incidence, parallelism, perpendicularity, angles between linear space; enumerative geometry; analytical geometry from projective and metric points of view; polytopes; elementary ideas in analysis situs; content of hyper-spacial figures. Bibliography. Index. 60 diagrams. 196pp. 5⅜ x 8.
Paperbound $1.50

ELEMENTARY CONCEPTS OF TOPOLOGY, *P. Alexandroff*

First English translation of the famous brief introduction to topology for the beginner or for the mathematician not undertaking extensive study. This unusually useful intuitive approach deals primarily with the concepts of complex, cycle, and homology, and is wholly consistent with current investigations. Ranges from basic concepts of set-theoretic topology to the concept of Betti groups. "Glowing example of harmony between intuition and thought," David Hilbert. Translated by A. E. Farley. Introduction by D. Hilbert. Index. 25 figures. 73pp. 5⅜ x 8. Paperbound $1.00

ELEMENTS OF NON-EUCLIDEAN GEOMETRY, *D. M. Y. Sommerville*

Unique in proceeding step-by-step, in the manner of traditional geometry. Enables the student with only a good knowledge of high school algebra and geometry to grasp elementary hyperbolic, elliptic, analytic non-Euclidean geometries; space curvature and its philosophical implications; theory of radical axes; homothetic centres and systems of circles; parataxy and parallelism; absolute measure; Gauss' proof of the defect area theorem; geodesic representation; much more, all with exceptional clarity. 126 problems at chapter endings provide progressive practice and familiarity. 133 figures. Index. xvi + 274pp. 5⅜ x 8. Paperbound $2.00

INTRODUCTION TO THE THEORY OF NUMBERS, *L. E. Dickson*

Thorough, comprehensive approach with adequate coverage of classical literature, an introductory volume beginners can follow. Chapters on divisibility, congruences, quadratic residues & reciprocity. Diophantine equations, etc. Full treatment of binary quadratic forms without usual restriction to integral coefficients. Covers infinitude of primes, least residues. Fermat's theorem. Euler's phi function, Legendre's symbol, Gauss's lemma, automorphs, reduced forms, recent theorems of Thue & Siegel, many more. Much material not readily available elsewhere. 239 problems. Index. I figure. viii + 183pp. 5⅜ x 8.
Paperbound $1.75

MATHEMATICAL TABLES AND FORMULAS, *compiled by Robert D. Carmichael and Edwin R. Smith*

Valuable collection for students, etc. Contains all tables necessary in college algebra and trigonometry, such as five-place common logarithms, logarithmic sines and tangents of small angles, logarithmic trigonometric functions, natural trigonometric functions, four-place antilogarithms, tables for changing from sexagesimal to circular and from circular to sexagesimal measure of angles, etc. Also many tables and formulas not ordinarily accessible, including powers, roots, and reciprocals, exponential and hyperbolic functions, ten-place logarithms of prime numbers, and formulas and theorems from analytical and elementary geometry and from calculus. Explanatory introduction. viii + 269pp. 5⅜ x 8½. Paperbound $1.25

A SOURCE BOOK IN MATHEMATICS,
D. E. Smith
Great discoveries in math, from Renaissance to end of 19th century, in English translation. Read announcements by Dedekind, Gauss, Delamain, Pascal, Fermat, Newton, Abel, Lobachevsky, Bolyai, Riemann, De Moivre, Legendre, Laplace, others of discoveries about imaginary numbers, number congruence, slide rule, equations, symbolism, cubic algebraic equations, non-Euclidean forms of geometry, calculus, function theory, quaternions, etc. Succinct selections from 125 different treatises, articles, most unavailable elsewhere in English. Each article preceded by biographical introduction. Vol. I: Fields of Number, Algebra. Index. 32 illus. 338pp. 5⅜ x 8. Vol. II: Fields of Geometry, Probability, Calculus, Functions, Quaternions. 83 illus. 432pp. 5⅜ x 8.
Vol. 1 Paperbound $2.00, Vol. 2 Paperbound $2.00,
The set $4.00

FOUNDATIONS OF PHYSICS,
R. B. Lindsay & H. Margenau
Excellent bridge between semi-popular works & technical treatises. A discussion of methods of physical description, construction of theory; valuable for physicist with elementary calculus who is interested in ideas that give meaning to data, tools of modern physics. Contents include symbolism; mathematical equations; space & time foundations of mechanics; probability; physics & continua; electron theory; special & general relativity; quantum mechanics; causality. "Thorough and yet not overdetailed. Unreservedly recommended," *Nature* (London). Unabridged, corrected edition. List of recommended readings. 35 illustrations. xi + 537pp. 5⅜ x 8. Paperbound $3.00

FUNDAMENTAL FORMULAS OF PHYSICS,
ed. by D. H. Menzel
High useful, full, inexpensive reference and study text, ranging from simple to highly sophisticated operations. Mathematics integrated into text—each chapter stands as short textbook of field represented. Vol. 1: Statistics, Physical Constants, Special Theory of Relativity, Hydrodynamics, Aerodynamics, Boundary Value Problems in Math, Physics, Viscosity, Electromagnetic Theory, etc. Vol. 2: Sound, Acoustics, Geometrical Optics, Electron Optics, High-Energy Phenomena, Magnetism, Biophysics, much more. Index. Total of 800pp. 5⅜ x 8.
Vol. 1 Paperbound $2.25, Vol. 2 Paperbound $2.25,
The set $4.50

THEORETICAL PHYSICS,
A. S. Kompaneyets
One of the very few thorough studies of the subject in this price range. Provides advanced students with a comprehensive theoretical background. Especially strong on recent experimentation and developments in quantum theory. Contents: Mechanics (Generalized Coordinates, Lagrange's Equation, Collision of Particles, etc.), Electrodynamics (Vector Analysis, Maxwell's equations, Transmission of Signals, Theory of Relativity, etc.), Quantum Mechanics (the Inadequacy of Classical Mechanics, the Wave Equation, Motion in a Central Field, Quantum Theory of Radiation, Quantum Theories of Dispersion and Scattering, etc.), and Statistical Physics (Equilibrium Distribution of Molecules in an Ideal Gas, Boltzmann Statistics, Bose and Fermi Distribution. Thermodynamic Quantities, etc.). Revised to 1961. Translated by George Yankovsky, authorized by Kompaneyets. 137 exercises. 56 figures. 529pp. 5⅜ x 8½.
Paperbound $2.50

MATHEMATICAL PHYSICS, *D. H. Menzel*
Thorough one-volume treatment of the mathematical techniques vital for classical mechanics, electromagnetic theory, quantum theory, and relativity. Written by the Harvard Professor of Astrophysics for junior, senior, and graduate courses, it gives clear explanations of all those aspects of function theory, vectors, matrices, dyadics, tensors, partial differential equations, etc., necessary for the understanding of the various physical theories. Electron theory, relativity, and other topics seldom presented appear here in considerable detail. Scores of definition, conversion factors, dimensional constants, etc. "More detailed than normal for an advanced text . . . excellent set of sections on Dyadics, Matrices, and Tensors," *Journal of the Franklin Institute*. Index. 193 problems, with answers. x + 412pp. 5⅜ x 8. Paperbound $2.50

THE THEORY OF SOUND, *Lord Rayleigh*
Most vibrating systems likely to be encountered in practice can be tackled successfully by the methods set forth by the great Nobel laureate, Lord Rayleigh. Complete coverage of experimental, mathematical aspects of sound theory. Partial contents: Harmonic motions, vibrating systems in general, lateral vibrations of bars, curved plates or shells, applications of Laplace's functions to acoustical problems, fluid friction, plane vortex-sheet, vibrations of solid bodies, etc. This is the first inexpensive edition of this great reference and study work. Bibliography, Historical introduction by R. B. Lindsay. Total of 1040pp. 97 figures. 5⅜ x 8. Vol. 1 Paperbound $2.50, Vol. 2 Paperbound $2.50, The set $5.00

HYDRODYNAMICS, *Horace Lamb*
Internationally famous complete coverage of standard reference work on dynamics of liquids & gases. Fundamental theorems, equations, methods, solutions, background, for classical hydrodynamics. Chapters include Equations of Motion, Integration of Equations in Special Gases, Irrotational Motion, Motion of Liquid in 2 Dimensions, Motion of Solids through Liquid-Dynamical Theory, Vortex Motion, Tidal Waves, Surface Waves, Waves of Expansion, Viscosity, Rotating Masses of Liquids. Excellently planned, arranged; clear, lucid presentation. 6th enlarged, revised edition. Index. Over 900 footnotes, mostly bibliographical. 119 figures. xv + 738pp. 6⅛ x 9¼. Paperbound $4.00

DYNAMICAL THEORY OF GASES, *James Jeans*
Divided into mathematical and physical chapters for the convenience of those not expert in mathematics, this volume discusses the mathematical theory of gas in a steady state, thermodynamics, Boltzmann and Maxwell, kinetic theory, quantum theory, exponentials, etc. 4th enlarged edition, with new material on quantum theory, quantum dynamics, etc. Indexes. 28 figures. 444pp. 6⅛ x 9¼. Paperbound $2.75

THERMODYNAMICS, *Enrico Fermi*
Unabridged reproduction of 1937 edition. Elementary in treatment; remarkable for clarity, organization. Requires no knowledge of advanced math beyond calculus, only familiarity with fundamentals of thermometry, calorimetry. Partial Contents: Thermodynamic systems; First & Second laws of thermodynamics; Entropy; Thermodynamic potentials: phase rule, reversible electric cell; Gaseous reactions: van't Hoff reaction box, principle of LeChatelier; Thermodynamics of dilute solutions: osmotic & vapor pressures, boiling & freezing points; Entropy constant. Index. 25 problems. 24 illustrations. x + 160pp. 5⅜ x 8. Paperbound $1.75

CELESTIAL OBJECTS FOR COMMON TELESCOPES,
Rev. T. W. Webb
Classic handbook for the use and pleasure of the amateur astronomer. Of inestimable aid in locating and identifying thousands of celestial objects. Vol I, The Solar System: discussions of the principle and operation of the telescope, procedures of observations and telescope-photography, spectroscopy, etc., precise location information of sun, moon, planets, meteors. Vol. II, The Stars: alphabetical listing of constellations, information on double stars, clusters, stars with unusual spectra, variables, and nebulae, etc. Nearly 4,000 objects noted. Edited and extensively revised by Margaret W. Mayall, director of the American Assn. of Variable Star Observers. New Index by Mrs. Mayall giving the location of all objects mentioned in the text for Epoch 2000. New Precession Table added. New appendices on the planetary satellites, constellation names and abbreviations, and solar system data. Total of 46 illustrations. Total of xxxix + 606pp. 5⅜ x 8. Vol. 1 Paperbound $2.25, Vol. 2 Paperbound $2.25
The set $4.50

PLANETARY THEORY,
E. W. Brown and C. A. Shook
Provides a clear presentation of basic methods for calculating planetary orbits for today's astronomer. Begins with a careful exposition of specialized mathematical topics essential for handling perturbation theory and then goes on to indicate how most of the previous methods reduce ultimately to two general calculation methods: obtaining expressions either for the coordinates of planetary positions or for the elements which determine the perturbed paths. An example of each is given and worked in detail. Corrected edition. Preface. Appendix. Index. xii + 302pp. 5⅜ x 8½. Paperbound $2.25

STAR NAMES AND THEIR MEANINGS,
Richard Hinckley Allen
An unusual book documenting the various attributions of names to the individual stars over the centuries. Here is a treasure-house of information on a topic not normally delved into even by professional astronomers; provides a fascinating background to the stars in folk-lore, literary references, ancient writings, star catalogs and maps over the centuries. Constellation-by-constellation analysis covers hundreds of stars and other asterisms, including the Pleiades, Hyades, Andromedan Nebula, etc. Introduction. Indices. List of authors and authorities. xx + 563pp. 5⅜ x 8½. Paperbound $2.50

A SHORT HISTORY OF ASTRONOMY, *A. Berry*
Popular standard work for over 50 years, this thorough and accurate volume covers the science from primitive times to the end of the 19th century. After the Greeks and the Middle Ages, individual chapters analyze Copernicus, Brahe, Galileo, Kepler, and Newton, and the mixed reception of their discoveries. Post-Newtonian achievements are then discussed in unusual detail: Halley, Bradley, Lagrange, Laplace, Herschel, Bessel, etc. 2 Indexes. 104 illustrations, 9 portraits. xxxi + 440pp. 5⅜ x 8. Paperbound $2.75

SOME THEORY OF SAMPLING, *W. E. Deming*
The purpose of this book is to make sampling techniques understandable to and useable by social scientists, industrial managers, and natural scientists who are finding statistics increasingly part of their work. Over 200 exercises, plus dozens of actual applications. 61 tables. 90 figs. xix + 602pp. 5⅜ x 8½.
Paperbound $3.50

PRINCIPLES OF STRATIGRAPHY,
A. W. Grabau
Classic of 20th century geology, unmatched in scope and comprehensiveness. Nearly 600 pages cover the structure and origins of every kind of sedimentary, hydrogenic, oceanic, pyroclastic, atmoclastic, hydroclastic, marine hydroclastic, and bioclastic rock; metamorphism; erosion; etc. Includes also the constitution of the atmosphere; morphology of oceans, rivers, glaciers; volcanic activities; faults and earthquakes; and fundamental principles of paleontology (nearly 200 pages). New introduction by Prof. M. Kay, Columbia U. 1277 bibliographical entries. 264 diagrams. Tables, maps, etc. Two volume set. Total of xxxii + 1185pp. 5⅜ x 8. Vol. 1 Paperbound $2.50, Vol. 2 Paperbound $2.50,
The set $5.00

SNOW CRYSTALS, *W. A. Bentley and W. J. Humphreys*
Over 200 pages of Bentley's famous microphotographs of snow flakes—the product of painstaking, methodical work at his Jericho, Vermont studio. The pictures, which also include plates of frost, glaze and dew on vegetation, spider webs, windowpanes; sleet; graupel or soft hail, were chosen both for their scientific interest and their aesthetic qualities. The wonder of nature's diversity is exhibited in the intricate, beautiful patterns of the snow flakes. Introductory text by W. J. Humphreys. Selected bibliography. 2,453 illustrations. 224pp. 8 x 10¼. Paperbound $3.25

THE BIRTH AND DEVELOPMENT OF THE GEOLOGICAL SCIENCES,
F. D. Adams
Most thorough history of the earth sciences ever written. Geological thought from earliest times to the end of the 19th century, covering over 300 early thinkers & systems: fossils & their explanation, vulcanists vs. neptunists, figured stones & paleontology, generation of stones, dozens of similar topics. 91 illustrations, including medieval, renaissance woodcuts, etc. Index. 632 footnotes, mostly bibliographical. 511pp. 5⅜ x 8. Paperbound $2.75

ORGANIC CHEMISTRY, *F. C. Whitmore*
The entire subject of organic chemistry for the practicing chemist and the advanced student. Storehouse of facts, theories, processes found elsewhere only in specialized journals. Covers aliphatic compounds (500 pages on the properties and synthetic preparation of hydrocarbons, halides, proteins, ketones, etc.), alicyclic compounds, aromatic compounds, heterocyclic compounds, organophosphorus and organometallic compounds. Methods of synthetic preparation analyzed critically throughout. Includes much of biochemical interest. "The scope of this volume is astonishing," *Industrial and Engineering Chemistry*. 12,000-reference index. 2387-item bibliography. Total of x + 1005pp. 5⅜ x 8. Two volume set, paperbound $4.50

THE PHASE RULE AND ITS APPLICATION,
Alexander Findlay
Covering chemical phenomena of 1, 2, 3, 4, and multiple component systems, this "standard work on the subject" (*Nature*, London), has been completely revised and brought up to date by A. N. Campbell and N. O. Smith. Brand new material has been added on such matters as binary, tertiary liquid equilibria, solid solutions in ternary systems, quinary systems of salts and water. Completely revised to triangular coordinates in ternary systems, clarified graphic representation, solid models, etc. 9th revised edition. Author, subject indexes. 236 figures. 505 footnotes, mostly bibliographic. xii + 494pp. 5⅜ x 8.
Paperbound $2.75

A COURSE IN MATHEMATICAL ANALYSIS,
Edouard Goursat
Trans. by E. R. Hedrick, O. Dunkel, H. G. Bergmann. Classic study of fundamental material thoroughly treated. Extremely lucid exposition of wide range of subject matter for student with one year of calculus. Vol. 1: Derivatives and differentials, definite integrals, expansions in series, applications to geometry. 52 figures, 556pp. Paperbound $2.50. Vol. 2, Part 1: Functions of a complex variable, conformal representations, doubly periodic functions, natural boundaries, etc. 38 figures, 269pp. Paperbound $1.85. Vol. 2, Part 2: Differential equations, Cauchy-Lipschitz method, nonlinear differential equations, simultaneous equations, etc. 308pp. Paperbound $1.85. Vol. 3, Part 1: Variation of solutions, partial differential equations of the second order. 15 figures, 339pp. Paperbound $3.00. Vol. 3, Part 2: Integral equations, calculus of variations. 13 figures, 389pp. Paperbound $3.00

PLANETS, STARS AND GALAXIES,
A. E. Fanning
Descriptive astronomy for beginners: the solar system; neighboring galaxies; seasons; quasars; fly-by results from Mars, Venus, Moon; radio astronomy; etc. all simply explained. Revised up to 1966 by author and Prof. D. H. Menzel, former Director, Harvard College Observatory. 29 photos, 16 figures. 189pp. 5⅜ x 8½. Paperbound $1.50

GREAT IDEAS IN INFORMATION THEORY, LANGUAGE AND CYBERNETICS,
Jagjit Singh
Winner of Unesco's Kalinga Prize covers language, metalanguages, analog and digital computers, neural systems, work of McCulloch, Pitts, von Neumann, Turing, other important topics. No advanced mathematics needed, yet a full discussion without compromise or distortion. 118 figures. ix + 338pp. 5⅜ x 8½. Paperbound $2.00

GEOMETRIC EXERCISES IN PAPER FOLDING,
T. Sundara Row
Regular polygons, circles and other curves can be folded or pricked on paper, then used to demonstrate geometric propositions, work out proofs, set up well-known problems. 89 illustrations, photographs of actually folded sheets. xii + 148pp. 5⅜ x 8½. Paperbound $1.00

VISUAL ILLUSIONS, THEIR CAUSES, CHARACTERISTICS AND APPLICATIONS,
M. Luckiesh
The visual process, the structure of the eye, geometric, perspective illusions, influence of angles, illusions of depth and distance, color illusions, lighting effects, illusions in nature, special uses in painting, decoration, architecture, magic, camouflage. New introduction by W. H. Ittleson covers modern developments in this area. 100 illustrations. xxi + 252pp. 5⅜ x 8. Paperbound $1.50

ATOMS AND MOLECULES SIMPLY EXPLAINED,
B. C. Saunders and R. E. D. Clark
Introduction to chemical phenomena and their applications: cohesion, particles, crystals, tailoring big molecules, chemist as architect, with applications in radioactivity, color photography, synthetics, biochemistry, polymers, and many other important areas. Non technical. 95 figures. x + 299pp. 5⅜ x 8½. Paperbound $1.50

THE PRINCIPLES OF ELECTROCHEMISTRY,
D. A. MacInnes

Basic equations for almost every subfield of electrochemistry from first principles, referring at all times to the soundest and most recent theories and results; unusually useful as text or as reference. Covers coulometers and Faraday's Law, electrolytic conductance, the Debye-Hueckel method for the theoretical calculation of activity coefficients, concentration cells, standard electrode potentials, thermodynamic ionization constants, pH, potentiometric titrations, irreversible phenomena. Planck's equation, and much more. 2 indices. Appendix. 585-item bibliography. 137 figures. 94 tables. ii + 478pp. 5⅝ x 8⅜.
Paperbound $2.75

MATHEMATICS OF MODERN ENGINEERING,
E. G. Keller and R. E. Doherty

Written for the Advanced Course in Engineering of the General Electric Corporation, deals with the engineering use of determinants, tensors, the Heaviside operational calculus, dyadics, the calculus of variations, etc. Presents underlying principles fully, but emphasis is on the perennial engineering attack of set-up and solve. Indexes. Over 185 figures and tables. Hundreds of exercises, problems, and worked-out examples. References. Two volume set. Total of xxxiii + 623pp. 5⅜ x 8. Two volume set, paperbound $3.70

AERODYNAMIC THEORY: A GENERAL REVIEW OF PROGRESS,
William F. Durand, editor-in-chief

A monumental joint effort by the world's leading authorities prepared under a grant of the Guggenheim Fund for the Promotion of Aeronautics. Never equalled for breadth, depth, reliability. Contains discussions of special mathematical topics not usually taught in the engineering or technical courses. Also: an extended two-part treatise on Fluid Mechanics, discussions of aerodynamics of perfect fluids, analyses of experiments with wind tunnels, applied airfoil theory, the nonlifting system of the airplane, the air propeller, hydrodynamics of boats and floats, the aerodynamics of cooling, etc. Contributing experts include Munk, Giacomelli, Prandtl, Toussaint, Von Karman, Klemperer, among others. Unabridged republication. 6 volumes. Total of 1,012 figures, 12 plates, 2,186pp. Bibliographies. Notes. Indices. 5⅜ x 8½.
Six volume set, paperbound $13.50

FUNDAMENTALS OF HYDRO- AND AEROMECHANICS,
L. Prandtl and O. G. Tietjens

The well-known standard work based upon Prandtl's lectures at Goettingen. Wherever possible hydrodynamics theory is referred to practical considerations in hydraulics, with the view of unifying theory and experience. Presentation is extremely clear and though primarily physical, mathematical proofs are rigorous and use vector analysis to a considerable extent. An Engineering Society Monograph, 1934. 186 figures. Index. xvi + 270pp. 5⅜ x 8.
Paperbound $2.00

APPLIED HYDRO- AND AEROMECHANICS,
L. Prandtl and O. G. Tietjens

Presents for the most part methods which will be valuable to engineers. Covers flow in pipes, boundary layers, airfoil theory, entry conditions, turbulent flow in pipes, and the boundary layer, determining drag from measurements of pressure and velocity, etc. Unabridged, unaltered. An Engineering Society Monograph. 1934. Index. 226 figures, 28 photographic plates illustrating flow patterns. xvi + 311pp. 5⅜ x 8. Paperbound $2.00